The Resurrection of the Roman Catholic Church

◆

A guide to the Traditional Catholic community

Griff Ruby

Writers Club Press
New York Lincoln Shanghai

THE RESURRECTION OF THE ROMAN CATHOLIC CHURCH
A GUIDE TO THE TRADITIONAL CATHOLIC COMMUNITY

Copyright © 2002 Griff Ruby.

All rights reserved. No part of this book may be used or reproduced by any means, graphic, electronic, or mechanical, including photocopying, recording, taping or by any information storage retrieval system without the written permission of the author except in the case of brief quotations embodied in critical articles and reviews.

Writers Club Press
an imprint of iUniverse, Inc.

iUniverse books may be ordered through booksellers or by contacting:

iUniverse
1663 Liberty Drive
Bloomington, IN 47403
www.iuniverse.com
1-800-Authors (1-800-288-4677)

Because of the dynamic nature of the Internet, any web addresses or links contained in this book may have changed since publication and may no longer be valid. The views expressed in this work are solely those of the author and do not necessarily reflect the views of the publisher, and the publisher hereby disclaims any responsibility for them.

Any people depicted in stock imagery provided by Getty Images are models, and such images are being used for illustrative purposes only.
Certain stock imagery © Getty Images.

ISBN: 978-0-5952-5018-9 (sc)
ISBN: 978-0-5957-7149-3 (hc)
ISBN: 978-1-4620-8710-5 (e)

Print information available on the last page.

iUniverse rev. date: 12/23/2019

The Resurrection of the
Roman Catholic Church

Imprimatur: None available, until an authentic Traditional Roman Catholic Bishop discovers his own actual authority to issue such.

All rights reserved by Author, © 1998, 1999, 2002, 2004, 2019.

No book, under my name or any other, may utilize extracts of my book in any manner which alters, diminishes, or expands the actual text as provided here, without written and notarized permission by the Author. Short extracts may be utilized in any sort of book, speech, or article for review purposes, provided that the quote is correctly given, in suitable context (as applicable), and that proper credit is given to the author by name, Griff Ruby, and that the title, "The Resurrection of the Roman Catholic Church," either with or without the subtitle, "A Guide to the Traditional Catholic Community," is given.

The Epilogue, Afterword, and Appendix A, created by this Author, are hereby donated to the public domain. Copies of any of these in full, together or separately, may be freely copied, printed, and even sold for profit by anyone, providing that no changes are made to the text and that this Author, "Griff Ruby," is credited as its author.

"A STATEMENT OF PRINCIPLES IN A TIME OF CRISIS," © 1988 by The Roman Catholic Association, Inc. Used by permission.

To Saint Francis de Sales, whose logic and argumentation served as an intellectual sledgehammer which totally smashed the false idol of Protestantism in Geneva; may my logic and argumentation defend the Catholic Church from Her enemies as well as his did.

To the older generations of cradle Catholics who still remember the miraculous saint-making Church, and who long for the return of that Church into their lives.

To the younger generations, some now entering middle age, who have grown up and lived their whole lives without seeing or even knowing of such a miraculous saint-making Church, but who long for something more than the dry gains of this world, the empty aphorisms of imitation Christianities and religions, and a so-called "Catholic Church" which takes its place among all the other Protestant sects and is no more distinguishable from any of them than they from each other.

Change is usually the result of lethargy; and stability, of intense effort.
—Edwin Faust

Contents

PROTEST OF THE AUTHOR xiii

ACKNOWLEDGMENTS xv

PREFACE xvii

INTRODUCTION. xxi

Chapter 1	HOW TO IDENTIFY THE ROMAN CATHOLIC CHURCH 1
Chapter 2	SOME BIBLICAL AND LITURGICAL PARALLELS TO THE CURRENT CRISIS 7
Chapter 3	THE JURIDICAL IMPACT OF THE SECOND VATICAN COUNCIL, OR HOW THE HIERARCHY LOST CONTROL. 12
Chapter 4	THE ROMAN CATHOLIC CHURCH VERSUS THE PEOPLE OF GOD 36
Chapter 5	OLD CATHOLICS, EAST ORTHODOX, AND THE SPECTRE OF DISSENT 71
Chapter 6	THE BEGINNINGS OF TODAY'S STAND FOR THE FAITH. 92
Chapter 7	THE ADVANCE OF MARCEL LEFEBVRE AND THE SSPX. 113
Chapter 8	THE BISHOPS OF PIERRE MARTIN NGÔ ĐÌNH THỤC 132
Chapter 9	THE ADVANCE OF THE SEDEVACANTISTS 163

Chapter 10	THE BISHOPS OF MARCEL LEFEBVRE AND ANTONIO DE CASTRO MAYER	192
Chapter 11	THE ADVANCE OF THE INDULT PRIESTS AND THE FSSP	211
Chapter 12	THE CATHOLIC LAY APOSTLATES AND FURTHER EVENTS	246
CONCLUSION		267
EPILOGUE		297
AFTERWORD		299
Appendix A	THE POPE CONDEMNS VATICAN II	313
Appendix B	QUESTIONS AND OBJECTIONS	341
GLOSSARY		429
SOURCES AND BIBLIOGRAPHY		441

PROTEST OF THE AUTHOR

In obedience to the decrees of Urban VIII, of holy memory, I protest that I do not intend to attribute any other than human opinion to any theory as to the ultimate cause of the fall of the Vatican organization from Catholic teaching, including my own. Such theories can only be regarded as such until confirmed by the Holy Roman Catholic Church and by the Holy Apostolic See, when better times return and Catholic leadership of an Apostolic and Papal character returns among those Apostolic successors for whom the condemned heresy of Modernism holds no appeal. I profess myself to be an obedient son, to the reliable Popes from Peter to Pius XII, and yet to come, and therefore I submit whatever I have written in this book to their judgment.

ACKNOWLEDGMENTS

Although mine is the only name appearing on this book as the author, a great many other persons have helped me with its contents in a great many ways. First thanks go to God of course and His holy angels and saints who guided me through this study, and then to the many reliable popes who came to my defense at the last moment, Paul III, Julius III, St. Pius V, Benedict XIV, Gregory XVI, Pius IX, Leo XIII, St. Pius X, Pius XI, and Pius XII, and next the great theologians who have expounded on the ecclesiological doctrines and dogmas in depth whose writings also give such clear support to my work, plus those who encouraged me to become familiar with their works. After that I would like to thank every bishop who either gave me useful advice, heard my ideas, reviewed my book, or offered suggestions, Bishops Richard Williamson, Clarence Kelly, Mark Pivarunas, Robert McKenna, Joseph Marie, Salvador Lazo, Thomas Curry. I would like to thank the many priests who did the same, Fathers M. E. Morrison, Kevin Vaillancourt, Dominic Radecki, Thomas Zapp, Anthony Cekada, Benedict Hughes, Benedict Van der Putten, Daniel Couture, Frederick Schell, and Ray Steichen. I do not claim to have followed anyone's advice slavishly, but their inputs have all been valuable. I would also like to thank Brian Walker whose enthusiasm kept me going, Richard Jamison, Madonna Fantz, Sisters Mary Cecilia and Mary Gemma who pointed me to some good books, P. C. Morantte who has long encouraged my writing, Mary Truong who filled in some details, Christopher Pullé who reviewed the manuscript in detail and provided much helpful advice and proofreading, Judith Sharpe who brought about the creation of the impressive "What We Have Lost" video, John Gregory who challenged me and brought to my attention every discoverable objection ever raised, and all the writers and publishers of the traditional books listed in my Bibliography, and the Editorial staffs of all the traditional periodicals listed there as well. Finally, I wish to thank all other traditional Catholic bishops and priests who have also labored to preserve the Faith and Church through these trying times.

PREFACE

Less than one in a hundred persons who regard themselves as Catholics are even aware that the Church as they knew it before Vatican II still exists, alive and well, utterly unchanged, and if anything, healthier (albeit much smaller) than it has been in centuries. Out of that "less than one in a hundred," less than one in ten of those realize that it is their absolute right, indeed their duty as faithful Catholics, to attend, support, and receive their sacraments from those still as yet very few traditional priests who function largely as if Vatican II and all of the chaos which followed it just never happened.

It is rare enough that one even finds any mention in the modern "Catholic" media that there are still any Catholic priests who say the Latin Mass and teach sound doctrine. Even those rare mentionings are of little help since they provide no real understanding as to what any of these priests, bishops, religious, and lay Catholics are doing or why. Invariably they are mischaracterized as "extremists," "fringe," "rightists," "schismatics," or any of several other non-complimentary terms, all carefully calculated to be very effective in discouraging any interest in them on the part of the reader or hearer, or else as some odd choice, a rare privilege to be extended to some peculiar few who seem to prefer it.

A case in point: The first time the Traditional Catholic community seems to have caught the international eye was the time when Archbishop Marcel Lefebvre consecrated those four new bishops back in 1988. Abp. Lefebvre was repeatedly spoken of as a "rebel," "schismatic," or "excommunicated," even by the secular press and news media. Many continue to sing that tired old song even though it has since been canonically proven and acknowledged at the highest levels that none of those vacuous charges ever applied to him at all. The second time comes fully fifteen years later when celebrity actor-turned-producer Mel Gibson decided to produce and release a truly Traditional Catholic film about the Passion of the Christ, and again, though what the man did is without fault, the press did everything in its power to deceive the public into thinking of the man and the film as if they were "anti-Semitic" (of all things!).

Thanks be to God that many souls have been graced with the perception and the bravery to find and attend Tridentine Masses despite that cloud of silence, calumny, confusion, and infamy with which Satan has shrouded the truth. The purpose of this book is to acquaint the reader with this Traditional Catholic community, who the key players are, how they relate to each other, what they are doing and why they are doing it, and finally to put to rest any fears devout Catholics might have about involving themselves and their families with it. It is amazing just how many supposedly devout Catholics will cheerfully attend an East Orthodox mass, an Episcopalian or Lutheran "Mass," or even a Baptist or fundamentalist service without a qualm, but invite them to a Tridentine Mass said by a Traditional Catholic priest and all of a sudden they get all concerned about being schismatic!

As one should be able to gather by now, this book is primarily intended for those who believe that Jesus Christ started a Church, that the Catholic Church is that Church, that union with the successor of Peter is essential to following Christ, and in all of the other things that the Catholic Church teaches. This is not an introduction to Roman Catholicism. For that, I would recommend the Baltimore Catechism, Bp. Morrow's catechism: *My Catholic Faith* (no edition later than 1960), the Catechism of the Council of Trent, or any other standard approved catechism which has been around over at least the last sixty years and which has proven itself. I am quite sure that any reader who does not believe in God, or that Jesus Christ is God Incarnate, or anything else which follows from those two facts would find themselves at quite a loss to appreciate the points brought out in this book.

Although a tremendous amount of research has gone into this volume, including the reading of numerous books, articles, periodicals, newsletters, fact sheets, booklets, brochures, tracts, and even Parish Bulletins, as well as live interviews with many of the persons mentioned herein, this is not intended to be a scholastic work. If I had, the footnotes alone would have been larger than the book itself. Only in Chapters 3, 4, and the Appendices is there any direct reliance on scholastic data. In a number of places, I refer the reader to various books listed in the Bibliography. The reason I list merely the book instead of citing some page or paragraph number within it is that I have summarized and described the contents and theme of the book as a whole rather than providing some selected quotation. Such references do constitute an endorsement on my part of the books so named as reputable sources of further information on the subject being discussed at that point. To that extent, therefore, this book is also a guide to the standard Traditional Catholic literature.

The remainder of the statements made here are for the most part either common knowledge to most educated Catholics, biographical and historical narrative, or else at least expected to stand on their own as being self-evident. In many cases, additional clarification can be found in Appendix B, Questions and Objections.

The first version of this book, written from 1996 to 1998, was revised in 1999 to accommodate the HTML format for being made available on the web, and for providing numerous internal and external links and pointers, and to fix a very few small typographical errors.

The second revision of this book corrected a few small details regarding the history of Abp. Thục, a number of minor grammatical errors, and also provided more up to date information regarding current events in the Church. An additional Appendix was added to round out the possible questions one might have, which is simply the file that used to be the "Miscellaneous Questions" file, previously listed under Traditional Catholic articles. Finally, the HTML had been cleaned up quite a bit, and important topics in it have been indexed for easy topical subject searching. For the printed version, all HTML links and indexing do not apply of course.

The third revision fixed a few more typographical errors, factored in a bit more of an international (read "Universal" or "Catholic," not "Internationalist" or "Globalist") outlook, and added a few more recent events of interest to the account. There are those who think of Traditional Catholicism as only, or primarily, an American phenomenon, but this is simply not the case. I apologize in advance if my book seems to contain something of an accidental Amerocentrism. That would only be the result of the fact that I live in the United States and that I have drawn primarily from English language and English-speaking sources. However, Traditional Catholicism is fully as international as Catholicism itself has ever been, and the struggles of Traditional Catholics in all parts of the world bear far more similarity than difference.

This fourth revision incorporates further events since the previous edition, breaks the eleventh chapter into two chapters, one to be for the Indult and the 2007 Motu Proprio (itself a significant additional event), the other for Catholic lay activities and current events, and serves as a corrective regarding certain key figures of this account and the background canonical and theological information which I previous editions failed to report accurately. To that extent this edition is a retraction and correction of things stated in the previous editions. The details of what is changed for this edition are addressed in the Afterword; for those reading this for the first time and learning about the life of real Catholicism for the first time it need merely serve as an interesting historical footnote as to the history of this book, and would best be read in sequence. But those familiar with this book, or with the issues and history and personalities described herein, whether from this book's previous editions or indirectly from others who have read it, or even more directly from the various flawed sources also responsible for the corresponding flaws in my account, the Afterword may be jumped to at this point before continuing on, if desired.

Some may not like the conclusions which logic invariably leads one to, but the search for truth, which I believe to be a defining characteristic of any person who is truly of good will, is what I hope will compel the honest reader to continue clear through to the end. If anyone objects to your reading of this book, please take note of the fact that no such person can ever point to any claim made in this book and prove that there is anything wrong with the logic or scholarship of that claim. If any "Catholic" wants to take issue with anything in this book, he must do so by taking a position which is not only heretical, but patently absurd.

"If anyone labors under the delusion that the (Novus Ordo) claptrap he hands out is real science (knowledge of the Faith) and wishes to dispute anything I have written, let him oppose this treatise if he dare. And let him oppose it publicly; not by whispering in corners or by refuting it before youngsters who have not the mental maturity necessary to be judges in questions of this kind. If he is looking for antagonists he will be able to find many, and not only myself (who am the weakest of the lot) but a host of others who love truth and who will be only too willing to refute his errors and instruct his ignorance."—St. Thomas Aquinas

INTRODUCTION

What could I, or anyone possibly mean by such a title as "The Resurrection of the Roman Catholic Church?" Did it ever die? Could it? These questions, shocking as they would be to any truly devout Catholic, lie at the heart of understanding the true nature of the Traditional Catholic community, what it is, why it exists, and what is certain to become of it. There is no easy way to introduce so profound and radical of a concept as Traditional Catholicism, at once as ubiquitous and invisible as the air we breathe, and yet also so startlingly and eternally new that no dust has never been able to gather on it.

Adding greatly to the complexity of trying to explain and introduce Traditional Catholicism is the widely varied experiences people have had with it as they have encountered it in their families, their chance encounters, their memories, the things told to them, and even in their own hearts. Many have been gravely hurt by the seeming death of the Roman Catholic Church. Many more have been baffled at how things could have gone so very wrong in the Church. "How is it possible that the Holy Spirit would allow such madness to prevail over the Church," Catholics have wondered.

In their anger, confusion, hurt and pain, many have resigned themselves to a life without moorings or guideposts. Many others have clutched onto any straw they could find: "Jesus," following one's own conscience, Do-Goodism, "Bible" Christianity, New Age Mysticism, prophetic speculation, Pentecostalism (both "Catholic" and "Protestant"), and so forth. Still others rationalized to themselves that the New Religion of the Post-Vatican II "Roman Catholic Church" must somehow yet be Catholic, and out of a sadly misplaced sense of obedience have gone along with the teachings and practices of the New Religion.

A small remnant of truly faithful Catholics has clung tenaciously to the Pre-Vatican II Roman Catholic Church, somehow sensing way deep down that it still exists despite all seeming evidences and even their own conscious belief that it doesn't. (By the phrase "Pre-Vatican II Roman Catholic Church" I refer not to chronology, since we all now live in a Post-Vatican II era, but to character. Previous to the Second Vatican Council, the Roman Catholic Church had a certain very distinctive "character" or "quality" or "nature" to it which anyone who knew it well would at once recognize. It is a character quite alien from the character of a certain new "Church" whose nature is most charitably illustrated with the exhortation, "Now, my brothers and sisters, let us gather around the table of the Lord, hold hands, and together sing a song of brotherly flaccidness.")

The following cases should illustrate that wide variance:

Bob is a young man who was raised in a devoutly Catholic home whose parents have attended an "independent" chapel, sponsored by one of the Traditional Catholic orders described in this book, for as long as he can remember. For him, the order that runs his parish speaks for all truly traditional Catholics and all other such orders are counterfeits or crackpots. He is now at the crossroads, trying to decide whether to become a priest for the Traditional Catholic order which sponsors his parish, or to simply get married. Either way, he knows what he believes and why.

Susan is a middle-aged widow whose truck-driving husband was recently killed in an accident. She and her family had long been regular parishioners at a "Catholic" parish affiliated with the local diocesan bishop in union with Francis I. While attending the funeral of her husband at the Church of his parents, she is at first impressed with their old-fashioned Catholic ways which remind her of what she vaguely remembered of the Church from when she was a little girl. She is shocked when she learns only after the funeral that the Catholic parish her husband's family attend and at which the funeral was held is not recognized by the local diocese. She is not sure whether to be more shocked over the fact of a parish operating without authorization, or over the diocesan "authorities" for neither recognizing nor accepting any friendly overtures from what is obviously the most Catholic parish she has seen in ages.

Gary doesn't go to Church anymore. With a wistful nostalgia he fondly remembers the Catholic Church he grew up in and loved and cherished. But when "the changes" had come through, to him it was as if the Church was done away with and no longer exists. Sunday morning now finds him on the lake with a fishing rod or else on the long green fairway of a golf course.

Like Gary, Mike also had his heart broken when everything he thought rock solid about the Church all got changed. What that (rather logically if mistakenly) caused him to conclude is that "Romanism" is just a man-made false religion, so now he devoutly attends a small Fundamentalist church and passes out Chick tracts.

Agnes is also a widow whose husband died some years ago after 58 years of marriage. She and her husband were raised Catholic and did their utmost to raise their 6 children in the Faith. She has been active in her parish for most of her life, even to having taught catechism for 20 years. When "the changes" came to her parish, it almost broke her heart, but out of a fierce and unswerving loyalty to

Francis I, she has remained in her parish and put up with, however grudgingly, all of the "new ways of doing things." She is very outspoken at parish council meetings where she is always pushing for a return to the old ways. She reads *The Wanderer*.

Sam converted to the Catholic Church some years ago when it became clear to him that the Protestant denomination he had been brought up in was not telling him the entire Gospel. When the Catholic parish he joined came to be even more Protestant than the Protestant church he had abandoned, he went on a quest to find the Catholic Church he intended to join and soon found an "independent" chapel and promptly became a member of it and never looked back. The fact that the priest of this "independent" chapel believes that Francis I is an antipope and possibly an antichrist as well does not disturb him in the least.

Rose was an avid reader of religious literature who had long been in various independent parishes for a number of years, before concluding that all of them lack official canonical status, and that all parishes which have such status lack the Faith. She now stays at home and prays her rosary every day, along with the prayers from her old missal. She makes a "spiritual communion" every day but it has been years since she has literally consumed a consecrated host. Since she believes the Catholic Church has been done away with, she is convinced that the world is about to end and that the Antichrist now stalks the earth. She often goes on and on about the "Three days of darkness" as though they are happening right now.

James just happened to encounter a bit of Catholic literature which mentions that Traditional Catholic worship and practice exist somewhere. He is intrigued by the idea and wonders just how he can find out if any such worship exists anywhere near his home, or at least at all. The literature he read could have been a traditionalist periodical such as *The Latin Mass*, *The Remnant*, *Catholic Family News*, *The Angelus*, *Catholic Family*, *The Roman Catholic*, *Catholic Restoration*, *Sacerdotium*, *Fortes in Fide*, *The Reign of Mary*, *Salve Regina*, *The Catholic Voice*, or *The Fatima Crusader*, or even any of quite an assortment of foreign language journals. Alternatively, it might have been a book, published by TAN books, or The Angelus Press, or any one of many much smaller traditional presses, or any of a large and growing number of Traditional Catholic informational websites springing up all over the net. Such authors as Dietrich von Hildebrand, Michael Davies, Christopher Ferrara, Atila Guimarães, Fr. Anthony Cekada, Dr. Rama Coomaraswamy, or the books of the twin Fathers Radecki have done much to document events which have spurred the traditionalist cause. What James read could even have been some non-traditionalist publication critical of Traditional Catholic worship or some Traditional Catholic priest or order, and yet even so he wants to see it.

Marlene was so impressed when she saw passing by a large group of Traditional Catholics making a pilgrimage on foot to Chartres that she joined them and finished the pilgrimage with them.

Bruce has heard of people who insist on Latin for their Mass and Altar rails and statues and incense, and he thinks it's totally stupid for anyone to make a fuss over such small details. To him, if someone wants that, and if his parish priest is willing to give them that, fine, and if not then they should just shut up. He even wrote a letter to a "Catholic" magazine about this viewpoint, and they promptly published it.

Jill had been raised in a Protestant family and regarded herself as a "Bornagain" believer. At the encouragement of her Protestant pastor, she attended with her church a screening of Mel Gibson's movie "The Passion of the Christ" and was so moved by His holy sufferings for her sins that she sought to know more of this aspect of Christ, and also to know more of the spirituality of the man who produced this film. She then decided to become a Catholic, and after spending some few months in disappointment attending her local "Catholic" parish, she finally learned about the Traditional Catholic community and the Latin Mass. Now she attends the Latin Mass exclusively.

Bessie and Janice are two gossipy old biddies who are just full of scandalous but unverified (and unverifiable) stories about the "horrible" new priest who has just set up an "independent" chapel in their neighborhood and the "evil schismatics" who go there. Everyone they know has heard from them all sorts of horrible things about "that" chapel and "that" priest which they would rather have not. Neither one of them have ever bothered to check it out for themselves to see whether or not the rumors they spread are true.

Bernard was a paid and respected apologist for the Vatican II changes and Church. Without realizing it over the years he had subtly gone from defending the changes as real progress and a good thing to more like excusing them as being "not enough beyond the pale as to justify jumping ship and joining those "rad trads," especially sedevacantists, which is to say, Catholics who believe the current Vatican leadership to be non-papal. But when Francis I asked, regarding homosexuality, "who am I to judge?" that was the last straw. Seeing his own theologians accuse him of heresy only confirmed for him that maybe the sedevacantists might be right after all. At least he realized he could no longer judge them as non-Catholics. But their literary productions had failed to impress him, long on detailed evidences on how unpapal Francis and the others before him were, but short on theological exposition and explanation for how all that could be and "where the Church is"; If only he could find such, and with any substance to it, he would jump in with both feet.

Diocesan Bishop Thomas Beezlebub wrote a pastoral letter condemning as schismatic and excommunicated some local retired priest who says private Tridentine Masses for some of his parishioners and friends without having obtained the permissions now "needed" to do so. Fortunately, the only reaction on the part of the faithful in "his" diocese has been to attend that Mass in considerably larger numbers than before.

Diocesan Bishop Frank Tryhard granted every permission to a priest from the Fraternity of Saint Peter to say a Tridentine mass in one of his parish churches every Sunday. Without any publicity it soon became the most popular mass in his entire diocese. Catholics who haven't been to Church in years are coming to these Tridentine Masses in droves and giving generously of their time and money. Bishop Tryhard is contemplating having more Tridentine masses and perhaps maybe even a Tridentine parish or two.

Traditional Catholic Bishop John Blessed was consecrated bishop without a mandate from the Pope, and has no conventional territorial jurisdiction, no diocese. Nevertheless he is in much demand at the seminary where he trains, forms, and ordains truly Catholic priests, as well as many "independent" chapels where he administers the sacrament of Confirmation and speaks at Traditional Catholic Conferences and heads up a community of a few Traditional Catholic priests.

Such experiences could be multiplied endlessly. In addition to that, there are many different understandings of what Traditional Catholicism actually is. To some it is the Latin Mass; to others it is the One True religion which must never pretend that other religions are just as good or right; to still others it is the religion which tolerates no error, even in one elected to be pope.

I say that it is all of these things and vastly more besides. Really, for reasons which will become clearer as one continues reading through this book, the best way to understand the Traditional Catholic community is to understand just what the Roman Catholic Church itself actually is. I therefore devote the next chapter to that subject.

1

HOW TO IDENTIFY THE ROMAN CATHOLIC CHURCH

Classic Catholic teaching states that the Church is "One, Holy, Catholic, and Apostolic." A few other principles of importance here are the Church being the mystical body of Christ, the indefectibility of the Church, the infallibility of the pope, and finally the promise of Christ that "the gates of Hell shall not prevail over it." Let us review each of these features in detail:

ONE: Unity of government is one essential quality which it must possess, even in such cases where there clearly was more than one group. At the top of course is God, Father, Son, and Holy Ghost, but after that (and first in the earthly realm) there comes the pope, who succeeds Peter in the role of using the keys of the kingdom. There can never be more than one pope at any given time. Even in those times where there were two or even three persons claiming to be the Pope and recognized as such by substantial portions of the Church during their lifetimes, only one of them at any time (at most) could have ever been the Pope and the others, regardless of how holy and pious and good and capable as leaders of the flock some of them in fact had been, were only "antipopes."

The only other alternative is to have no pope, such as what happens when each Pope dies and until the next Pope accepts his office. Just as having two and three claimants to the papal throne did not destroy the unity of government of the Church, neither does that frequent but usually brief period without any living pope. Were such a popeless period to be sustained over a period of several years, even that would not pose a major problem for the Church. In 1268, when Pope Clement IV died, the Church allowed nearly three entire years to pass before electing Pope Gregory X in 1271.

In those times when the Church is between popes, She is united in government by being submitted both to the popes which have already gone and to the popes yet to come and to the faithful and legitimate bishops who lead particular flocks to eternal life. We unite ourselves to the popes of the past by obeying the things they commanded while they were alive as if they were still alive and still there ruling and expecting us to obey them. We unite ourselves to the popes of the future by having resolved that we shall obey them as the final authority when they come to exist. It is quite proper for any Catholic to speak of being "in union with the Pope" and not merely "the Papacy," even when the Church is between popes, or uncertain as to which (if any) of more than one claimant is the Pope. Yet it is also proper to speak of "being united to the Chair of Peter" regardless of whether that Chair is occupied at a given time or not.

Another part of that oneness is the unity of the believers themselves who all live in the same framework of right and wrong. Even in those rare cases where there is a gray area where the Church has not as yet ruled as to what is right or wrong, believers continue to stand up for each other and help each other. We Catholics are all united to the Truth and since there is only one real Truth, we are therefore united to each other in that one same Truth.

This oneness does not necessarily imply personal friendship or even an absence of petty rivalries and even animosities. Paul even fought with Barnabas in the Bible (Acts 15:39), and Church history is littered with great Saints, such as Augustine and Jerome, or Epiphanius and John Chrysostom, who were at odds with each other for various personal reasons. That is still very much the case today, all the more so in the relative absence of any clearly and authoritatively Catholic leadership who might be concerned with resolving these difficulties.

One essential part of that unity is that every Catholic is united in submission to the Pope. Every true Catholic with a right to the name MUST be united to the Pope, but as will be seen from the foregoing chapters this is not necessarily the same thing as a union with Francis I.

HOLY: Holiness, piety, reverence, a life led in accordance with the Ten Commandments of God and the Six Commandments of the Church, is an essential quality unique to the Roman Catholic Church. It is true that we are all sinners, and that no one walking about on the surface of the earth other than Jesus and Mary is known to have completely avoided sin. The Church serves and must always serve the purpose of drawing people to God and separating them from their sins.

Holiness is not just being "nice people." Many false churches and religions and even non-religions are filled with "nice people." Holiness most certainly does include kindness as a basic component, but a supernatural kindness which goes the extra mile and even to death, willing to sacrifice all the usual compensations of life, and even fit to merit the occasional rare miracle, and yet can also be as firm and stern as death itself, tolerant of sinners but intolerant of the lies that deceive them and keep them in their sins. The standard is not human but Divine. "Nice people" may do well by human standards, but Holiness is a Divine standard. The Founder of our Church, our Lord Jesus Christ, is perfectly holy, and what holiness the Church possesses is all drawn from, and fed by, the holiness of Her Founder.

What better way is there to be Holy than to behave as if God is watching, since in fact, He is. With great humility, the holy priest approaches the sacred altar of God where also resides the tabernacle containing the consecrated hosts, actual pieces yet whole and entire of His sacred body and blood and soul and divinity. The congregation kneels in awe and silence as the priest prays, both as one of them by facing the altar as they do, but also on their behalf as only a priest can do.

If a priest should puncture this sacred action with irreverent jokes or other mannerisms or behaviors which mock God, or act as though He is not there or He doesn't exist, that priest is not a holy priest, regardless of how moral his private life may be. It is written that God does not tolerate mockery. If a priest's worship is not holy, then his is not the worship of the Roman Catholic Church, and any real Catholic, if present, should leave. It does not make the slightest bit of difference whether the priest does this on his own initiative, as any priest potentially could have done in the "good old days" before "the changes," or if he does so at the direction of his equally irreverent diocesan Post-Vatican II bishop. The moral implications of both situations are identical.

Irreverent worship scandalizes, offends, and disgusts those who attend it, or else it teaches them to mock God. Irreverent worship therefore provokes sin, the very opposite of what reverent worship does. What is said of worship also applies to all actions of the Church and its representatives. A man of the cloth must always live as though God were watching. By so doing he causes those he meets up with to abandon their sins, and therefore he so does the work of the Church in his community.

CATHOLIC: That is simply an ancient word meaning "universal." The Catholic Church is universal primarily in the sense that it is for all persons all around the world and throughout all of time from that of Christ to the End of the World, and not merely just people of this or that particular nation, or language, or race, or economic level, etc. The moral standards promulgated by the Church are binding on all life which is capable of moral choice. The truths of the Church are not merely "true" for some people and not others any more than the notion that 2 + 2 = 4 is only "true" for certain persons. In particular, the Catholic Church is not owned or controlled by any secular power. The East Orthodox churches are controlled by the Russian government, the Greek government, and so forth. The Protestant churches were also founded (and many are still run) by secular rulers in Germany, England, Switzerland, Norway, Sweden, and so forth.

Another sense of the word Catholic is that this Church teaches universally all things which are to be believed. There can be no picking and choosing in Faith or Morals. A Catholic must believe **all** that the Church teaches.

"Catholic" has also come to imply that certain "character" distinct from "Protestant" or "Jewish" or "Atheist." There are many numerous tiny details and actions and an outlook of the Faithful, such as making the sign of the cross, or grace both before and after meals, or veneration of the Catholic saints, which reflect Catholic belief and alliance. When an individual employs or exhibits all of these details in his life he is saying through them, whether hypocritically or genuinely, "I am a Catholic."

Finally, "Catholic" is how our Church is referred to generally. Only one Church retains that title by right out of all of the other "Christian" churches. Those who separated from it in the East take the title of "Orthodox." Those who separated from it during what some call the "Reformation" take the title of "Protestant," or at closest, "Anglo-Catholics" in reference to the Protestant Church of England. Those who separated from it to follow Dollinger and the Jansenists approximate it by taking the title of "Old Catholic." the nearest anyone else has come to keeping the title of "Catholic," yet still obviously different. Only in particular historical circumstances is the Church obliged to add a temporary appellation, such as "Homoousian," "Ultramontane," "Roman," or "Traditional" to set itself apart from some then (or now) overgrown heretical rival.

APOSTOLIC: The Catholic church has the interesting distinction of having been started by Jesus and the Apostles back in the first century. All others come along on the scene much later. The Catholic bishops of today have a line of succession which can be traced all the way back to the original twelve Apostles. It is also the only church which has exactly the same faith as those Apostles. The connection between then and now is something I call continuity of which more will be said in later chapters. What this amounts to is the fact that the Catholic Church is not merely similar to the early Church (indeed, some differences might be found, particularly in a number of procedural and disciplinary details), but in fact the identical self-same thing.

Another aspect of being Apostolic is the authority to send missionaries and the exercise of that authority so as to bring people from every tribe and nation into the Church, and to organize and coordinate all the internal affairs of that Church. The Church is never content to leave anyone in spiritual ignorance or darkness, but desires that all should come to a true knowledge of Christ. Many missionary priests, bishops, monks, and nuns have suffered much to bring the Gospel to hostile nations and tribes.

THE CHURCH IS THE MYSTICAL BODY OF CHRIST: As such it is a living thing which grows and changes over time even as a person does from infant to child to adolescent to mature adult. This growth occurs naturally and organically, every new detail firmly rooted in other details which have been long established and logically derived from them. For example, much follows from the fact that Jesus Christ is one person, the Second Person of the Trinity, yet with two natures, His human nature and His divine nature, namely that it is therefore proper to refer to His mother as the Mother of God, or the Theotokos (Greek for "God-bearer") and so many other things we must believe as Catholics. With this status comes the divine authority to represent God's interests in this world. As Canon Francis J. Ripley writes in *This is the Faith*, "[the Church] is the continuation of the work of a divine person; she will never fail simply because Christ can never fail. Her every Calvary will be followed by a fresh Resurrection." Even so, to speak of a "resurrection" of the Church is not to imply that the Church has died, such that it must be brought back to life—though for billions, so long bereft of the teachings and guidance and sacraments of that Church, finding it still alive must indeed seem like such a resurrection from the dead—but rather that the mystical body possesses within itself that powerful Resurrection life as was sufficient to bring Him back from His literal death on Calvary.

THE POPE IS INFALLIBLE: It is often said that the Church is infallible, but this pretty much boils down to the fact that the Pope is infallible. The "infallibility of the Church" is the product of the Church echoing what is taught by Her popes. This mainly means that a Catholic pope, as pope, and when speaking to the entire Church, could never teach error in matters of faith or morals. As an individual, or private theologian, even after his election to the papacy, he may still be subject to error, as would any other theologian of comparable wisdom and knowledge. Also, outside the realms of Faith and Morals (and most particularly in those Disciplinary matters not connected to doctrine), his competence may be a matter of his own personal prudence as a man and as a leader. Everything taught by the Church has been either taught or at least affirmed and approved by some pope at some time.

Finally, Infallibility is intrinsically intertwined with universal sovereignty. All Christians, in order to follow Christ, must be subject to the Supreme Pontiff. Conversely, the Supreme Pontiff (Pope), in order to exercise such authority, must be able to claim jurisdiction or authority over all Catholics. That is part of the definition of his office. For him to claim that his jurisdiction is over any less than all the Church is for him to relinquish the universal authority of the papacy and take on a status similar to that of the other bishops. The jurisdiction or authority of all bishops other than the Bishop of Rome is limited, in most cases to a certain diocese or archdiocese, or to the members of the religious order to which one is Abbot. For precisely that reason they are not, in and of themselves, infallible, but derive what limited infallibility they do have from their pope.

THE CHURCH IS INDEFECTIBLE: This means that the Church could never be changed from the perfect society which Jesus Christ made into anything else. If ever it should seem that it has, either the changes are not substantial or else it means that that which is changed has ceased to be the Church, and (since the Church could never be done away with) something else has become the Church. It also means the Church can never disappear, even until the End of time, meaning it must always be somewhere, even if somewhere unexpected, but always identifiable by its marks. The only eternal society in the earthly realm is that perfect and eternal society created by God Himself in the person of Jesus Christ. All other societies will eventually disappear.

THE GATES OF HELL SHALL NOT PREVAIL AGAINST IT: No matter what evil Satan and his demons throw at the Church, the Church suffers but is never destroyed. It is unpublicized yet widely known, scourged but not killed (2 Corinthians 4:7–12). Even Christ, while dead on the cross, or in the grave, was spiritually alive and preaching to the spirits in prison. So too, even the Church, if mortally attacked, is in fact very much spiritually alive, and as it turns out, bodily alive as well.

With all of that in mind, I now introduce the startling and shocking but true thesis which resides at the heart of the remainder of this book and is to be demonstrated throughout it:

**THE TRADITIONAL CATHOLIC COMMUNITY
<u>IS</u>
THE ROMAN CATHOLIC CHURCH!**

When we of the Traditional Catholic community use such words as "Traditional" to describe ourselves or "Tridentine" to describe our Roman Rite worship, these terms are strictly descriptive adjectives, on par with "short" or "tall." We are simply Catholics – Homoousian Ultramontane Roman Traditional Catholics. Our right, and indeed our duty before God, to cling tenaciously to that title is absolute and irrefutable! It may take a large number of martyrs to prove this, but if that is necessary then **so be it!** God forbid that we should ever become a "Traditional Catholic" Church that is not also one and the same as the Roman Catholic Church. We simply are **that** Church; it is the Modernists who have set up a brand-new rival Church, and we must not allow them to continue using the names of "Catholic" or "Roman" to describe themselves.

2

SOME BIBLICAL AND LITURGICAL PARALLELS TO THE CURRENT CRISIS

Catholics for Pro-Choice! Call To Action seeking to legitimize priestesses! Unheard of compromises with the leaders of other religions in the name of ecumenism! The total disappearance of all remaining Catholic nations! Pathetic liturgies, complete with clowns, hand puppets, a live donkey, and priest-in-a-box! A huge decline in mass attendance, baptisms, marriages, and religious vocations! A huge increase in marriage annulments! Huge numbers of Catholics using birth control, or even abortion, and imagining that they commit no sin in doing so! Huge numbers of Catholics either converting to other religions or losing all interest in religion! An almost complete disappearance of Latin liturgy! Financially strapped parishes forced to spend enormous amounts of money on such renovations as destroying the altar and the altar rail, installing a funky wooden table, removing the statues of the saints, replacing the crucifix with abstract designs (or worse!), and removing the tabernacle to some place of dishonor! Deals made with Communists, Socialists, and Liberation Theology! Collectivism (Communism by another name) openly advocated in church! "Catholic" schools which pride themselves in having so many non-Catholic students who attend them owing to their suppression of "school prayer" or any Catholic teaching! "Catholic" schools incorporating "sex education" classes which encourage the little children to explore the various sensations which result from fondling different parts of their bodies! Laypeople being put in charge of teaching catechism with absolutely no training other than a one-day workshop in "Building a Community" and no idea what they are supposed to teach! Scandals of priests and bishops molesting little children! Homosexuality tolerated or even encouraged in the seminaries! Pope accused of heresy by his own theologians! Masons trading secret oaths with the Knights of Columbus so that each can vouch for the essential innocuousness of the other! Catholics no longer forbidden to become Masons! Priests who teach unity and community, but not Christ, forgiveness, and the Four Last Things!

All of those things are related to each other. All of them are symptoms of the same malady. "The abomination of desolation now stands in the holy place!" Hard to believe as it is, the above described organization actually still calls itself the Roman Catholic Church! From a human standpoint one would think that the gates of Hell **have** prevailed over the Church. Many today even think that we have come to the End of the World, or else that God must have died.

When Jesus, who was and is Everlasting Life, died on the cross, the disciples then must have felt the same way very much. The Devil had won. Jesus was dead, and soon buried as well. What was there left to do but scatter in utter humiliation and fear for their lives? The Shepherd had been struck and the sheep were scattered. As it was then, so it is now.

Just as the death of Christ in many ways prefigures the present crisis in the Church, so also did other events leading up to that crucifixion. The events I wish to discuss briefly here and in more depth in the next couple chapters are comparable to the Trial of Jesus before Pontius Pilate, the Way of Sorrows and Crucifixion, and the death and burial of Christ.

When Jesus Christ appeared before Pontius Pilate, He admitted that He was a King, and after a short explanation of the nature of His Kingdom, never spoke to Pilate again. Like the sacrificial lamb He went peacefully to the slaughter, never once defending Himself against those who spoke ill of Him.

Christ, who lives and moves in the Sacraments and power of the Roman Catholic Church, was similarly silent at the Second Vatican Council where they put Him in the Dock and decided His fate. All previous councils from Nicea to Vatican I had been expressions of Christ to His Church and also the World. Vatican II on the other hand put Christ and His Church on trial before the Court of the World.

There was also at that time an insurrectionist, a robber and murderer named Barabbas who had been condemned to die for his crimes. Pilate gave the World a choice of either letting Christ go free or letting Barabbas go free. They chose Barabbas and so Barabbas was set free and Christ was crucified.

At the time of the calling of the Second Vatican Council there was present in the world a huge and growing political force which like Barabbas had murdered many in advancing its cause, had robbed many of their hard-earned wealth, and which promoted insurrections wherever it was not already established. That political force was called Communism. It was not only a political force but also a heresy. Had a council been convened to condemn the heresy of Communism, that would have been a very good and proper use for a Council.

When given a choice between judging Communism and judging the Church, Communism was allowed to go scot free while the Church and all of its functioning came under opposition scrutiny. John XXIII had ruled early on that Communism and its evils would never be discussed at the new Council they were convening. In all fairness to John XXIII, negotiations were underway with several communist nations for the release of certain Catholic clergymen whom they were holding prisoner, so John XXIII may have felt that condemning communism might in some way impede those negotiations. Be that as it may, he did what he did; Pilate no doubt had his reasons too.

Pilate had no intention of condemning Christ to be crucified. He even acquitted Him four different times. Likewise, most of those bishops, cardinals, and other prelates who came to the Second Vatican Council had no intention of injuring the Church, (even though most of the pet theories many of them wished to inflict on the Church would prove gravely injurious). What they intended was one thing; what they ended up doing was quite another.

Pilate literally washed his hands of the matter, thus relinquishing his authority in the matter and allowing the angry mob to decide what was to be done. Vatican II likewise did not so much directly mandate the chaos described in this chapter's first paragraph as cause many (or all) members of the hierarchy to relinquish the authority they had previously possessed and exercised to **prevent** such madness from happening.

After that trial of course comes the Way of Sorrows and the Crucifixion. All of the post-conciliar changes to the liturgy, to the way Vatican politics works, and to such basic documents of the Church such as the New Code of Canon law, all together constitute the Way of Sorrows. Even as so many Jews then turned their backs to their long-promised Messiah, the Novus Ordo presider similarly "turns his back" to the tabernacle, to Christ Himself. The presence of His Body, Blood, Soul, and Divinity in the Sacrament of an irreverent, anti-Catholic, and yet still valid Mass was his Crucifixion and sufferings on the Cross. The frequent invalidity of the new sacraments (as shall be explained in later chapters) maps to His death and burial.

Just as the veil of the temple was torn from top to bottom (Matthew 27:51) thus desacralizing it by granting free access between the more and less holy areas, the once Catholic Churches are similarly desacralized by the removal of altar rails, rood screens, or even the iconostasis (in the Eastern rite churches). Likewise, the stripping bare of the altar, the consigning of the Eucharist to a side chapel or other location totally outside the sanctuary, and the absence of Mass on Good Friday all correspond to the present expulsion of the Church with its true and valid Sacraments to, at best, some sort of side chapel, and at worst, some place totally outside the previous sanctuary, and the absence of valid sacraments in the New Rite. Just as Christ died outside Jerusalem after being condemned to die by Pontius Pilate, nearly all the resources of His Mystical Body, the Church, die after being condemned to die by Vatican II.

Jesus taught, "A servant is not better than his master. If they persecuted Me they will also persecute you." (John 15:20) Indeed, they crucified Christ. "They" were the religious authorities of His day. Although numerous individual Christians have been persecuted, tortured, even crucified just like Christ, the amazing thing is that such a thing never happened to the Church as a whole, the Mystical Body of Christ, until Vatican II during the twentieth century.

An individual can usually tell when he is being persecuted. The Church is made up of numerous individuals who may or may not be persecuted at any given time, but it is often only their own individual persecution or that of a few close friends or relatives which they are aware of. The Church, being the Mystical Body of Christ, is actually a societal, but also mystical, extension of Christ's Body, and it is therefore He that is being persecuted this time. He is the one who is directly aware of the persecution His Mystical Body is receiving which is distinct from the persecution which individual Christians experience, which He as God and knowing all is also aware of.

He, being aware of His persecution, is the One who is really taking action, not us. Since He lives and moves in us, His Mystical Body, there are ways in which many Catholics often (even unwittingly) have participated in what He is doing, even though it is He who is really doing it in those of us who cooperate with His grace. He has no need for any of us to be doing what we have been doing in keeping alive the Faith and the Church, but He chooses to invite us to share in His work even as He invites us to share in His sufferings. As He said, "If these should keep silent, the stones would cry out." (Luke 19:40)

The trial of Christ before Pontius Pilate has many similarities to the Second Vatican Council. Just as no one has the right to judge God, or put God in the Dock so to speak, neither does anyone have the right to judge the Church God made. While it may be fair to point out a few scattered individuals here and there who have abused their authority, or had to be deprived of it, owing to some weakness or decadence or laziness on their part, no such criticism may ever fall upon the Church which is by definition a perfect society, created by God and sustained by His Will.

Pilate started out with no intention to kill Christ. He even exonerated him four times. However, he did give the order to have Him scourged, and when he later gave permission to have Him crucified, the order to scourge Him still stood. Likewise, most of the bishops, prelates, and pope who gathered at the Vatican for a council in October of 1962 had no plan to destroy the Church or turn it into something else, but they did give the order to scourge the Church with an aggiornamento of "updates" to the Church to bring it in line with current worldly thought. As will be seen in the next chapter, that council also generated the permission given to the enemies of the Church to crucify it; when Vatican II opened the windows of the establishment ruled from Vatican City, not only did the some of the smoke of Satan leak in, but some of the Church leaked out.

3

THE JURIDICAL IMPACT OF THE SECOND VATICAN COUNCIL, OR HOW THE HIERARCHY LOST CONTROL

The Church was shocked and devastated by the sudden and major changes made in nearly all the Catholic churches all around the world, back in the 1960's and 1970's. The shockwaves of this enormity then reverberated throughout the whole world, which had thereby lost its divine lampstand. Whether one loved the Catholic Church or hated it, its divine and miraculous stability brought stability to everyone; love it or hate it, the "it" was ever and eternally the same. No one could destroy it; no one could change it; no one could overthrow it. Thousands of years of historical events of every possible nature and character had sought to destroy or modify it, all to no avail. And then one day, somewhere in the 1960's, all of that ceased to be true. It was as if the blessing and protection of God had simply and silently evaporated.

It is this sudden and dramatic loss of divine protections which lies at the heart of that great mystery of what happened. Many explanations have been ventured: Perhaps the doctrines and dogmas of the Church were never meant to have the permanency that Catholics had long expected; perhaps the massive defection (for there is no other way to describe it) somehow remains, by some tortured line of rationalization, within the pale; perhaps God really has withdrawn His blessing as the means of bringing about prophecies about that Great Apostasy to take place just before the world ends; perhaps the Church had long possessed an inherent weakness, an "Achilles' heel" so to speak, which had only just now been finally discovered and exploited by the Enemy of souls; or perhaps, as many speculated to the point of its being itself a newsworthy event in 1966, God had died.

THE JURIDICAL IMPACT OF THE SECOND VATICAN COUNCIL, OR HOW THE HIERARCHY LOST CONTROL

Are the dogmas about the Church being eternal and indestructible false, or at least were they scheduled to expire at some point? Or might those dogmas mean so little as to allow for literally any scenario one might imagine? Or might there be a different solution, one which retains the dogmas in full, showing the Church to have continued to shine as She always has, albeit as a tiny remnant somehow no longer to be identified with that now-fallen society? This last has always been possible, and even has precedent in history. What has happened certainly CAN be explained, at least by what is to follow here if by no other means, regardless of whether this explanation is one day confirmed by the Church, or some other explanation (whatever that would be) yet to be developed and so confirmed.

The Mystical Body of Christ is identical to the Roman Catholic Church. That is Catholic Dogma. The Roman Catholic Church has long been identical to the Vatican organization. (By Vatican organization I refer to the visible, "hierarchical" organization, or institution, headed in Vatican City, ranging from its leader who had for all this time been the pope, clear through his cardinals, archbishops, bishops, priests, religious, all the way down to its lay members.) Such an arrangement, this practical identity, was a historical fact going back clear to the beginning. But this historical arrangement has been sustained and enforced by what amounts to Church Discipline, not Faith or Morals.

To illustrate how that is true, for example, suppose a pope and his cardinals were to sell off the facilities of Vatican City and go and establish his See somewhere else. While such a thing is gravely ill-advised and will almost certainly never happen quite that way, there is no clear law on the books which absolutely forbids it, or at least none which a pope might not have the authority to set aside, if he so wills and presses for it hard enough.

However, were such a thing to happen, the people buying the facilities of Vatican City and moving in and taking over would quite properly be spoken of as the Vatican organization, which would therefore no longer be the same thing as the Roman Catholic Church, nor even the center of its hierarchy. There would also have to be on public record a "bill of sale" of the properties.

Admittedly, this particular thing has not happened, but the bare possibility, no matter how remote or hypothetical it may be, establishes beyond all doubt that there is no dogmatic reason to identify the Roman Catholic Church with the Vatican organization. Nevertheless, that separation would have to take place formally and legally in order for the identities of the two to become distinct. Such a thing simply cannot be done secretly.

Some other obvious consequences of such a thing happening would be that the Vatican organization would lose Catholic authority along with the Charisms of infallibility and indefectibility. That authority and those Charisms wouldn't disappear, but only remain attached to the Roman Catholic Church, whatever has become of it and wherever it is subsequently to be found. They have been historically attached to the Vatican organization only because, and insofar as, the Catholic Church and the Vatican organization had so long been identical.

It is an article of Faith that those Charisms (and all other details of the Catholic Church) can never be done away with while the world still exists, but it has always been possible for the Vatican apparatus to relinquish its role as the custodian of those Charisms by legally signing their "title" to them over to another corporate entity. Why is this important? As will be demonstrated later this chapter, something somewhat of the sort has happened which would not have been possible were it a Dogmatic Truth that the Roman Catholic Church is always and eternally to be absolutely identical to the Vatican organization. Before I get there however, several other points must be clarified first.

The first point regards the Catholic Dogma that "There is no salvation outside the Church." However, salvation has never been identical to being Catholic. For one thing there exist Catholics, baptized, and knowing fully well what is right and wrong, yet choosing for whatever reason to live and die in a state of mortal sin. Not surprisingly, such Catholics are not saved. Being Catholic is no guarantee of being saved, only persevering as a practicing Catholic clear to the end.

More difficult to understand is the fact of those who are not Catholic and could be saved. It has been defined as a Catholic doctrine that "there is no salvation outside the Church." If they are not Catholic, they are outside the Church. That being the case, how is it that they can be saved? The Church, having wrestled with this problem, has come to the following conclusions:

Water baptism is what ordinarily defines who is Catholic. All persons who are validly baptized belong to the Roman Catholic Church and are bound by its laws, even if they were baptized by non-Catholics, until upon attaining the use of reason they reject the Church or its teachings. If, through invincible ignorance, their refusal to enter the Church does not spring from any culpability on their part (whether baptized or not), they are often spoken of as being united to the "soul" of the Church, but still outside the "body" of the Church. Even though such Persons might be counted as spiritual members of the Catholic Church they are still spoken of as "outside the Church" because they reject Catholic truths and disciplines or even the Church itself.

Baptism comes in three basic forms, namely the normal way with water, by desire, or by blood (martyrdom). Baptism of Desire is the most difficult to define since there are many who think it is enough to wish to "do good" or even to "ask Jesus to be their Lord and Savior" but these things, fine as they may be in some sense, do not qualify as a valid Baptism of Desire any more than wishing to take a bath would. The absolute bare minimum is a commitment to keep the precepts of the natural law that have been written by God in the hearts of all men, be prepared to obey God, and lead a virtuous and dutiful life, avoiding all voluntary fault, and being truly sorry for any faults they have nevertheless.

It is theoretically possible that some small number of especially virtuous unbaptized pagans might be graced with a vision God gives which enables them to know what a baptism is and to desire it, and thereby gain a valid and explicit Baptism of Desire. Alternatively, if a soul should be genuinely seeking the truth and at the time of death have true and perfect contrition for its sins, that soul would have an implicit Baptism of Desire which is also valid. While the Church admits the likelihood that some souls have done this, it has never been dogmatically defined as to whether or not any particular soul has done it. In other words, no non-Catholic but sincere and well-meaning "saint" has ever been canonized by the Church. There are however a small number of canonized saints whose baptism is in their own blood.

Only those who are baptized either by water, blood, or desire can ever enter Heaven and see God, but it is reasonable to hope that other virtuous persons might be at least spared the positive misery of Hell even though they can never enter heaven and see God, because they might go to a place the Church calls Limbo. Theological opinion is less than universal on this point since the Church has never formally defined dogmatically whether or not such a "Limbo" exists. In any case, this last option would only apply to those who don't have the use of reason (such as small children or those with severe mental handicaps).

When the Church teaches that there is no salvation outside the Church, by "salvation" it is talking about the power to save, to confer sanctifying grace unto salvation. This power resides exclusively within the Church. Anyone who is outside the Church in the normal understanding of that expression yet gets saved would be saved by virtue of their spiritual union to the Catholic Church and not by whatever rival religion they mistakenly hold instead. The Grace of salvation flows to that soul from Christ through the Roman Catholic Church. Also, all saved persons go to the Catholic Heaven; if in some case a Moslem or a Buddhist were somehow to obtain a Baptism of Desire unto justification and eternal life, the Moslem would find no houris in Heaven and the Buddhist would not go to Nirvana.

While I am on this point, I must mention the strange case of Father Leonard Feeney. One occasionally hears of him in the Traditional Catholic community, and so therefore some introduction to the man and the controversy which surrounds him bears mention. Fr. Feeney was a priest in the Boston area who was one of the first to see the doctrine of Baptism of Desire being increasingly abused by certain

"Catholic educators" in his area, including even his own Archbishop Cardinal Cushing, as an excuse for all sorts of false religion.

Fr. Feeney heroically fought against that heresy but in doing so he went just a little bit too far by rejecting the Catholic doctrines of Baptism of Desire and of Blood. While he was correct that the dogma "no salvation outside the Church" was soon to be compromised, and with devastating effects, he mistook the doctrines of Baptism of Blood and Desire as being themselves exceptions to that dogma, and he was afraid that they might serve as precedents for many more such "exceptions." So, he took the drastic step of removing the doctrines of Baptisms of Blood and Desire from his own theological outlook. While it is vital to continue to fight the evil which he fought, one does not help the situation any by denying the Catholic doctrines of Baptism of Desire and Blood, nor by denying any other Catholic dogma. Heresy is a poor weapon to use against heresy.

I bring up that first point in order to clarify an important detail: "The set of those who are saved" are <u>not</u> and never have been exactly the same as "The Mystical Body of Christ," "The Roman Catholic Church," nor "The Vatican organization." Within the scope of this book I am not concerned with "who is or isn't to be saved," (apart from the conclusion where I give some practical advice about that) but rather with the nature of the Visible Church, that which itself alone saves, as She survives this current bizarre and difficult period.

Most importantly, I ask the reader to keep in mind that when I later mention Catholics operating as such outside the Vatican organization, I am most emphatically **NOT** referring to those saved by a Baptism of Blood, or Desire, or a non-Catholic Baptism coupled with invincible ignorance! All real Catholics, as such, saints and sinners alike, are simply "in" the Catholic Church as members, regardless of any other category of saved or lost, or any other affiliation. Conversely, others such as sincere but ignorant Protestants, and regardless of any other affiliation of any kind, may well be justified such that God may excuse their ignorance and save them, but we cannot be certain as to the status of any particular individual as that is up to the mercy of God.

The second point is the importance of understanding the nature of a what might be speculatively called a "partial abdication," or a "partial relinquishment" of authority. On many levels and in many ways, the membership of the Vatican hierarchy have resigned, abdicated, and relinquished a great many of their former prerogatives and offices, collectively speaking. Once upon a time they exercised a direct and personal authority over what Catholics could read, who (or what parties) they could vote for, how Catholic nations are to be run with respect to the role of the Church in such states, and so forth. The Pope had direct and unilateral authority over the bishops, who in turn had the same over their priests and so on down to

THE JURIDICAL IMPACT OF THE SECOND VATICAN COUNCIL, OR HOW THE HIERARCHY LOST CONTROL

lay children in the parishes. For the most part, something has disrupted that order. However, it is not any mere individual or particular relinquishments of power which I intend to speak of in this chapter, but something far more general and broad-based which swept many from their former places. Even more insidiously, all bishops were invited and encouraged to vacate their sees, and a great many did, resulting in a partial, yet substantial, loss of much of their collective authority.

Such a thing has no clear precedent in all of human history and so people may find it hard to picture. Neither does any clear, full, and direct precedent exist in all of theological study to account for or explain such a situation; this is something with some entirely new aspects. Also, a certain tendency on the part of many to see everything in terms of black or white may also make it more difficult understand this concept.

A partial abdication means to relinquish some, but not all aspects of authority of one's office, in such manner that one is permitted to doubt whether or not they retain that office, or at least an office, at all, and if so, then in what sense or of what sort. As an absurd but physically possible example, suppose what would happen if some pope were to wake up one fine day and suddenly declare and teach that he personally was infallible in Faith but not in Morals. Obviously in that case he would be relinquishing all moral authority, but what about his doctrinal authority? Without any advance ruling on precisely that question, Catholics would be quite at liberty to disagree with each other as to whether or not that pope retains his doctrinal infallibility.

Another such absurd but physically possible example would be if a bishop were to declare that all Catholics in his diocese (along with all of those already not Catholic but living in the geographical territory of his diocese) whose last names start with the letters M through Z are not Catholics while also declaring that all persons in his diocesan territory whose last names start with the letters A through L are Catholics, including even those who are obviously not.

Since his recognition of someone in his diocese being Catholic or not has absolutely no connection with whether or not that person is really a Catholic, his place, if any, becomes quite unclear. At the very least he has relinquished jurisdiction over Catholics M through Z. It remains to be seen whether he retains jurisdiction over Catholics A through L and if so what kind. Let us even allow that he might even win a sort of non-Catholic "authority-like" power over the non-Catholics A through L by teaching that all their religions count as "Catholic" no matter what they believe, but only for those whose names do not begin with M through Z.

Catholics A through L might still be reasonably well managed by this bishop, but what about Catholics M through Z? Let us suppose that even some parish priests are M through Z. Should these priests stop saying mass or stop catering to the spiritual needs of Catholics M through Z? Obviously not, since the charge against him and them is false and unfounded, and furthermore one can reasonably expect that the next bishop to take over that diocese would reverse the bizarre legislation of his predecessor. One could even reasonably argue that Catholics A through L ought to join Catholics M through Z and have nothing to do with such a weird bishop.

The third point consists of a few general considerations regarding the Second Vatican Council itself. Much of what was written comes across as something that could have been written by a layman for which no one would have challenged his orthodoxy. Coming from the level of Pope, Cardinals, and Bishops all gathered at an Ecumenical Council, the statements as given lack by far the required precision and care in their drafting that documents of such a status must have. If one were to read the rather schizoid documents which emerged, one would find a great many things in it which are quite orthodox and, in some cases, extremely well said. Perhaps about 60% of them (by volume) are absolutely fine and perfectly Catholic, or at least well within acceptable limits. The balance of them however is quite another story. At least some degree of interpretation is needed to reconcile them with Catholic teaching, and some points are so contrary as to be impossible to reconcile with Catholic teaching at all, unless one is willing to resort to some serious rationalization, extremely complex and convoluted mental gymnastics, logic stretching, and special pleading. It only takes a single drop of poison to render the whole drink deadly, but in Vatican II, the drops of poison were numerous. In all of Catholic existence there have been twenty ecumenical Councils (and any number of lesser councils and synods) all directed at counteracting some heresy, error, schism, dissent, or to increase disciplinary order. Vatican II stands alone as an ecumenical Council directed towards yielding to heresy and integrating worldly patterns of thought then fashionable into the Faith so as to create a new Faith, and decrease disciplinary order.

The following statement best summarizes the quality of those documents: That in them which is good is not original; that in them which is original is not good. Guess which parts have been given legal force in the subsequent practice of the Vatican organization! The good parts proved to be nothing but padding with which the "Council Fathers" were lulled to sleep in order that they would not notice those other parts in which they were signing away their authority and at least partially resigning from their respective sees.

Some people might contend that it all must somehow be "good" since Vatican II was simply the twenty-first of a series of ecumenical councils the Church has held, of which it has to be said that they are all part of the Church's Infallible

Magisterium and therefore absolutely right and perfectly good throughout, but this will not stand with regards to Vatican II.

Apart from John XXIII's unprecedented move of declaring this to be a "Pastoral" council, one which would define no dogmas and pronounce no anathemas, there is an even more basic, obvious, and irrefutable reason why the documents of Vatican II simply cannot be any part of the Infallible Magisterium of the Church: They are not a part of the Infallible Magisterium of the Church because they are not a part of the Magisterium of the Church, which in turn is because they are not Magisterial. The Catholic Church uses the terms "Magisterium" and "Magisterial" to refer to its authority to teach, or its teaching authority. In order to be Magisterial, a document must teach. The Vatican II documents do not teach.

The story of every Ecumenical Council of the Church from Nicea to Vatican I is always the same: One or more serious questions having been raised, the council is convened, the Council Fathers argue the question(s) back and forth until they come to some consensus most in line with the history and tradition of the Church, the Pope gives his formal approval to that consensus, and the consensus is promulgated in a document or group of documents which answer the question(s) definitively and authoritatively.

A key point here is that such documents promulgated under such circumstances are in absolutely no need of interpretation. They are clear, unambiguous, and teach the reader what the Church has to say about the specific question(s) raised. The documents of a council as promulgated are intended to be, themselves, the Church's interpretation of Her teaching as well as the last word on the subject(s) and therefore <u>not</u> subject to further interpretation or debate. "Rome has spoken; the cause is finished."

What a striking contrast between that and what happened at Vatican II! No significant questions had been raised since the previous council. Any questions which might potentially have been raised (e. g. Communism) were ruled out from the outset. Most of the bishops, cardinals, and other leaders in the Church who gathered in Vatican City back in the 1960s had no idea why a council was being convened. The documents it produced answered no questions and solved no problems. Indeed, the very reverse happened: Those there then who read those documents (namely the Council Fathers themselves who signed them) might have started out as Catholics, but by the time they were done reading them (and signing them), and thereby having committed themselves to everything in them, they ended up not knowing <u>what</u> to believe anymore.

The documents, owing to their ambiguous and even heterodox or heretical implications, do not stand on their own but are in dire need of further interpretation. The problem here is "Which interpretation do we go by, the almost

Catholic interpretation Cardinals Ratzinger, Gagnon, Stickler, or Oddi gave them, or the utterly uncatholic interpretation given to them by Cardinals Noe, Villot, Bernardin, or Mahoney?" The documents themselves provide absolutely no basis for preferring one interpretation over the other: They don't teach. With the documents of Vatican II, one could properly say "[Modernist] 'Rome' has spoken; can anyone even make out what they said?" "If the trumpet gives an uncertain sound, who will rise to battle?"—1 Corinthians 14:8

Nevertheless, the documents of Vatican II (or some of them anyway, as shall soon be explained) by any conventional evidences seem to be official documents of the Roman Catholic Church and as such would carry with them a certain legal and disciplinary force, and therefore have a certain juridical impact. The crucial fact about this "certain juridical impact" is that its result is not what the Council Fathers thought, intended, wanted, nor expected, but something else entirely, something unforeseen. Just as computer programs are famous for having "bugs" in them which perennially cause unexpected problems, the Council documents also have a certain "bug" in them which I am about to point out.

To be properly understood, the Council must be divided into three phases, roughly corresponding to the 1963, 1964, and 1965 sessions. (The 1962 session generated no documents.) It is the middle phase which is by far the most crucial since it is in that 1964 session that the Council Fathers and Pope Paul VI together jointly signed away a crucial portion of their authority and thereby decreed into existence a new Vatican organization detached (in law) from the Roman Catholic Church.

In the first phase (1963), only two documents were promulgated and both of them merely established committees in order to carry out their directives. The less important of these did nothing more than set up a committee to "study the problem of social communication and the media" and for all I know they may very well still be off somewhere doing just that. Or perhaps they prepared some report or two, regarded their job as fulfilled, and then disbanded. The more important of these is the Constitution on the Liturgy, which set up a committee to "reform" the Liturgy, and then provided rather detailed guidelines as to what sort of reformed liturgy the committee was meant to come up with. Since that committee thus set up never has come up with any liturgy (Catholic or otherwise) which conforms to those guidelines, none of the monstrosities they have come up with (such as their new "Mass" I will describe next chapter) could possibly be the object of any obligation generated by the Council. Maybe they too continue to invent further changes, or else maybe they too have regarded their job as finished and have therefore disbanded as well.

The third phase (1965) comes after this unforeseen juridical effect, and therefore can only be described as official documents of a new Vatican organization, but not of the Roman Catholic Church itself. By virtue of that fact, these documents could not have been infallible, even if they were to have "taught" anything (which they didn't). Before one can really understand the nature and purpose of the documents of the third phase, it is necessary to know and understand just exactly what happened in that crucial second phase (1964), and so I will focus primarily on that from here on.

The fourth and last point concerns the application of any formal document of legal or disciplinary force. People often have the mistaken notion that a law, a canon, or a decree, etc. should be "interpreted" in the light of its author. Certainly, there have been many unfortunate results from such documents saying something other than what was intended owing to such ridiculous things as typographical errors, and which nevertheless had to be applied as written and not as intended. Going back to the "computer bug" example just above, a computer does exactly what the program literally states, not what the programmer intended.

The Church always "interprets" (more properly, "applies") its official documents (e. g. Canons, Decrees, Constitutions, the Code of Canon Law) in no way but as literally stated. The only legitimate room for some "adjustment" or "interpretation" is that which is specified within the text of the law itself, or those things obviously based on generally accepted jurisprudence. That is because if one gets away from that into the world of the intention of the author, or "what's best for everyone now," or anything else, one ends up entering a very murky, spooky territory of having to second-guess the intentions of a lawgiver who may not be available for comment, or of having to predict precisely what the consequences or side-effects of a certain implementation of a Law might be. If a Law as written fails to function well in some particular case, the procedure is to revise the Law so that it will handle the new situation properly. Only in some rare and exceptional case in which the law works an evident injustice, as obvious to all concerned, might it be set aside under a principle known as Epikeia.

A few paragraphs ago, I stated that the documents of the Second Vatican Council don't teach. That does not mean that nothing in them can be understood or taken at face value. Many of the good portions state their meaning quite clearly and unambiguously. It's just that what they say has already been stated in terms every bit as clear by some previous Magisterial document of the Church, if it is of any legitimate value. An example of that happening could go as follows:

A member of the board of some club could submit a document to the president of the board and the rest of the board to the effect that "Gobbledygook. Gobbledygook. The sky is blue. Gobbledygook. I quit. Gobbledygook." The gobbledygook's, that is to say, all the ambiguous things that don't teach, can be interpreted any way as suits one's fancy and therefore mean nothing, the sky being blue is simply an unremarkable statement of fact typical of the perfectly orthodox statements found scattered throughout the documents of Vatican II and therefore of no practical significance or implications, and so therefore the only part that matters is the part which says "I quit."

If the president accepts this document for consideration and the board approves of it in a formal vote, said document becomes yet another official document of that club. However, its only real consequence is to affect the resignation of that member.

Even if the "I quit" portion is written in such complex legalese that neither the person who submitted it, the president of the board who accepted it, nor the board members who voted for its approval, were to have understood it, as long as there is no other legal interpretation grammatically possible it has legal force and that member has resigned from the board. Furthermore, if a board member questions the new document asking "Might this not imply the resignation of So-and-so?" and everyone else (president included) says "We don't care about that; we think the document should stand as written," then its legal force is only all the more binding, and couldn't be argued away, even in the courts, short of abrogating or revoking the document in full.

Another example: If a man signs a document with which he gives his house over to a con artist, no matter how complex and confusing the legalese in it may be, his house now belongs to that con artist. It won't do him any good to sit on top of what was once his house with a gun and attempt to shoot anyone who tries to take it from him. He signed it away; it's no longer his. Perhaps if he fights it out in the courts, he might win it back, but not until and unless the courts should so rule in his favor and invalidate the said document.

With all of the foregoing four points in mind, we are now in a position to address the crucial statement within the 1964 documents of the Second Vatican Council. In the interest of sound scholarship and in order to avoid any charge of taking something out of context, I present the chief one of the relevant paragraphs in full, along with a short discussion of its context.

> This is the sole Church of Christ which in the Creed we profess to be one, holy, catholic and apostolic, which our Savior, after his resurrection, entrusted to Peter's pastoral care (Jn. 21:17), commissioning him and the other apostles to extend and rule it (cf. Matt. 28:18, etc.), and which he raised up for all ages as "the pillar and mainstay of the truth" (1 Tim. 3:15). **This Church, constituted and organized as a society in the present world,** *subsists in* **the Catholic Church,** which is governed by the successor of Peter and by the bishops in communion with him. Nevertheless, many elements of sanctification and of truth

are found outside its visible confines. Since these are gifts belonging to the Church of Christ, they are forces impelling towards Catholic unity. *(Dogmatic Constitution on the Church (Lumen Gentium), Chapter I, section 8, second paragraph, translated by Fr. Colman O'Neill, O. P., Austin Flannery, O. P. General Editor, 1988 Revised Edition)*

This quote comes after a long, detailed, and generally rather admirable, if amateurish, description of the Catholic Church. But then there comes this paragraph in which the Catholic Church, having been so described, is said to "subsist in" the "Catholic Church" (**bold** emphasis mine). Let's take a good hard look at what that language means, precisely. Classically, and correctly, it had long been the case that the Church, as described, simply is the Roman Catholic Church. Saying "subsists in" instead of "is" conveys a substantially different message. To subsist basically means to exist, or to persist. It is a special theological term that refers to a special mode of existence, namely one that necessarily and intrinsically exists eternally, and cannot be eliminated. This is exclusively a characteristic of the Church, though other than that special meaning it is grammatically equivalent to "exist." It is the use of the word "in," grammatically introduced by the word "subsists," which carries with it an implicit denial of identity. To state that "a thing is (exists, or subsists) in a thing," is to imply that the first thing is not the same thing as the second thing.

Had the phrase used at that point been "subsists as," it would have been rendered harmless, since identity would still be implied. But "subsists in," like "subsists on" (as for example one could properly say that John the Baptist "subsisted on" locusts and wild honey, and again no one would claim that John the Baptist was one and the same as these foods—i. e. he was not eating himself), amounts to a declaration of distinction. One could say that a man "subsists in" (or more exactly, "exists in") a given house, but obviously the man is not the house and the house is not the man. Who would ever say that a man "exists in" himself?

As with all illustrations, the "man existing in the house" illustration is not a perfect one to use here however since it has some serious differences from the current situation. No part of the man comprises any part of the house, and furthermore, that relationship is non-symmetrical. The house does not exist in the man, only the other way around. When applied to groups of individual persons, be such groups the Church, the Rotary Club, the Jewish Community, or all of humanity at large taken as a whole, to say that one exists in the other implies an intersection between the two groups. If group A of persons exists in group B of persons, then at least some members of group A must also be members of group B. However, since the groups are different, there can be members of group A who do not exist in group B and vice versa. In this case, there would necessarily be persons comprising a part of group A who themselves also comprise a part of group B. Furthermore, such a relationship is symmetrical. Group A exists in group B, so therefore group B exists in group A.

One here (and as it turns out, throughout all the Vatican II documents) has to judge from context just what is referred to by each instance of the phrase "Catholic Church." In the barest handful of cases in these documents (hangovers from earlier drafts in which this whole "subsists in" language had not yet been invented), it is clearly a reference to the Mystical Body of Christ, the Church of Christ, the visible and eternal Roman Catholic Church, "entrusted to Peter's pastoral care." However, in all other instances such as the one in this quote, it clearly refers to an entity which is distinct from that Church. This other distinct entity, now newly created and "governed" by the men who had formerly served as "Peter" and "the bishops in union with him," had no distinct name for itself with which to set it apart from the Roman Catholic Church. For clarity and convenience of nomenclature in this book, I have referred to that distinct entity as "the Vatican organization." Up until the creation of that distinction the Vatican organization simply WAS the Roman Catholic Church, as a visible society, absolute identity.

The intent of those who used the phrase "subsists in" where the word "is" belonged was to have a "Catholic Church," in fact this new Vatican organization, that could affirm other (rival) religious organizations to be sources of Divine Grace. At the Council, many conservative fathers intervened to have that passage (and others like it) corrected, but they were simply outvoted. The question was officially raised in an intervention, and then simply ignored by a room full of people who simply wanted to get it over and done with so they could move on to other things and return home, rather than seriously consider opening the document for further wrangling yet one more time.

Here is another example to demonstrate the significance of that key operative phrase in this quote, namely "subsists in." Having one thing merely subsist "in" another instead of subsisting "as" implies that, while having an overlap, connection, or relationship between the two, there may now be portions of each which do not belong to the other. In that vein, one can properly say that the Catholic Church "subsists **in**" the United States of America (or any other mere secular nation one may care to name). That is right and proper since there are Catholics who are Americans, but also Catholics who are not Americans, and also Americans who are not Catholics, and finally those who are neither Catholics nor Americans.

But imagine what blasphemy it would be to state that the Catholic Church "subsists **as**" the United States of America! Clearly that would be to equate the two, which would be ridiculous, for in that case one would be saying that every American is a Catholic and every Catholic is an American. Such an identity can

THE JURIDICAL IMPACT OF THE SECOND VATICAN COUNCIL, OR HOW THE HIERARCHY LOST CONTROL 25

only be rightly applied to the Church, the Church's own visible society (itself) in which case it is proper to say that the Church subsists "as" its own visible society, and correspondingly improper to say that the Church would only subsist "in" its own visible society, as that puts its own visible society exactly on par with any and every merely human and secular society, such as the United States.

What this new distinction does is declare it possible that ministers who do not belong to the Vatican organization would nevertheless possess and lawfully exercise (in a normal and routine manner) the salvific authority which exclusively belongs to the Catholic Church. The fact that this document was mandating that some Catholic priests and bishops could be recognized as truly Catholic (and therefore in union with Rome) while being outside the organization ruled by Paul VI and his successors is further clarified in that quote where it states that "many elements of sanctification and of truth are found outside its [the "Catholic Church's," or more exactly and properly, the Vatican organization's] visible confines."

That would contradict (if the Catholic Church were still identical to the Vatican organization) the *De fide* teaching that all Grace (including sanctification and truth) comes from Christ to mankind <u>through</u> the Roman Catholic Church, that is, by means of priests and bishops in union with the Successor of Peter. By saying that a part of the source of sanctification and truth (i. e. some percentage of the truly Catholic priests and bishops in union with the Successor of Peter) is outside the Vatican organization, what they have really said is that part of the Roman Catholic Church is outside the Vatican organization.

That conciliar statement, from what is called a "Dogmatic Constitution" of the Church (not truly Dogmatic since the entire Council was purely disciplinary and juridical, not doctrinal or dogmatic), states quite clearly and unequivocally that the Roman Catholic Church, the Mystical Body of Christ, the Church of Christ, the visible and eternal Church headed by the Pope, and against which the gates of Hell would not prevail, now only "subsists in," not "as," the Vatican organization. The two entities are now made distinct, overlapping but no longer identical. As an interesting aside, it is also true that something of the Vatican organization exists "in" the Catholic Church. There is something of a symmetry here; each has a portion which is in the other and another portion which is outside the other: Some of the Vatican organization could be (and rapidly would be) outside the Roman Catholic Church.

There can be Roman Catholics who are part of the Vatican organization and Roman Catholics who are no part of the Vatican organization. And there can be non-Catholics who are part of the Vatican organization (namely those who adhere to a new religion I will talk about in the next chapter) and as always, non-Catholics who are no part of the Vatican organization (Moslems, Buddhists,

Animists, Atheists, Satanists, Hindus, Jews, Protestant Fundamentalists, etc.). As I have just shown, it is the Second Vatican Council itself which has decreed and "dogmatically constituted" such a state of affairs!

Such a distinction also decrees into existence what might be described as a "Great Detachment," or a "Roman Schism." By its own formal declaration here (which has legal force of law), the Vatican organization is not the Roman Catholic Church anymore. As such, it therefore cannot be held bound to the rules of the Catholic Church. Paul VI and the members of the hierarchy signed a formal document, a "Constitution," which makes it so that they are not answerable to the teachings, doctrines, precepts, or disciplines of their Catholic predecessors. This is no less schismatic than it was for King Henry VIII of England to make it so that in 1534 that he and his new church were not answerable to the teachings, doctrines, precepts, or disciplines of the Pope in Rome. The only difference is that where the boundaries of the English Schism were geographical, separating England from Rome, the boundaries of the Roman Schism are aterritorial, separating that true Church which exhibits the marks and attributes of the Church from what amounts to a new Vatican organization in parallel to it which need not, and ultimately doesn't, exhibit such marks and attributes. As with many schisms, a false continuity is created by which the resources (human and physical plant) are appropriated by a new schismatic body, while keeping almost everything precisely where it was, the same bishops and priests in the same cathedrals and parish churches, the same tribunals and even continuing to use the same books of baptismal records. No wonder nearly all English back then (or so very many all around the world this time around) stuck with their now Anglican parishes etc., seeing them as still "the Church."

By defining into existence a new Vatican organization in parallel to the Roman Catholic Church and partially overlapping it, and themselves collectively as being the new society's new officers and leadership, this Constitution created a nonhierarchical doppelgänger office for each member of the Vatican hierarchy. Where before each official of the Church only had one job, they now had two. Since few had the personal stamina to do the work of two different persons, and one job came to be increasingly at odds with the other job, most had to choose between continuing to function within their previous office as hierarchical bishops of the Church, or else functioning within their new office as local functionaries within this newly created Vatican organization, effectively resigning their former post by acceptance and entry into their new offices. Needless to say, the heroes of this account are those few bishops who devoted at least the brunt of their efforts to their original posts as bishops of the Catholic Church, and also their lawful successors.

THE JURIDICAL IMPACT OF THE SECOND VATICAN COUNCIL, OR HOW THE HIERARCHY LOST CONTROL

The Catholic Church cannot be changed because She is the Mystical Body of Him who is the same yesterday, today, and forever (Hebrews 13:8), but the Vatican organization did get quite substantially changed by Vatican II and many other official actions of its leadership in the years following. Such a thing only became possible when the two were detached from each other in Law. That had the effect of making the Vatican organization at most like merely a secular power, political party or faction, or club. Whatever "authority" they can be properly said to wield is strictly that of any merely human and secular authority.

Given all this, the proper response or attitude regarding the Vatican organization is neither allegiance ("I must join it or else I am out of the Church") nor hostility ("It is evil; I must have nothing to do with it") but indifference. "I am going to go about my Catholic business. The Vatican organization is free to accept that or not; for my own sake, I honestly don't care which choice they make about me." When the Vatican organization, in the person of those who comprise its hierarchy and leadership, chose to become indifferent as to what is Catholic and what is not, that was the cue for Catholics to become indifferent as to what is or is not approved by the Vatican organization.

Creating such a distinction between the Catholic Church and the Vatican organization by signing those documents directly resulted in a partial abdication. Like that bishop in the above example who relinquished jurisdiction over Catholics M through Z, the leaders of the Vatican organization relinquished jurisdiction over those Catholics who remain faithful to the Roman Catholic Church while functioning outside the Vatican organization's now questionable structures.

Important to note is the distinction between "being in the Catholic Church independent of being in the Vatican organization" and "being in a state of Grace." Some of those in the non-Catholic portion of the Vatican organization, even though they are materially "ex-Catholics" are not formally so (owing to invincible ignorance) and therefore not in a state of sin, but they are visibly outside the Church. Such ones are honestly not conscious of having left the Church even as those who remained in the Anglican parishes were not conscious of having left the Church. By the same token, it is reasonable to expect that there are some "tares" among the "wheat" of the Roman Catholic Church, both the parts inside and the parts outside the Vatican organization, although it must be pointed out that the proportion of "tares" within the Catholic Church is dramatically lower than it has ever been in many centuries.

A big question which now needs to be asked is "What are we to make of the laws and decrees of the Vatican organization?" In particular, are they binding on the Catholic Church? Some things they promulgate are practically incompatible with the teaching of the Roman Catholic Church. Obviously, anything promulgated which is contrary to Catholicism must be ignored. For example, if a bishop orders all of the married couples in his diocese to go on the Pill or use condoms the Catholic couples must disobey their bishop and obey Catholic teaching. Any Catholic should be able to figure that one out. That is not the difficult case.

Somewhat trickier are those cases in which certain laws and decrees of the Vatican organization contradict only more subtle teachings of the Church that the typical layman cannot be expected to be familiar with. In this has lived the gradualism with which the new Vatican organization has subtly and insidiously introduced further and further aberrations, gradually and quietly transferring souls out of the Church, and reducing that intersection between itself and the real Catholic Church from something almost its whole size to only the barest fraction thereof.

The difficult case would be those orders given which are compatible with the Church's teaching. Do we allow the reasonable commands of the Vatican leaders to be binding on Catholics as well? Can there exist such a thing as a part-time bishop or pope? If such a thing does exist, how does one determine which times belong to their functioning as a Catholic bishop or pope and which times do not? These are challenging questions put before Catholics today and many truly devout and serious individuals have argued quite persuasively for many sides of this question.

The Church has not, and cannot, rule officially on this question because we are still waiting for a pope to come along who will hand down a decision. Therefore, it is quite proper that I also do not here insist upon any particular answer to this difficult question.

One thing I do know is that the Pope we do get this answer from cannot be Francis I because he is one of those under suspicion and therefore not in a position to exonerate himself, nor those five predecessors of his who are also under the same (or similar) suspicion. For reasons which will become clearer in later chapters, a small but growing contingent of faithful Catholics had come to be of the opinion that the conciliar and post-conciliar Popes (John XXIII, Paul VI, John Paul I, John Paul II, and Benedict XVI as of this writing) have so seriously failed to function as Catholic popes that they ought not be considered as bearing the title of pope at all. With the extreme failures of Francis I to act anything at all like a Catholic pope, this Sede Vacante finding has advanced from being a small but substantial current within the Traditional Catholic community to attaining a mainstream status in the whole world at large. The antics of Francis I have forced billions to begin asking themselves, "Does the Catholic Church still have a Pope?"

Within this volume, in the interest of speaking to, for, and on behalf of all true Catholics, I shall use the phrase "reliable popes" to refer to all popes from

Peter to Pius XII, excepting only such popes as have been judged "unreliable" by later popes within that sequence. By the phrase "reliable pope," I refer only to that pope's doctrinal trustworthiness. Some few whom I must refer to as reliable popes would be most charitably described as scoundrels owing to the nature of their private lives, but at least God guarded their public teaching so as to keep it authentically Christ's. They are "reliable" to the extent that the Church has ever since relied upon their dogmatic teachings

A good example of what I herein call an unreliable pope would be Pope Honorius I. In 634, Honorius, in response to a query by Sergius, Patriarch of Constantinople, made an ambiguous statement which opened the door for a heresy known as "Monothelitism," which denies that Christ had a human will. Pope Leo II (the Great) confirmed the conclusion of the Third Council of Constantinople to the effect that Honorius I was "incapable of enlightening this apostolic Church by the doctrine of Apostolic Tradition." Therefore, Pope Honorius I's ambiguous declaration cannot be authoritatively quoted against the more definitive and precise declarations of reliable popes. Even whatever perfectly good things he might have said on any number of other occasions fail to balance this conclusion of the Church that he was unreliable.

Any and all Catholics with a right to the title of "Catholic" MUST accept all teachings of the reliable popes as legitimate authority within the Church. However, the leaders of the Vatican organization from John XXIII to Francis I (as of this writing) have not proven to be "reliable popes." Either they are unreliable popes (in fact more untrustworthy than Pope Honorius I), or else they are not popes at all. It is even possible that some will be in one category and others in the other. I leave it to the Church, in the person of the next truly Catholic and reliable pope or any council he convenes for this purpose, to rule authoritatively on that matter. For this work I shall herein call those last six "doubtful" or "unreliable" popes.

One position which is truly untenable is to regard the six doubtful popes as being reliable popes who can safely be followed blindly the way one with one's eyes closed could have safely followed any of the many reliable popes who preceded them. That is because their leadership is of the Vatican organization and not necessarily of the Roman Catholic Church. If the leader of the Vatican organization makes an official teaching or action, that teaching or action cannot be binding on those Roman Catholics outside the Vatican organization, and therefore cannot be construed as teaching or ruling the entire Church. Remember also, the Charisms of infallibility and indefectibility rest upon the Roman Catholic Church but NOT necessarily upon the Vatican organization.

One effect of this, and a major part of what had once been the hierarchy of the Church losing control (and any real hierarchical status beyond that of bare secular power) is how "ministers who do not belong to the Vatican organization would nevertheless possess and lawfully exercise the salvific authority which exclusively belongs to the Catholic Church." Since such ministers do not belong to the Vatican organization, then neither do they answer to it nor to any part of it. As a result, where before Catholic bishops truly ruled their dioceses exclusively such that without the consent of the local parish priest or bishop no one may exercise spiritual functions therein, such as marrying, baptizing, preaching, burying, giving extreme unction, etc., **now no minister, even an admittedly Catholic one or not, can be excluded from performing these functions within anyone's parish or diocesan territory, even altogether without permission or consent.** As it turns out, this would affect not only the Vatican organization's local functionaries, but also the real bishops of the real Catholic Church as well.

The change wrought by "subsists in" has an even more extreme impact on the papacy of the last five doubtful popes. In supporting (in the case of the John Paul's, Benedict XVI, or Francis I) or even in signing (in the case of Paul VI) the documents of Vatican II they admit to a lack of authority or jurisdiction over part of the Catholic Church which is the source of grace. They are therefore, in that action, relinquishing their universal sovereignty. With that goes their infallibility, since a teaching must be not only in Faith or Morals but be authoritatively applied to the entire Church.

Any bishop whose jurisdiction is anything short of universal (i. e. any bishop other than the pope) lacks the authority to address the entire Church and for that reason is incapable of proclaiming any teaching infallibly. In 1964, Vatican II deprived the leader of the Vatican organization of universal jurisdiction and therefore of the charism of infallibility. The only way such a leader might possibly teach infallibly would be to do so completely outside the parameters set by Vatican II, that is, as if the Council simply did not exist.

In a strangely parallel mirror reflection of what happened in England when most members of the "Church **in** England" defected to the schismatic (and soon to be heretical as well) "Church **of** England," the "Bishop **of** Rome" reduced himself to being merely a presiding "Bishop **in** Rome." All Catholics must be subject to the Pope, but not all Catholics need be subject to the Vatican leadership; Vatican II itself has so decreed.

It was in the few weeks leading up to that time that Paul VI gave away the Papal Tiara, an action which itself was merely symbolic of the fact that he was relinquishing his papal authority. Few Catholics have ever read the Vatican II documents; fewer still have ever reflected upon their true meaning, but even the simplest Catholic could read the symbolic meaning of that action.

A fair question to ask is, "Where have the authority, marks, and Charisms of the Catholic Church gone; who possesses them today?" The answer is surprisingly simple: They can only be held by real Catholics. They are the ones who have taken the Anti-Modernist Oath as instituted by Pope Saint Pius X in 1910 and who live by it, adhering to, and teaching, the entire magisterium of the Church as promulgated by the reliable popes and councils. It is these priests and bishops of whom more will be said in later chapters.

The doubtful popes can and have said many fine and Catholic things, but regrettably one is permitted to doubt whether or not they exercise Catholic authority even when they do. Even when they intend to do the right and Catholic thing their authority to do so has been greatly curtailed. John Paul II could and did rightly condemned abortion, and yet within the structures of the Vatican organization even he lacks the authority to shut down "Catholics for Pro-Choice," to everyone's lasting regret. Also, because of their detachment from the Catholic Church, the Charism of Infallibility no longer rests upon them. This makes it at least theoretically possible for them to promulgate heresy in the name of their own authority.

Even if one allows that they may be part-time popes, all that means is that they have a "part-time infallibility," which is the same as not having infallibility at all. In the absence of any clear and objective delineation as to which periods are portions of the "part-time" and which are not, "part-time infallibility," like part-time reliability, is the same as fallibility or unreliability. Perhaps a future reliable pope, if he chooses, might comb through the teachings of these unreliable figures and designate certain orthodox statements they have made as being worthwhile.

A key point of the case being made here is that I am not judging individuals. One could quite reasonably make the case for some of these leaders (such as John Paul II) to have been personally, as men, the moral superiors of some actual reliable popes of centuries gone by. What ruins them is their adherence to Vatican II, especially beginning with the Vatican II schema *Lumen Gentium* in which the official detachment of the Vatican organization from the Catholic Church was decreed. That adherence places them each at least partly outside the visible structures of the Roman Catholic Church, and in many cases entirely outside those structures (even while retaining full membership in the Vatican organization), and that is what enables them, like so many other well-meaning religious teachers who are outside the visible unity of the Church, to descend into heresy and schism and to promulgate error.

That is also why the documents from the third phase of the Second Vatican Council were free to contain error. The actual (if not intended) result of these documents was to put a certain distance between the Vatican organization and the Roman Catholic Church. Although formally and legally detached from each other, the Vatican organization and the Roman Catholic Church still substantially overlapped each other when the 1964 session of the Council closed. By voting for the third phase documents, and particularly the one on Religious Liberty which is almost unavoidably heretical, the bishops and cardinals (and pope?) who did so thereby remove themselves entirely from the Roman Catholic Church. Only Marcel Lefebvre and 73 other faithful Roman Catholic bishops voted against the schema on Religious Liberty. All the rest embraced heresy, something they were free to do only now that they were simply leaders of an organization which is no longer identical to the Roman Catholic Church.

The first document of the third phase was the infamous schema on the Episcopacy, in which the new structure of "collegiality" was imposed. The practical effect of this document was to reduce in rank the leader of the Vatican organization from Monarch or King to that of mere President. In the Catholic Church, the authority of the Pope is final; his authority surpasses the authority of everyone else put together. Within the scope of the Vatican organization, according to this document, even though the "pope" still retains more authority than any particular "bishop," he can now be outvoted by any Congress of his "bishops." Where before he was "Monarch," now he becomes "President." The limitations, strictures and partial loss of jurisdiction (namely over those Catholics outside the Vatican organization) so severely constrain the Leader of the Vatican organization that if he is to function as a pope at all, he must do so outside the channels established and defined as "normal" by Vatican II. Since the documents of the third phase are official documents of the Vatican organization, but not of the Church, they need not be further discussed here.

Whether one filters the words and official actions of the Vatican leadership for orthodoxy or rules them out completely as non-popes may not matter much until the Church rules on that, but it is essential that the test by which any statement/teaching/official action/papal candidate is either accepted or rejected MUST be conformance to the Universal and Historic Magisterium of the Church as promulgated by the reliable popes, the twenty Ecumenical and Dogmatic Councils of the Church from Nicea to Vatican I, by all standard and approved catechisms prior to the Second Vatican Council, and the clear teaching and practice of the Church as expounded upon by the doctors of the Church and other approved theologians, and documented in Sacred Scripture and in the ancient Church Fathers. Any other standard is the mark of a non-Catholic.

One other point which must be made is that the Catholic Church did not cease to be a visible Church. Rather, what happened is that where we once had one visible Church, we now have two visible Churches (that is, insofar as the Vatican

THE JURIDICAL IMPACT OF THE SECOND VATICAN COUNCIL, OR HOW THE HIERARCHY LOST CONTROL

organization as it exists now, detached from the Roman Catholic Church, is to be regarded as a "Church" at all). The main problem people often have in recognizing the visible Roman Catholic Church is that their eyes have been fixated on the visible Vatican organization as though it were still the Catholic Church. Train your eyes to be simply oblivious to the Vatican organization, and once your vision adjusts you will see the true, visible, Roman Catholic Church standing out in sharp relief. There is a well-known "optical illusion" that shows a picture of a brick wall with what looks like a small rock stuck into the mortar between two rows of the bricks. But once one can see the cigar in the picture (many seem to need to have it pointed out) the cigar becomes something which cannot be "unseen," forever changing one's perspective of the picture to what is in fact its true nature. Once you see the truth, while you can turn your back to it and even lie about it, one thing you cannot do is unsee it.

The visible structures of the Catholic Church no longer coincide with the visible structures of the Vatican organization. Due to their activity being at least primarily or in many cases exclusively devoted to running the Vatican organization, the visible Catholic hierarchy has been gutted, many members having entirely deserted their sees, and the rest would be most charitably described as seeing to the Catholic needs of their dioceses on at most a part-time basis. What very few truly and consistently Catholic bishops there are have no conventional territorial dioceses and limited recognition, but spread thin as they are, they do their best to see to the Catholic episcopal needs of many of the thousands of remaining Catholic parishes worldwide. Even if virtually the entire membership of the Church hierarchy were to be destroyed (for example by the detonation of a nuclear bomb in the middle of a council), the hierarchical structure of the Church remains a reality mandated and divinely protected by Christ Himself. The Church is a great deal smaller than before, but like Gideon's valiant 300 men, She is only all the more powerful and all the more ready to take on all the evils of the World.

While one would feel much better having intact an entire conventional hierarchy so as to enforce the existing teaching and to answer the questions which have been raised by the current crisis, the fact is that God Himself has been standing in that gap. All one has to do is look in on any of the thousands of traditional parishes and Mass centers all around the world and one can only be amazed by what an incredibly good job God does at this. These parishes overflow with such Graces and Virtues as only God through His Church could ever possibly provide. Just as in any chain of command, if an officer in that chain of command goes AWOL, the next ranking officer fills in for him, and life goes on only slightly disturbed.

The final proof of any "independent" priest's unity with Eternal Rome is the fact that at all times he conducts himself as if he were answerable to a truly Roman Catholic hierarchy. The fact that he may go for a very long time with no access to such a hierarchy is no different than in the old days when any Catholic priest would go for months or even years between visits from his diocesan bishop. The holy priest just functions as if he knows he could be visited at any time. Just because the Catholic hierarchy may not be present at a specific time or place does not preclude one from being in union with that hierarchy.

Even questions as might become something of a bone of contention between bishops do not pose a real problem to church unity since the parties of each side are united in submission to the Pope. One day the question will be put to the Pope and he will answer it and both sides will accept that judgment. Like missionaries of rival religious orders with differing approaches to their work, having to go all the way back to Rome in order to settle their differences (such as the Jesuits and Dominicans as they evangelized China), we too must travel a long and difficult way to visit the next reliable Catholic Pope in order to settle our differences. The only way in which today differs from the old days is that the large distance was merely geographical then, but now it is a travel through time as we journey towards the time of the next reliable Catholic Pope at a rate of sixty minutes per hour.

Let me now dispel one more myth. While it is true that we often hear the phrase "independent Catholic priest," even from these priests themselves and even in reference to themselves, the fact is that no Roman Catholic priest is ever truly "independent." All we can ever validly mean by that is that the priest has no current relationship with the Vatican organization, or else in addition, that neither is that priest currently a member of any of the Traditional Roman Catholic priestly orders (such as I will introduce in the chapters which follow). The fact is that <u>all</u> Roman Catholic priests are members of the visible and hierarchical Roman Catholic Church regardless of whether they may or may not be members in good standing, or at all, of the Republican party, the National Rifle Association, the Coalition for Literacy, the Rushlight Club, or most pointedly, the Vatican organization.

So that's the view or theory by which the things which have happened can all be explained, could even in fact have been predicted back during Vatican II itself. To summarize, where before there was one subsisting society the Church, *Lumen Gentium* legislated that there would now be two parallel and overlapping societies, one still being that subsisting Church, expressly described as possessing the Marks of the Church, and the other being ontologically separate and distinct and explicitly possessing different boundaries than the first, such that the first could "subsist" partially in the other (and partially outside), while the new and second could exist partially in and partially outside the first. Since the divine promises would apply only to the one which is the subsisting Church, the other society would be no longer entitled to such promises, would be free to drift into error and even heresy, would in fact, in seeking a new and distinctive identity of its own, come to differ from the first in its teachings, morals, and to reflect its changed law of belief, also come to differ in its liturgical expressions, and lose the Marks of the Church, as a society. The overlap would mean that at least some few small corners of the new society would continue to tolerate the presence of the first and subsisting society, but not by and large. Nevertheless, the first and subsisting Church would also continue, despite being seriously stripped of so many of its former resources, unchanged other than in size, recognizable, and still possessing the Marks that inform us where and what the Catholic Church is.

All of this could have been trivially foreseen and deduced by the close of that fateful day of November 21, 1964, had anyone back then actually bothered to look closely and carefully at the text of what they promulgated that day. In the chapters to follow, we will see the outworking in history of this new arrangement, first within that majority part of the new organization which is no part of the Church, then with those who properly belong to neither society, and then most importantly with those remaining faithfully with the first and subsisting Church, both those outside the new society and those within it. We will see just how well the view, or theory, stacks up against actual historical events to follow.

4

THE ROMAN CATHOLIC CHURCH VERSUS THE PEOPLE OF GOD

Having lost a substantial portion of their Catholic authority and hierarchical status, the leadership of the Vatican organization gained in exchange the power to teach heresy and propagate evil. This is because Catholic infallibility is always inextricably interwoven with Catholic authority. Where Catholic authority goes away, Catholic infallibility goes with it.

While it is true that quite a number of those in leadership positions in the Church before Vatican II had harbored certain heretical beliefs and all manner of "pet" theories or even formed unholy alliances with Masons, Communists, and others hostile to the Church, the authority they possessed in the Church and the guarantee of the guidance of the Holy Spirit prevented them from making any public acts, mandates, or teachings which were contrary to the Faith. Ambiguous, ill-conceived, unfortunate, dangerous, or even malicious, yes; explicitly contrary to Faith and Morals, no. Once they signed away a portion of their authority at Vatican II however, they also signed away the guidance and protection of the Holy Spirit, and in that gained the power to make evil mandates, teachings or public acts. The Charism of infallibility is not like the priesthood which is an indelible mark on one's soul which one retains throughout eternity, but only something which one has while they function as pope. When Pope Celestine V signed his resignation, he lost not only the Catholic authority he had exercised as Pope, but also the Charism of infallibility which he had enjoyed as Pope. Though a portion of the Church continues to subsist within a tiny portion of the Vatican organization, the remainder of the Vatican organization, the many portions not subsisted in by the Church, has defected starkly from the religion of Christ. This chapter is the story of that major portion of the Vatican organization which is no longer Catholic.

Most Catholics first became aware that something was amiss when they started noticing changes being made to the Mass as they knew it and architectural changes began to defile the Church buildings. Conservative Catholic publications such as *The Wanderer* would repeatedly say of each new aberration that it would be the last ever to be tolerated, only to be later contradicted by further outrages. Tabernacles were torn out from their place and installed in nearby closets and basements; altars were separated from the wall and then later smashed into powder and replaced with funky wooden tables; altar rails were removed, along with statues and historic stained-glass windows. Even Protestants, brought in to do the actual altar and stained-glass smashing (after promises that the stained-glass windows would simply be shined up by having the dust and grime cleaned off), have been known to shed tears over having been hired to perform such a blatant sacrilege. Communion began being handed out by "helpers," and then "in the hand," and finally altar girls have made an appearance. Parishioners soon came to sense that their obedience was being exploited for nefarious purposes, and they then realized that they have no power to stop it from happening. For those reasons it can properly be said that "Malaise is a defining characteristic of the new Vatican organization."

In some parishes, the problem becomes more obvious as the "priest" presiding might dress like a clown or even use hand puppets, stuffed animals, balloons, or food fights. Admittedly, those more extreme things have come to be categorized as "abuses" which go beyond the liturgical norms, even for their new "Mass." To their credit, the highest echelons have even spoken out against some of the more extreme abuses, but no real authoritative action has been taken to stop them from happening. How can it, since they regrettably no longer have the authority to do any good anyway? I will not elaborate with further instances of abuses and horror stories, such as "liturgical dance," etc. since practically everyone in contact with the new Church has been exposed to many such abuses and, I am sure, could tell quite some stories of them. However, even the craziest abuses contrived by some "Father Bozo" have a place in the plan of God. Their purpose is to draw attention to the fact that they are not Catholics anymore.

Even in certain traditional countries where such abuses have not as much occurred, there has nevertheless been a tremendous power shift. Where before, the high standards of the Catholic traditions not only inspired people to convert, but even inspired many of those who didn't convert to at least seek the highest and noblest expressions of their own traditions, now it is whatever nobility remaining in the surrounding culture's pagan traditions that sustains the local "Catholic" parish's reverence. Directly, the problem is that the new "Mass" has such loosely defined rubrics and forms that practically anything is allowable from clown and puppet masses to Voodoo and Santeria masses clear to reverent, dignified masses even said in Latin, serenaded by Gregorian Chant and having an almost fully Catholic flavor to them. In promulgating such a loosely defined service, the formerly Catholic leadership has once again, in yet another way, abdicated any right or power to tell Fr. Bozo that he can't clown around.

There are some who think that it's enough to push for more reverence to be used, but that thinking has two basic fallacies (not counting the false assumption that the Vatican organization promulgating it was still the Catholic Church). The first fallacy is that as long as the new "Mass" is the official worship of the Vatican organization, clown, puppet, Voodoo, and balloon masses will also be every bit as "legitimate" as reverently said Latin Novus Ordo masses complete with Gregorian Chant, and perhaps even more so in some cases. One Bishop Beezlebub somewhere in Canada was even known to have forbidden the Franciscan Friars of a monastery in "his" diocese to say <u>any</u> mass in Latin, even the Novus Ordo! Clearly, that diocesan See is entirely vacant!

The second fallacy with the approach of pushing for "more reverent masses" in the Novus order format pertains to how the General Instruction included with the original promulgation of the new Mass actually defines "the Mass" as "7. The Lord's Supper is the meeting or congregation of the people of God assembled, the priest presiding, to celebrate the memorial of the Lord. For this reason, Christ's promise applies eminently to such a local gathering of holy Church: 'Where two or three come together in my name, there am I in their midst' (Mt. 18:20)." (official ICEL translation; the Latin original states the same, except that the opening phrase reads "The Lord's Supper, or Mass, is the sacred meeting…" By contrast, the Catholic Mass is the re-presentation of the crucifixion of Christ, a propitiatory sacrifice in which the priest, as Alter Christus, transubstantiates the bread and the wine into the Body and Blood and Soul and Divinity of Christ. So heretical was that original definition that within a year a new definition had to be substituted, one less obviously heretical, though still rather weak and blended with Novus Ordo-isms.

The third fallacy is the seriously flawed nature of the Novus Ordo prayers themselves, along with their associated rubrics. Under the acid rain of Annibale Bugnini and his committee, nearly every prayer of the Catholic Mass was simplified, watered down, and deprived of any and all pith. For example, prayers for the intercessions of the Saints which not only reiterate but draw their strength and effectiveness from the truth of the Catholic doctrine of the Communion of the Saints are replaced with prayers which vaguely ask for a nice day.

The most central and basic prayer is called the Canon of the Mass. This prayer, which grew gradually and organically over the course of the first six centuries, was officially canonized at the Council of Trent after remaining utterly stable and unchanging for nearly a thousand years. Acts of canonization are by definition acts of the Infallible Magisterium of the Church and as such irreversible and irreformable. Canonization is that process by which the Church formally recognizes that something or someone is perfected, be it a saint, a scripture, or a prayer. That which is canonized <u>cannot</u> be officially changed, since any such change would put it in an imperfect and therefore inferior state and would furthermore carry with it an implicit denial that it had ever been canonized in the first place. A canonized saint cannot be "de-canonized," even though the observance of that saint's feast day can be displaced from the calendar by some new saint or feast day. To come out with even one let alone several new "canons" of the Mass is every bit as nonsensical as coming out with one or more new "canons" of Scripture. The four "canons" of the new "Mass," along with all other new canons, official or improvised were all ruled out for all time when the prayers of the consecration of the Mass were canonized at the Council of Trent.

Even Canon Number One, which is sometimes called the "Roman Canon" because it retains at least a somewhat similar sequence of prayers to the prayers of the Catholic Canon of the Mass, is no good because of something done at the very core, with the words that Jesus Christ Himself said, as repeated at every Catholic Mass of every Rite, Eastern and Western, and as recorded in Scripture. But the same manner of damage (or worse) is found in the alternate "canons."

At so sacred, precious, and intimate a moment of the Mass, He is speaking not to the "all" for whom He died in order to offer them an opportunity to be saved, but only the "many" who would actually accept His gift. What has been said from the days even before the Bible was written is as follows: "Who the day before He suffered took bread into His holy and venerable hands, and with His eyes lifted up to Heaven, to Thee, God, His almighty Father, giving thanks to Thee, He blessed, broke, and gave it to His disciples, saying: <u>Take and eat you all of this.</u> **FOR THIS IS MY BODY.** In like manner, after He had supped, taking also this glorious chalice into His holy and venerable hands, again giving thanks to Thee, He blessed and gave it to His disciples saying: <u>Take and drink you all of this.</u> **FOR THIS IS THE CHALICE OF MY BLOOD OF THE NEW AND ETERNAL TESTAMENT: THE MYSTERY OF FAITH: WHICH SHALL BE SHED FOR YOU AND FOR MANY UNTO THE REMISSION OF SINS.** <u>As often as you do these things, you shall do them in memory of Me.</u>" The underlined words here are the exact words of Christ Himself during that famous "last supper." The rest, describing what actually took place on that momentous occasion, have been recited by the Church ever after.

It is worth pointing out that the four descriptions of this event in the Bible not only differ somewhat from this, but from each other as well. Since they are describing a single event which happened only one time, and not several or many separate events which could have varied from each other, it is clear that each is telling only part of the story, and indeed different parts. Each biblical account as given is a proper subset of the prayers and words of Christ as given above. The only phrase missing from all four biblical accounts is "the mystery of faith." Although slight variations to the above have been shown in the Eastern Liturgies, all such variations are clearly traceable to characteristics of the distinctive grammar of the various languages in which the Eastern Rite Mass is said. The point of all of that is that it must be believed that the above underlined words are exactly what Christ said on that occasion and the remainder, an accurate description of what He did while saying these things.

Perhaps you have noticed that the way this prayer is was said at the Novus Ordo "Mass" from 1985 clear until 2011 is not quite what you just read. In fact, it read (in Canon One), "The day before He suffered, he took bread in His sacred hands and looking up to Heaven, to you, His almighty Father, he gave you thanks and praise. He broke the bread, gave it to His disciples, and said: <u>Take this, all of you and eat it:</u> **THIS IS MY BODY** <u>which will be given up for you.</u> When supper was ended, He took the cup. Again, He gave You thanks and praise, gave the cup to His disciples, and said: <u>Take this, all of you, and drink from it:</u> **THIS IS THE CUP OF MY BLOOD, THE BLOOD OF THE NEW AND EVERLASTING COVENANT. IT WILL BE SHED FOR YOU AND FOR ALL SO THAT SINS MAY BE FORGIVEN.** <u>Do this in memory of Me.</u> Let us proclaim **THE MYSTERY OF FAITH**: Christ has died, Christ is risen, Christ will come again/ Dying you destroyed our death, rising you restored our life, Lord Jesus, come in glory/ When we eat this bread and drink this cup, we proclaim your death, Lord Jesus, until you come in glory/Lord, by your cross and resurrection you have set us free, you are the Savior of the world" (emphasis and capitalization mine).

Here, I shall chart out just what happened to one phrase of this in going from the original to the Novus Ordo:

- **FOR YOU AND FOR MANY**—Jesus' original words as said at the original "last supper" and said by the entire Church, East and West alike, until the coming of the Novus Ordo Missae

- **FOR MANY**—Jesus' words as recorded in Matthew and Mark (see that only part of it is there, "for you and" was omitted)

- **FOR YOU**—Jesus' words as recorded in Luke (see that only part of it is there, "and for many" was omitted); Paul omitted the entire phrase from his first letter to the Corinthians, although he supplies a similar phrase in connection with the bread which has never been part of the sacramental formula in any part of the Church at any time.

- **FOR YOU AND FOR MANY**—Latin Novus Ordo Missae (at least they got that part right; nevertheless, the phrase "the mystery of faith" is displaced in the Latin as well as in the vernacular)

- **FOR YOU AND FOR ALL MEN**—Original ICEL vernacular translation of this phrase in the Novus Ordo (notice in English the use of the word "men" which phonetically sounds similar to "many"; that may have been done in an attempt to hide the word "all")

- **FOR YOU AND FOR ALL**—Leading ICEL vernacular translation of this phrase in the Novus Ordo used for decades; omission of the word "men" done in order to make it more "inclusive" of women as well

- **FOR YOU AND FOR MANY**—Current vernacular text, released in 2011 after 26 years of the previous leading ICEL translation and 43 years of "for all" in some form or another; but the "mystery of faith" is still transposed to the end as a separate sentence.

As you can see here, even the Bible shows up the deficiencies of the ICEL vernacular translation of the Novus Ordo Missae. Pope Benedict XIV in his encyclical *De Sacrosanctae Missae Sacrificio* confirmed the teaching of Saints Thomas Aquinas and Alphonsus Liguori and the *Catechism of the Council of Trent* that "for you and for many" is the correct understanding of Christ's words, and that "all" is not permissible. Even more obvious than that, one can see that Annibale Bugnini and Co. had no objection to changing Christ's words even as found in sacred Scripture. And though "all" has finally been replaced with "many" after all those years, so much else remains off kilter: the phrase "Mystery of Faith" remains transposed to the end, and in such a manner as to introduce such small verses as "Christ will come again" (as if He was not already present in the newly-consecrated Eucharist) and the whole thing still reads as a "narrative" rather than an enactment by the priest as "Alter Christus." The passage is still given in undifferentiated text as if the presider is merely giving a "narrative" rather than confecting transubstantiation, and the remainder of the ceremony continues to reflect no sacramental intention as the Catholic Mass does. So while the changes to the consecration formulae are the most crucial change in the new "mass," there are many other changes which I will here illustrate with just a few examples:

Early in the Mass, one encounters a prayer, known as the Confiteor, which goes like this: "I confess to almighty God, to blessed Mary ever Virgin, to blessed Michael the archangel, to blessed John the Baptist, to the holy apostles Peter and Paul, and to all the saints, that I have sinned exceedingly in thought, word and deed, through my fault, through my fault, through my most grievous fault. (strike breast three times here) Therefore I beseech Blessed Mary ever Virgin, blessed Michael the archangel, blessed John the Baptist, the holy apostles Peter and Paul, and all the saints, to pray to the Lord our God for me." Experienced altar boys will remember saying also "and to you, Father" after each occurrence of the phrase "all the saints." Priests insert the phrase "and to you, my brothers" at this point. It has long been known, taught, and understood that the setting for this prayer is that one has just been brought into Heaven, before the very throne of God. Aware of one's own unworthiness to appear there, one begs His angels and saints seen there to "pray to the Lord our God for me."

Contrast that with the Novus Ordo version of this prayer: "I confess to almighty God, and to you, my brothers and sisters, that I have sinned through my own fault, (Strike breast once here) in my thoughts and in my words, in what I have done, and in what I have failed to do; and I ask blessed Mary, ever virgin, all the angels and saints, and you, my brothers and sisters, to pray for me to the Lord our God." Notice the change in position. Where the first takes place in Heaven, in the very sight of God, His angels, and His saints, the second clearly takes place only on the earth, in the Novus Ordo church building, and surrounded by "you, my brothers and sisters." God, His angels, and His saints are only addressed in the Third Person since they are not present. Why would they be?

I take the next example from the Libera Nos, a prayer which is said after the Pater Noster (Our Father). "Deliver us, we beseech Thee, O Lord, from all evils, past, present and to come, and by the intercession of the blessed and glorious ever Virgin Mary, Mother of God, together with Thy blessed apostles Peter and Paul, and Andrew, and all the saints, mercifully grant peace in our days: that through the bounteous help of Thy mercy we may be always free from sin and secure from all disturbance. Through the same Jesus Christ Thy Son our Lord, who liveth and reigneth with Thee in the unity of the Holy Ghost, one God, world without end." The Novus Ordo equivalent of this reads as follows: "Deliver us, Lord, from every evil, and grant us peace in our day. In your mercy keep us free from sin and protect us from all anxiety as we wait in joyful hope for the coming of our Savior, Jesus Christ."

The original ICEL translation of the preface (the prayer which goes right before the Sanctus (Holy, Holy, Holy)) which was to be used whenever the fourth of the new "canons" was to be used began like this: "Father in Heaven, it is right that we should give You thanks and glory. You alone are God, living and true." That reading clearly and indisputably affirms the Arian heresy that Jesus and the Holy Spirit are not God, but only the Father "alone." This translation was fixed in late 1985 (the same time that "all men" was replaced with "all" in the words of Christ) to read, "Father in heaven, it is right that we should give you thanks and glory: you are the one God, living and true." That current ICEL translation, like the Latin original of that Novus Ordo prayer is now rendered ambiguously open to an orthodox interpretation, but the explicitly heretical version had been allowed to stand for nearly seventeen years.

One can chart a logical progression from the slight, trivial changes made in the few years preceding the Council, and even the tolerable if strongly ill-advised changes recommended at the Council in the Constitution on the Sacred Liturgy (1963), on into the more serious changes made more or less in accordance with that conciliar document during the mid-1960s, and clear through the seriously deficient Latin Novus Ordo Missae which itself is in turn less harmful than the vernacular translations (such as ICEL's) which introduce yet more problems, and finally culminating in Fr. Bozo's "mass" littered with abuses of every sort.

Even the smallest and seemingly harmless changes at the top have a way of becoming drastically serious by the time they work their way to the bottom. It is just like the rotation of a wheel: a small movement near the hub results in a large movement at the rim. Now, the hub is Christ, who cannot be moved even slightly from His place at the right hand of the Father, but the rim is where the faithful reside. Popes and Ecumenical Councils are so very close to the hub that a change (movement) which would have been of no noticeable consequence if done at the rim would be catastrophically significant if done so very close to the hub. Such a continuum from the slight evil at the top to the significant evil at the bottom has been an excellent tool for Satan to use to divide the Catholic faithful who seem to be at a loss to agree as to where to draw the line.

For those who might be curious as to what liturgical changes took place in the Mass before the Council, here is a thumbnail sketch: In 1955, Pope Pius XII in a sadly mistaken attempt (no doubt at the advice of his secretary Giovanni Battista Montini, the future Paul VI) to gain control over a growing faction which ostensibly existed to study the origins of the details of the Liturgy, but which was actually nothing but a group of subversives bent on destroying the Church, decided to give his approval to a small revision they had proposed to the Holy Week Liturgies, namely the omission of a couple readings and the shortening of a couple other readings, and permission to do the Saturday (before Easter Sunday) Liturgy in the evening. As a diplomatic gesture, that should have been enough to satiate any reasonable requests that anyone could ever care to make, but after his death it was used as a precedent for further changes.

In 1960, John XXIII gave approval to a not quite so slightly revised Missal in which the opening prayers at the foot of the altar and the closing reading from the Gospel of John were suppressed on certain days, several Collects (the distinct daily prayer before the readings) were deleted, ember day lessons became optional, several Feasts, such as Saint Peter's Chair in Rome or the Finding of the Holy Cross or the Seven Sorrows of Our lady were either abolished or downgraded, and many Octaves and Vigils of Feasts were also abolished. In 1962, John XXIII added the name of Saint Joseph to the list of Saints in the Canon, changing a prayer which had been utterly unchanged for over 1400 years, and which had furthermore already been canonized (perfected) in the form it had enjoyed all those years.

So far, I have only used portions in the unchanging parts of the Mass to illustrate the difference between the Catholic (Tridentine) Mass and the Novus Ordo (the "New Order") "mass." There are also the prayers which change from day to day over the course of the Church's liturgical year and seasons, namely the prayers of the feasts and the saints on the calendar. First of all, one finds many saints shuffled around from one day to another, many more deleted, and their prayers changed, in the Latin as well as the vernacular. I provide here as an example only one of the prayers in honor of Saint Gertrude the Great. "O God, who didst prepare for thyself a pleasant home in the heart of the holy virgin Gertrude…" both old and new start, as translated directly from the Latin. The old continues, "by her merits and intercession do Thou mercifully wash away from our hearts the stains of sin and grant that we may rejoice with her in heavenly fellowship." The new continues, "by her intercession do Thou mercifully enlighten the darkness of our hearts that we may joyfully experience Thee working and present within us."

In the book *The Problems With the Prayers of the Modern Mass* (See Bibliography) the changes in the prayers for specific days, feasts, and saints are covered in far more detail. A specific set of themes of the changes has also been noted, namely that certain topics are changed or deleted wherever they occurred:

1. "Negative Theology," which has to do with the less pleasant features of our faith such as Hell, Purgatory, Sin, Repentance and Penance, and so forth, is cut way back,

2. Detachment from the World, which means that where the hope of the Catholic used to be on Heaven, is no longer mentioned; in the new religion it is in an earthly hope,

3. Prayers for the Departed, in which we help the dead and are therefore reminded that they need our help, now deleted,

4. Ecumenism, which here means that anything which Jews or Protestants would rather not hear is cut way back, and some new things are even done out of deference to them, such as adding a couple prayers from the Jewish Shabbat service, or adding to the "Our Father" a doxology popular among Protestants owing to its presence in their King James Bible,

5. The Merits of the Saints, by which their sufferings and sacrifices can actually help us now and after death have been completely deleted, and

6. Miracles, by which God intervenes in human affairs, sometimes in a dramatic fashion, have been cut way back.

Furthermore, the very words, "Grace" and "soul" have been systematically eliminated from all the prayers, both common ("Ordinary") and daily changing ("Propers"). All of that is just yet another mark of the Novus Ordo being a soulless religious service devoid of any capacity to convey any Grace.

Some might allow that such changes were obviously ill-advised, but still claim that they were allowable on the basis that worship has always been subject to the discipline of the bishops, cardinals, and pope of the Church. Since discipline, unlike Faith and Morals, is always revocable and changeable, subject only to legitimate Church authority, many assume that the changes made to the worship in the Novus Ordo Missae are allowable. To some extent that is true, but only to a certain point. Worship contains some elements which are by their nature subject to discipline, but other elements of worship enter the arena of Divine Revelation which is higher than Faith and Morals. Tampering with Divine Revelation is such a serious thing that not even a pope has the authority to do that, let alone any mere "president" of the semi-Catholic Vatican organization. I point that out because some people tend to treat the exact manner of worship as being merely "disciplinary," which is <u>inferior</u> to Faith and Morals in roughly the same sense that Divine Revelation is <u>superior</u> to Faith and Morals.

This issue of discipline, Divine Revelation, and worship bears some discussion. There are some very superficial aspects of worship which do in fact fall merely into the category of discipline, such as what language is to be used or which particular saints will be honored in which geographical areas. The use of the Latin language itself is not really the issue. There are many good reasons that the Church has so long used Latin for all of its official documents, actions, and liturgies (excepting only the liturgies of the Eastern Rites). The problem is not the bare fact of using the vernacular (especially for those portions of the Mass which are spoken to the people, such as the homily or the Scriptural readings), but that a) the Latin text itself was corrupted, and b) a further corrupted translation was made into the vernacular languages. Most curious and interesting of all here is the fact that all the same further corruptions show up in virtually all of the vernaculars: The "for many" of the Latin (even the Latin Novus Ordo) became "for all" in nearly every vernacular translation (Polish is one of the minuscule handful of exceptions). And in every vernacular, the phrase "the Mystery of Faith" was transferred to a new sentence and introduced with the added words "Let us proclaim the…"

There are several reasons put forth by the Church as to why She has used Latin for Her official language for so long. From a theological standpoint, the main reason is that since Latin is what is commonly referred to as a "dead" language, it is no longer in a state of flux. This way the meaning can stay absolutely the same across all the many centuries the Church shall endure so that "the Faith which was once for all delivered to the Saints" may be perfectly preserved clear to the end of time. Furthermore, nearly all the great questions the Church has faced during Her history were settled in the Latin language, and so therefore these infallible declarations of popes and councils are kept alive in the exact words and meanings originally spoken. Finally, Latin is a source of global unity for the Church. The Mass is the same almost no matter where one goes and can also be attended by persons of differing vernacular languages and of equal value to all. Nevertheless, it must be conceded that the use of the Latin language is only a disciplinary concern, revocable at any point no matter how ill-advised that would be. The changes in what the prayers literally say and mean is quite another matter.

When the books of the Bible were being canonized, the criteria of this canonization was not only so much a matter of whether a book was written by this or that person or in this or that time, but whether the book had a good effect on the faith of those who were listening to it being read as part of the Liturgy. Several books which were prominent in the early Church and of indisputable authenticity, such as the Epistle of Barnabas or the Revelation to Peter, were ultimately excluded from the canon for that reason.

The books of the Bible were chosen by the pastoral leadership of the Church for their usefulness in Christian Liturgy. All of the things said in the course of Christian Liturgy (other than the Homily and any special announcements or prayers) fall into one of only three categories. There are prayers, such as the Canon of the Mass, which are always prayed at every Mass and which were pretty much fixed by the fifth century after they had been honed and refined to perfection under centuries of persecution (otherwise known as the "Ordinary" of the Mass);

there were prayers associated with particular saints or feasts which were written when these saints were canonized and these feasts were added to the calendar (otherwise known as the "Proper" of the Mass); and finally there was scripture to be read. The Bible is therefore actually a part of Christian Liturgy. It is therefore a part of a larger corpus of works which, collected and taken in their entirety, constitute the entire Liturgy, of which any actual Liturgy used in any actual time and place must be a proper subset.

The Bible falls under the category of "Scripture," and the rest of the Liturgy, since it was only orally transmitted in the beginning and committed to writing later, is a prominent portion (along with the other writings of the ancient Church Fathers) of what is called "Tradition," which the Protestants fault us Catholics for "adding" to Scripture. One fact from which we can see that the prayers of the most ancient Liturgies, like the Bible, must never be changed is that virtually all of the great Church Doctors and theologians have argued from details, even in the very turn of phrase, of various Liturgical prayers, as one would from Scripture, something one would not do if the content of those prayers were changeable at the whim of any Church authority.

There is a place for discipline in all of this, and that pertains to the selection of a particular subset of the totality of Liturgy which is to be used in this or that time and place. A good example of this would be the Gospel reading for Good Friday. It is a matter of Church discipline that the reading for the Passion should be taken from the Gospel of John instead of Matthew or Mark or Luke, but it is certainly not within the domain of Church discipline to rewrite the Passion as given in John (or Matthew or Mark or Luke), nor to substitute for it an excerpt from *The Last Temptation of Christ*.

Likewise, it is a matter of Church discipline to choose which (if any) saint is to be honored on a given day shared by two or more, or even to grant permission for one saint to be honored in one geographical area and the other saint in another, but it is beyond the domain of discipline to rewrite the distinctive prayers used to honor any of those saints. Another example: In certain Eastern Liturgies, there comes a point at which the priest exclaims, "The doors, the doors!" The doors he is talking about actually separate the altar from the faithful in attendance. The priest must go through these doors to the altar where he can confect the sacrament. Since Western Church architecture seldom has such doors, the discipline of the Western Liturgy has long been to omit that exclamation. Certainly, it is within the Pope's prerogative to apply or forbid the Eastern practice universally (either of which would be obviously imprudent), or to sustain the status quo, but he has no authority to change that exclamation to "The windows, the windows!" One more example: It is a matter of discipline as to whether or not the Gospel of John is to be read at the end of the Mass, but it is beyond the domain of discipline to rewrite John or to replace it with a reading from Shirley MacLaine.

What would one say of a Bible translation which was 99.95% accurate, but the remaining 0.05% error consisted of a systematic removal of each and every reference regarding what Christ's death on the Cross achieved for us? As Cardinal Ratzinger wrote (in his preface to Klaus Gamber's book, *The Reform of the Roman Liturgy* (see Bibliography), "in the place of liturgy as the fruit of development came fabricated liturgy. We abandoned the organic, living process of growth and development over centuries, and replaced it—as in a manufacturing process—with a fabrication, a banal on-the-spot product."

The new Mass as promulgated by Paul VI therefore goes well beyond the domain of Church discipline. It literally changed and negated a large portion of the Revelation as given to us by God, including Christ's words at that crucial point illustrated above, and is therefore a change of not only doctrine, but even the very source on which doctrine is based, namely Divine Revelation.

The changes in the prayers of the Mass are only the initial drops of poison into the recipe for saying the Mass. That poison has diluted and spread to many other areas of the Mass, including the rubrics and even the music. In the area of rubrics, the priest(?) faces his audience virtually the entire time, prays the Eucharistic prayers aloud, and in such a manner as to include the entire audience as participants and as if he is telling them a bedtime story, not enacting a sacrament. One sees in this no mark of the priesthood he should have received at his ordination. He faces the people; the people face him, and in some (more and more) places the people even face each other. Altar and sacrifice are reduced or even eliminated and replaced with table and supper. All of these actions positively reek of a Man-centered religion rather than a God-centered religion. Having communion placed in one's hands cannot help but impart the impression that it is after all merely ordinary bread or merely symbolic of Christ's Body instead of actually being Christ's Body. No wonder consecrated hosts have frequently turned up under the pews, or even in the pockets of clothing (found at laundry-time)!

Let us now take a look at what else of significance happened when they changed Christ's words. You might have noticed that some of His words as given above were in **BOLD CAPITALS** while the rest were not. In old (pre-Vatican II) Missals, the exact same words are usually made to stand out in a similar way. The clear consensus of the Church Fathers and Doctors, scholars and theologians, has long been that those words, no more and no less, constitute what is called the "form" of the sacrament. The "form" of a sacrament is one of the four essential ingredients necessary in order for a sacrament to be what the Church calls "valid."

The validity of a sacrament is strictly a measure of whether or not the sacrament actually took place, i. e. if it "worked." An invalid sacrament is one which did not work. In the case of the Mass, if the sacrament is valid, you no longer have bread and wine on the altar, but the actual Body, Blood, Soul, and Divinity of Christ our Lord and those present receiving that communion have literally eaten His flesh and drunken His blood as He commands (John 6:53). If it is not valid, you still have merely bread and wine and those who consume it have received what amounts to a continental breakfast. Four things must all be correct in order for any sacrament to be valid:

1. Form,
2. Matter,
3. Intention, and
4. the Minister of the sacrament.

The form is the words actually spoken, the matter is the physical material used, the intention is the intended purpose or goal of administering the sacrament, and the minister of the sacrament must have certain qualifications. If any of these things is not correct, the sacrament is not valid. The bread and wine do not become the Body and Blood of our Lord; the couple do not marry; the penitent is not absolved; the dying soul is not healed; the soul of the recipient receives no indelible mark of Baptism, Confirmation, or Holy Orders.

The Novus Ordo changes the words, thus gravely damaging the form of the sacrament of the Mass. Providing that the other three considerations are met, the priest who says Christ's words at this point does the work of Christ. God Himself has promised that. That is what is meant when the Church teaches that the priest (or bishop) is at that very moment "*alter Christus*," another Christ. On the other hand, there is no promise from God regarding a priest who says the words of Annibale Bugnini, ICEL, and Co. instead at that point, since they are not Christ's words. Even though the words "Thou shalt commit adultery" are each quoted from a passage in God's word, as presented in this sentence they are not God's words. It is the omission of one of God's words in the middle, namely "not," which alters the meaning and renders it mere Man's words.

One small objection worth addressing here is the claim made by some that changing "many" to "all men" or "all" and removing the phrase "the mystery of faith" to another place could be harmless to the validity of the sacrament because one can sometimes find respectable scholars who seem to list the words "this is my body; this is my blood" as being all that is needed, thus seeming to be able to do without the words, "of the new and eternal testament: the mystery of faith: which shall be shed for you and for many unto the remission of sins," from the necessary form of the sacrament. An example of this would be Ludwig Ott's *Fundamentals of Catholic Dogma* which says at one point "the words of instruction demonstrate… that at the Last Supper Jesus effected the transmutation with the words, 'This is my body; this is my blood.'" That may seem to threaten the case I am making here, but the solution is readily apparent. The phrase "this is my blood" as used by Ludwig Ott here and also by others as well is in fact a kind of shorthand for "for this is the chalice of my blood of the new and eternal testament: the mystery of faith: which shall be shed for you and for many unto the remission of sins." One can see that it would be quite tedious to write out the entire form every time one wants to refer to it. As another evidence that "this is my blood" is only a paraphrase with which one refers to the entire form, neither Jesus Himself at that Last Supper nor any priest in any Rite of the Church since then has ever said "this is my blood," but always "this is <u>the chalice of</u> my blood…" Finally, *De Defectibus*, a short papal decree printed in the front of all old Missals for priests to use in the Mass, states that the necessary and unalterable form of the sacrament is given as the long form as shown in **BOLD CAPITALS** in this chapter.

One is therefore permitted to doubt that the Novus Ordo is even a Mass at all. The validity of such a new "mass" is one of the more difficult questions put before Catholics today. For 43 years people had been expected to accept hosts consecrated "for all" (or "for all men"), expressly at variance with the words of Christ, and of itself gravely imperiling validity. While at least some more recent editions appear to have repaired this problem (but to what extent has that fix been rippled to all editions and also translations into the other languages?), the "Mystery of Faith" remains transposed and repurposed. At this point, I will only provide the reader with a relatively vague, general compromise consensus on this issue. At the one extreme, even the most hard-core traditionalists might grudgingly have to concede the possibility that maybe some very few Novus Ordo masses said under all of the most optimum conditions possible may have been valid. At the other extreme, many of those who are caught up in the new Novus Ordo religion, yet still retaining just enough orthodoxy to understand and believe in the concept of validity with respect to sacraments have in fact admitted that some instances of the new "mass" have definitely been invalid. The difficult question is, "Where, between those two extremes, does one draw the line between validity and invalidity?"

My own opinion here is to draw a distinction between intrinsic and extrinsic validity. Intrinsic validity applies to the Tridentine and Eastern Rite Masses which have God's promise resting upon them. Providing that all four conditions are met,

God <u>must</u> and <u>will</u> honor an intrinsically valid mass with His Divine Presence. That is like someone coming up to you and asking for some money which you have already promised to pay him. Justice obliges you to pay up regardless of your feelings about the person in question. Extrinsic validity is like someone coming up to you and begging for you to please give him some money. Out of charity or mercy you may see fit to give him some, but there is no obligation in justice for you to do so. Extrinsic validity applies to the Novus Ordo Missae. God never promised His Eucharistic Presence at such a man-made prayer meeting, but filled with compassion and mercy, especially for those misguided souls who are honestly trying their best to be good Catholics but who just simply don't know how to do it, one could speculate that it is up to the mercy of God whether to see fit to honor some of these meetings with His Presence, especially where deep reverence for the Blessed Sacrament remains, though obviously that has no reliability.

The next aspect of the validity issue is the matter, in the case of the mass, the bread and wine which are to be used. It is impossible for any priest to transubstantiate an Oreo cookie and Kool Aid into the Body and Blood of Christ, valid ordination, correct form, and intent notwithstanding. Here, at least, the official (albeit unenforced and unenforceable) rules of the Novus Ordo are still within acceptable bounds (providing we ignore the times that John Paul II allowed parishes in Africa to use cakes of millet or cassava root and corn wine).

Another aspect of the validity issue is one regarding the intention of the minister of the sacrament. The other three considerations are more or less external and readily verifiable as to whether or not they have been observed. Intention resides entirely within the mind and heart of the priest himself. Nothing he says or does can absolutely establish just what his intention is. That would seem to get a little spooky since it is always possible for a priest to say a perfectly correct Tridentine mass and yet have no intention to confect the sacrament and so thereby render it invalid. To prevent this issue from posing a danger to the Church, it has been ruled that a "presumption" must be made that provided the form if full and correct and the priest's own training and qualifications verified, the priest's intentions are to be taken as correct, unless there is clear, publicly available evidence to the contrary. For example, if a priest states at the beginning of the "mass" that he has no intention to confect the sacrament or "do what the Church does" then that "mass" is invalid, no matter in what Rite, Eastern or Western, Tridentine or Novus Ordo, it is said.

Given the heretical training one typically finds in Novus Ordo seminaries, the "presumption" concerning intent works the opposite way for the Novus Ordo Missae. One must presume that the intent is defective unless there is clear, publicly available evidence to the contrary. Where the intention is defective, whether by presumption or by clear, publicly available evidence, the "mass" is invalid. There is one other consideration regarding the validity of the new "mass." That is the fact that the new "mass" is clearly a parody or satire of the Catholic Mass, very sacrilegious at best and explicitly blasphemous at worst. In some cases there might even be a certain horror at the prospect that some instance of the new "mass" might actually be valid, sort of like encountering the warm carcass of a freshly mutilated animal; you can't help but hope that the poor creature is not still alive. Given the irreverence frequently shown towards the "Blessed Sacrament" (?) in Novus Ordo churches, any true Catholic might rather hope that it wasn't validly consecrated, at least in those instances.

Has any defense ever been put forth for the Novus Ordo? A remarkable and significant fact is that only one defense has ever been put forth for it. This defense boils down to the following: "Pope Paul VI promulgated it, giving it his fullest support and approval. Since the Pope cannot err concerning something so very important, it must somehow be perfectly fine, good, edifying, and valid. Who ever heard of a pope promulgating invalid or irreverent sacraments?" As I have already explained in the last chapter, Paul VI had already relinquished his and his hierarchy's exclusive claim over the Catholic Church, a significant portion of his authority, his papal role, his infallibility and the guidance of the Holy Spirit. Had he been able to promulgate the Novus Ordo "sacraments" while yet retaining all of that, it would have been the End of the Church and Christ's promises to be with the Church until the end of time would have been proven false right there.

Without that one argument, just imagine trying to defend the Novus Ordo Missae. You can't, and neither can I nor anyone else. In particular what is lacking is any semblance of a scholastic argument in its defense. All other masses said by the Church, both Tridentine as well as each and every Eastern Rite in their authentic and traditional forms (Byzantine, Malabar, Antiochian, Monrovian, etc.) have a clear and well documented pedigree. One can easily chart their valid change as they grew and achieved final stability into their current forms.

By contrast, the Novus Ordo Missae has absolutely no precedent anywhere within the entire history of the Church. One can find a pedigree for it which goes back a few centuries, but this pedigree is of no use to those who would want to defend it since this entire pedigree takes place *outside* the Catholic Church. In particular, this pedigree consists of several "masses" written by Thomas Cranmer, Martin Luther, John Calvin, and the like. The details of this pedigree have been admirably demonstrated in *Cranmer's Godly Order* (See Bibliography) for those who wish to know more about it.

If I have talked rather long and hard about the Mass, it is primarily because that aspect of the new Church is what has been most scrutinized and studied by the Traditional Catholic community. It is also because the Mass is the sacrament most often seen in the ordinary day-to-day life of the Church. The liturgy is also the means by which the Church, both as a whole and individually in every Catholic, professes our Faith in the worship of our God. A changed liturgy creates and reflects a changed Faith. *Lex orandi, lex credendi* ("the law of what is to be prayed is the law of what is to be believed"). The fact is that what was done to the Mass was also done, in varying degrees, to each of the other sacraments as well. Their validity (in many cases) is also imperiled. Allow me to step through the other sacraments and point out what was done to each:

While the form and matter of Baptism are retained, many of the new priests have been taught that Baptism "welcomes the recipient to the Community of the People of God," rather than that it washes away all sin and all penalties (in the next life) for such. Since they baptize with the wrong purpose in mind, it must be presumed that the intent is not valid unless that priest has clearly demonstrated that he knows and intends "to do what the Church does," namely to remove the stain and penalty of Original Sin together with any actual sins already committed by that soul. Furthermore, since there is often little to no concern as to the validity of sacraments, they may accept as valid a highly questionable Baptism given in some other church, or even skip Baptism altogether, as happened to a large number of souls in Canada a couple decades ago. Other (admittedly less serious) injuries to the Rite of Baptism are:

1. the elimination of the exorcisms,
2. the lack of a requirement for a Saint's name to be used,
3. the reduced role and relaxed qualifications for the Godparents.

The sacrament of Marriage, as performed in the wedding, is probably the least adversely affected sacrament as regards form, matter, intent, and minister. Where the current Vatican organization most errs with respect to this sacrament is the way that annulments have become so easy to obtain that there is never a question of whether or not you can get one. Of course, you can, no matter what. If only King Henry VIII of England could have come to the twentieth century, he could have gotten as many annulments as he wished.

Thankfully, to his credit Paul VI in *Humanae Vitae* managed to pick up the expected morality regarding the regulation of births precisely where Pius XI left it in *Casti Connubii*. On the level of the local parish however, one is doing well if he is able to know that only natural methods, based on periodic (or total) abstinence, may ever be used to regulate or limit births. The "something more" which one virtually never hears is that regulating births by total abstinence for any reason or even no reason at all is always allowable, providing only that both partners willingly agree to it, but regulating births by periodic abstinence is only allowable for "serious motives…which derive from physical or psychological conditions of husband and wife or from external conditions," all under the guidance of their confessor.

Unfortunately, the new Church has a new (and different) doctrine regarding the purpose of marriage. The Catholic Church teaches that the primary purpose of marriage is the children, not only making them but also raising them in the faith, and that the secondary purpose of marriage is the mutual help (practical assistance, companionship etc.) of the spouses. In the new Church, these two purposes are treated as being equal to each other, and several important documents even list them in the opposite order as if to imply that the secondary purpose of marriage is of more importance than the primary purpose.

The crucial words of the form and the matter of Confession in the new rites are still within acceptable limits, but intention is more of a gray area. There are two weaknesses in the New Religion's understanding of this sacrament which often injure or invalidate the intention for it, but it must be admitted that there are still some priests in the New Religion who can and do validly absolve from sin. These two weaknesses are 1) the tendency to minimize sin in a number of ways, such as by refusing to clarify its seriousness or distinction between venial and mortal sins, to educate the penitent in knowing which actions or omissions are serious and which are not, or in the denial that some serious sins are sinful at all, a refusal to take serious sins seriously, and finally an all too generous willingness to excuse it, and 2) the tendency to see sin in terms of breaking with the Community of the People of God rather than breaking with God Himself, thus causing them to be more intent on "restoring the penitent to society" instead of bringing peace between the penitent and God.

Another innovation is the introduction of a public "reconciliation service." This service is modeled quite loosely on the new "mass," complete with the new Confiteor, the Pater Noster (Our Father), scripture readings, and a homily. At best, such a service is only a tedious, useless, and time-consuming activity which precedes or surrounds a more or less conventional visit to the confessional. At worst, it is used as a replacement for the confessional, with only some vague and virtually always invalid "general absolution" being given.

The sacrament of Extreme Unction can only be described as having been completely done away with. In, roughly, its place the "Anointing of the Sick" lacks even the most rudimentary, crucial, and essential words of the form of Extreme Unction, "By this Holy Unction may God pardon thee whatever sins thou hast committed by the evil use of sight, smell, or touch." That is like having a "mass" which omits the words "this is my body." The matter of that sacrament (olive oil, blessed by the bishop, or in some rare and exceptional cases where granted permission by the Apostolic See, the parish priest) has been replaced with "any oil of plant origin," and again the standards as to how it is to be blessed have been significantly relaxed. Beyond that, the intention is totally changed. The intent was to prepare the soul of the dying to meet its maker.

Granted, on many occasions, especially in cases where the danger of death was from injury or disease instead of old age, the recipients would frequently get better and go on to need this sacrament again the next time they were in danger of death. Now, only earthly health is thought of, and this new "sacrament" (at best, mere "sacramental") is no longer given to those who are so old they are expected to die soon. Also, this is now often done in public settings, as an adjunct of their new "mass" or "reconciliation service" instead of privately, as indicated in Scripture: "Let him **call for** the elders of the Church, and let them pray over him, anointing him with oil in the name of the Lord." Such "services" are unquestionably invalid. The official priestly request for the forgiveness of sins, an essential component of every Rite's sacrament of Extreme Unction, and even mentioned in the Biblical account ("...if he be in sins, they shall be forgiven him..."), is entirely missing from the form of "the anointing of the sick."

The sacrament of Confirmation has been altered roughly on par with the mass, thus similarly imperiling its validity. The blow to the cheek with which the bishop symbolically prepared the recipient to accept blows of persecution as a soldier of Christ has been replaced, usually with a handshake or something similar. Gone from the new form of the sacrament is any clear concept of receiving the indelible mark of confirmation on the soul, but only a more vague reference to being "sealed with the Gift of the Holy Spirit." The old form read as follows: "I sign thee with the Sign of the Cross, and I confirm thee with the chrism of salvation. In the name of the Father, and of the Son, and of the Holy Ghost."

In 1972, with a stroke of the pen and nary an explanation, Paul VI deleted the following steps in the reception of Holy orders: Tonsure, Porter, Lector, Exorcist, Acolyte, and Subdeacon. The essential forms for the orders of priest and bishop which Pius XII clearly defined in *Sacramentum Ordinis* are, for the priesthood, "Grant, we beseech Thee, Almighty Father, to these Thy servants, the dignity of the priesthood; renew the spirit of holiness within them so that they may obtain the office of the second rank received from Thee, O God, and may by the example of their lives inculcate the pattern of holy living," and for the episcopacy, "Fill up in Thy priest the perfection of thy ministry and sanctify him with the dew of Thy heavenly ointment this Thy servant decked out with the ornaments of all beauty." The new forms are sufficiently different to again imperil the validity of the sacrament, and also the intention is often no longer to empower the recipient to go to the altar and offer sacrifice, to absolve from sin, and in the case of the bishop, to empower him to ordain priests to those ends.

Another point worth mentioning is the lack of any real training in the Catholic faith seminarians now receive in the seminaries. Where they used to learn Latin, Canon Law, Church History, Thomistic Theology, and how to perform the sacraments, they now learn Community building, psychoanalysis, and a new "theology" based on the writings of Teilhard de Chardin and many others who echo his ideas to a greater or lesser extent, and whose works have, for the most part, been censured by the pre-Vatican II Church while still under popes Pius XI and XII. In modern seminaries, discipline is lax, homosexuality is tolerated, and many seminarians have their faith completely destroyed. More about that can be read in the books *Are Today's Seminaries Catholic?* and *Goodbye, Good Men* (See Bibliography). The rampant spread of homosexuality among the Novus Ordo clergy has led to widespread "priest-abuse" scandals in which minor children (usually teenage boys) are seduced (recruited to homosexual behavior) by sexual predators in clerical garb. Even more scandalous has been a widespread policy of protecting these predators from the consequences of their bestial acts through various legal and other subterfuges.

The loss of reliable Holy Orders is especially dangerous since an invalidly ordained priest is no priest at all and therefore useless for all sacraments except Baptism and Marriage, no matter how well and traditionally he does them. Moreover, if a bishop fails to obtain a valid consecration to the episcopacy, he fails to gain not only the power to ordain priests and give the sacrament of Confirmation, but most importantly, he lacks the power to consecrate new bishops validly and continue the Church.

And, although the performance of an exorcism (something along the lines of the story and film, *The Exorcist*, but not quite so Hollywood dramatic) is not a sacrament, the vitally useful ceremony by which a Catholic priest performs this function was also gravely mutilated, so much that even the Vatican's own official "chief exorcist," Fr. Gabriele Amorth, quickly ceased to bother using the new ceremonial, and reverted to the classical (pre-Vatican II) ceremonial for exorcising demons, since he found the new to be altogether ineffective. The old continued to deliver as promised.

As you should now be able to see, the new "sacraments" pose a major threat to the very existence of the Church. Yet even all of that is only the beginning of the problem. There are two areas in which the present-day Vatican organization especially departs from the belief and practice of the Catholic Church and Faith:

Ecumenism and Religious Liberty. In addition to those two areas, common cause has been made with Masons and Communists, even at the highest levels. One repeatedly would hear bishops and cardinals say that we must join the Communists in their fight for "Social Justice," an ironic misnomer if ever there was one, since Communism has always been, and is by its very nature, Oppression Incarnate! Real, authentic Catholic social justice is of course poles apart from communistic "social justice" as each can quite properly and accurately be defined as a total negation of the other. Finally, the prohibition against involvement in Masonry was quite pointedly removed from the new (1983) Code of Canon Law.

In the area of ecumenism, there have been major compromises with those who reject Catholic teaching on many points. It may sound so very high minded and idealistic, the "hope that all nations, while differing indeed in religious matters, may yet without great difficulty be brought to fraternal agreement on certain points of doctrine which will form a common basis of the spiritual life," but see what a complete loss of faith hides behind those noble-sounding words! By the time you encompass all nations, there are no "certain points of doctrine" left with which to "form a common basis of the spiritual life."

Such an ecumenicalism, expressed in those very words, is exactly what was condemned by Pius XI in the Encyclical *Mortalium Animos*. In it, he states that, "such efforts can meet with no kind of approval among Catholics. They presuppose the erroneous view that all religions are more or less good and praiseworthy, inasmuch as all give expression, under various forms, to that innate sense which leads men to God and to the obedient acknowledgment of His rule. Those who hold such a view are not only in error; they distort the true idea of religion, and thus reject it, falling gradually into naturalism and atheism. To favor this opinion, therefore, and to encourage such undertakings is tantamount to abandoning the religion revealed by God." What the Pope rightly perceived is that certain teachings of the Church have to be regarded as "unimportant" in order to make common cause with those who reject those teachings. The very notion of allowing **any** Church teaching to be "unimportant" is a tacit permission to hold **any and all** Church teachings to be "unimportant," the deepest heresy.

In the Vatican II schema on ecumenism, *Unitatis Redintegratio*, it is recommended that there be a greater fidelity to the original calling of the Church, which if actually performed would be a good thing in any case. Unfortunately, the way things have worked out, this call to fidelity has been used only as a pretext for spreading a false and heretical primitivism in the Church. We would have reached other Christians best by being fully, completely, and honestly Catholic in the fullest sense of that word, and by defending rather than hiding or apologizing for the Church's impact on history. The Church has always and consistently been a force for good, for civilization, for honor, for the happiness, prosperity, and well-being of every society which has been guided by Her throughout history. We Catholics have nothing to hide, and nothing to apologize for.

Instead, the Vatican organization hides our heritage, our history, and our doctrine by gutting nearly all references to such from its liturgy and catechisms. It has become a Church without a past, all in the name of being ecumenical; it is now just like all of those other denominations which really have no past, except that it once had one and they never did. In addition, "dialogue" has come to replace evangelism all around the world in every mission field. The Vatican organization does not bother with trying to make Catholics, or even the pseudo-Catholics of the post-Vatican II variety, out of persons of other faiths. Now, it is considered enough to encourage Muslims to be "better" Muslims, Hindus to be "better" Hindus, Buddhists to be "better" Buddhists, Mormons to be "better" Mormons, Wiccans to be "better" Wiccans, and so forth.

The practical upshot of these ecumenical distortions to our Faith has been the removal of the doctrine of the Godhead of Christ. Though many Novus Ordo "catechisms" may still make passing mention of the doctrines of the Trinity or even the Deity of Christ or the Incarnation, in practice these doctrines have been systematically removed. These doctrines most of all are "offensive" to Jews who reject their Messiah, Moslems who relegate Jesus Christ to the mere status of "just another prophet," Hindus, Buddhists, and all other Pagans of all sorts to whom Jesus can only be (at most) just some other guru. The very idea of "God" becomes merely some sort of vague "great architect," at most an impersonal and ineffective balancing force of nature. That the Great God should be personal, and incarnated as the God-Man Jesus Christ at a particular time and place, and with all the authority to tell us what is right and how to live, has become just yet another part of faith to be rejected by the secular and godless world's "political correctness" which is tacitly accepted by the Novus Ordo "church."

It is fair to ask whether or not there is any possible ecumenism which might have validly been intended by a Conciliar document. The answer is yes, there is another kind of ecumenism, one that would exist strictly as a stopgap measure for enlisting the aid of souls who have serious doctrinal defects, but who do have a sincere attachment to Christ, in attainment of a goal. An excellent example of this would be the fight against the evils of abortion, a fight in which we are helped greatly by many Christians of other denominations. This is a stopgap form of ecumenism because the souls of those other Christians are still at risk of being lost due to their separation from the Church Christ founded. It can often take a great deal of time and effort to dismantle the errors which keep a soul out of the Church, or to win them over, sometimes without a word. During that time, in the interest of social order, or some other practical and secular concern of mutual interest, there certainly does need to be a protocol for getting along with them and working together for the common good. But at no point would it be right for such a protocol to confirm the heretic in their error, whether directly or indirectly, by action or omission.

The false ecumenism based on treating other religions as equals (not to be confused with treating the <u>people</u> of other religions as equals) is what lies at the heart of what went wrong at Vatican II. Those statements about the Church "subsisting in" the Vatican organization were intended to grant the non-Catholic religions the power to confer Grace. That is impossible, and in fact it is heretical to claim otherwise. Only the Catholic Church can have the power to confer Grace. Since the doctrinal impact of such a statement is heretical, the Holy Spirit could not have allowed such a statement to pass unless there were a way for that statement to be applied exclusively to the disciplinary realm, wherein it has always been possible to detach the Catholic Church from the Vatican organization.

Religious Liberty, defined as the loss of the Church's temporal authority over the secular rulers, has been taking place over the past five centuries, and so much of that cannot be blamed on Vatican II. What **can** be blamed on Vatican II is the fact that what few Catholic nations as still existed were de-Catholicized, even in many cases where they could have remained Catholic indefinitely. In view of the almost total secularization of the modern world, to say nothing of the American tradition of the separation of Church and State, it will not do here to cite the numerous popes and councils which have defined with great precision the role of the Church in relation to secular authority, which the new "Religious Liberty" of Vatican II contradicts. Many today have never seen how Catholic authority can function within Catholic nations, and might therefore imagine that the old popes were wrong to condemn Religious Liberty. Therefore, I prefer to defend and explain, rather than merely state, the teachings on this subject of the reliable popes such as Pius IX and Leo XIII.

First of all, the separation of Church and State is and always shall be a myth. Any state, that really is a state and not merely some stretch of land with total anarchy prevailing, must have an authority structure and laws. There can be no authority without a purpose for that authority, and the laws it maintains and enforces are its purpose. Likewise, laws are merely ideas unless maintained and enforced by some authority. The laws of a nation must "hang together" in some sense, be at least kind of a set, and not merely an assortment of meaningless *ad* h*oc* rules.

Philosophically, the basis of Law is Ethics. The ethics of a society define its purposes and its goals, its priorities and its concerns. Human life, of at least some sorts, is commonly a priority in most societies, but only some will include all human life, others might include animal life, others, the environment, etc. Some place a premium on art and literature, others on science and technology. Some value comfort and the avoidance of all pain while others value personal strength and integrity, or even ability to tolerate pain. The list of possible variations is endless. In any society, its Law and Ethics are its art and study of carrying out its objectives.

The basis of Ethics, however, is Religion. It is a culture's religion which defines its world view. In these modern times, many make the mistake that Atheism is not a religion, but it is. What a person believes is always a religion regardless of whether one believes in zero, one, two, or many gods. The world view, or religion, of any nation which takes "Religious Liberty" as some kind of philosophical axiom instead of merely an acknowledgment of its own particular incompetence in religious matters, is a world view in which "God" is merely an idea, a concept in people's heads, some figment of their imagination welling up from their unconscious. In such a world view, there is no "The God" out there in any objective sense, only people's subjective opinions about Him, Her, It, or Them. While the Church cannot be blamed if a state, on its own initiative, should elect to proclaim itself incompetent in religious matters, there is a great evil in the Vatican II mandate that "the Church" must teach that all states are necessarily incompetent in religious matters and so must therefore treat all religions on an equality basis. Such a teaching, if ever actually claimed by the Church (and not merely some ex-ecclesial Vatican organization), would make the Church (and Christ Himself) culpable and carry with it the logical implication that there is no objective standard for judging religious truths.

That world view is a religion, regardless of whether they call it such or not, which is quite alien to Catholicism, or even Protestantism or Judaism or Islam, for that matter. When the United States was being founded, its religion was primarily Protestant, with perhaps just a trace of Masonry, and in some localities such as Maryland, even Catholicism (hence the name). "America," as founded, was a Christian nation, at least in the Protestant sense of the word. Over the last forty years or so, it has gradually shed most aspects of its Christian heritage, the only remaining present vestiges being "In God We Trust" on our currency and the use of a Bible for swearing in a witness or inaugurating a president.

The American "State" has indeed separated itself from the Christian Church, but all that means is that the guidance provided to our government has gone from being that of the (Protestant) Christian Church to that of another Church. This other "Church" has no name, no parishes, no congregations, no preachers, no visibility, and no public recognition of its existence. This "Church" of the

American Mindset holds all (other) churches to be equally true, in other words equally false. This Church with no name also has no catechisms and no publicly acknowledged doctrines. One just has to try to infer its teachings from the actions of those who practice it in the arena of American Politics and Law. Perhaps there are elements of Atheism, Masonry, Humanism, Existentialism, Empiricism, Rationalism, and perhaps sometimes even a vague Theism, but that is probably just a start of the list. The key doctrine of this religion which concerns us here is its denial of any objective reality of God other than its own unknown and unknowable belief, the <u>real</u> basis of the teaching known as "Religious Liberty."

Even a pure Atheist, in the modern western European sense, will still believe that there exists an order in the Universe, parts of which are discoverable, and that a hypothesis can be either true, by being in line with that order, or false, by being at variance with that order. On that point at least, the Atheist and the Catholic and all degrees of unbelief and belief in between can agree, while certain Far Eastern modes of thought will allow that any hypothesis can be both true and false in exactly the same way and at the same time. People who use those Far Eastern modes of thought will allow that any hypothesis can be both true and false in exactly the same way and at the same time. People who use those Far Eastern modes of thought (and their revival in the West in the form of the "New Age") have no concept of the principle of contradiction, namely that if **A** is true, **A** cannot also be false at the same time.

What the Western thinker has, Atheist and Catholic alike, is at its heart a belief in an objective reality. Such a belief is essential in the physical sciences, such as Physics or Chemistry. In those, the fundamental criteria for judging any hypothesis is the science experiment. If you mix this chemical with that at a given temperature, such-and-such occurs. If your hypothesis predicted that it would, then that hypothesis is correct and graduates to the status of a "theory." If not, then it is false and we go back to the drawing board. The experiment can be repeated as many times as one likes, with always the same results, because the chemicals, the temperature, and the chemical processes which cause the such-and-such to occur are objective realities. If they were not, if they were mere subjective experiences or opinions, then one could cause the experiment to come out any way one likes without having to change the experimental procedure.

Just as matter and energy exist "out there" as objective realities, so do God, Angels, Heaven and Hell, Eternity, and all elements of a theology. Contrary to current popular opinion, theology has every bit as solid of a foundation in reality as any other science, one which even has objective, clear, and indisputable tests with which one can determine which theology is true and which are false. Indeed, theology used to be called the "Queen of the Sciences." This is the point at which the Atheist and the Catholic part company. While both resort to objective tests to ascertain the nature of material reality, the Atheist arbitrarily refuses to admit any objective tests to ascertain the nature of spiritual reality. Where the test in Physics or Chemistry is the science experiment, or in Mathematics, the mathematical proof, the primary test for theology (as it is for Law, History, and Literature) is scholarship. Few would argue that one of the greatest scholars of all time is Saint Thomas Aquinas. Through his intensive study of every known (in his lifetime) document which could have any bearing on the study of God, he managed to be able to explain every conceivable spiritual, moral, ethical, or other religious experience of all of Mankind with an understanding as simple and clean and elegant as "$E = mc^2$" and every bit as profound.

The American government still sets certain standards for certain things, even though such standards are being successively relaxed in many cases, and already entirely so in religious matters. For example, we still have a system by which schools are "accredited." This means that what they teach is compared against some objective standard and if found satisfactory, the schools are accredited, else they are not. (For the sake of argument I will here ignore the fact that such an accreditation process is often abused.) A school of Obstetrics which taught the "stork theory" of child delivery and which therefore sent its students out hunting birds or harvesting cabbages would understandably not get accredited, even with today's lax standards. A Catholic nation is merely a nation which applies the same standards to its religious education as this nation still applies to its medical education.

It is way beyond the scope and scale of this work to step through all of the documentation which proves that Jesus rose from the dead, and that Roman Catholicism is the only religion He founded, nor even to expose the many methods of pseudo-scholarship with which false doctrines are "proved" by non-Catholics of every sort, but rest assured that our Catholic Faith is supported by all reliable scholastic sources in a way which no other faith before or since has ever been. A good set of books to get started on for this is the three volumes of *Radio Replies* by Fathers Rumble and Carty, published by TAN Books and Publishers, Inc. These books contain summaries of Catholic apologetic arguments directed towards all manner of opposing doctrines. Also recommended is *Catholicism and Fundamentalism* by Karl Keating, published by Ignatius Press, and *The Catholic Controversy* by Saint Francis de Sales, though these books primarily address Protestant objections to Catholicism, and these books are also published by TAN Books (See Bibliography). The Catholic Church has the right and duty to teach that Her doctrines are correct and all contrary doctrines therefore incorrect. This prerogative is not a product of arrogance but rightly deserved because the Catholic

Church, unlike all others, has "done its homework." There is also the protection of the Holy Spirit of God who protects the Church from error, though that is a matter of faith rather than science or research. At the very least, it can never be the duty of the Church, or of individual Catholics, to support, sustain, or defend non-Catholic religious doctrines. Therefore, within this book, I take it as an assumed premise that Catholicism is true and all other belief systems, no matter how well meaning many who hold or even teach them might be, are false, and furthermore have been proven false where Catholicism has been proven true.

Given that Catholicism is true, it becomes crystal clear why a Catholic nation must prevent those of other religions from expounding their false doctrines. Even in the case where this effort requires the use of corporal force, such force is justified even though it may not be a pleasant thing to witness. A Catholic nation which has the guts to protect its citizens from dangerous and divisive ideas is a strong nation which can only have a bright prospect for its future. Contrary to popular twentieth and twenty-first century misconceptions, such censorship is a good and worthy thing to do. Teachings which advocate divorce are every bit as destructive to the family as pornography, if not more so. Why is it that so many Americans feel so free to censor one, but not the other? Censorship can only become ugly and dangerous by being used to silence the truth, never by silencing deceptive and stupid opinions. In any case, the Truth can never be silenced anyway. Again, do we like false advertising to be legal?

When the Council opened in 1962, there were still many Catholic nations: In Europe there were Ireland, Portugal, Spain, Italy, Belgium, and certain Cantons of Switzerland which were still Catholic in Law and not merely in demographics (such as France and Poland). In the far East, the Philippines and even Việt Nam were Catholic. Several nations in Africa, especially those which were still controlled by the Catholic European nations, such as Rwanda, were also Catholic. Finally, practically every nation of Central and South America was Catholic. Even granting that several of these nations, such as Spain, Belgium, and some of the African colonies were pulling away and might have ceased to be Catholic nations with or without the Council, there remains a great many others such as Portugal, Ireland, Columbia, Ecuador, and the Philippines which would cheerfully have remained Catholic nations to this day and well beyond it as well.

What, you may ask, did the Council have to do with the loss of those Catholic nations? In "obedience" to the Vatican II schema on Religious Liberty, the presidents of the Vatican Organization, or part-time "popes" if you will, approached the leaders of each of the few remaining Catholic nations and got them to sign new agreements (Concordats) to the effect that they shall cease to restrict the propagation of false doctrines. As a result, these countries are going

Protestant, Communist, Moslem, or whatever. They are even ceasing to have large numbers of Catholics in them. I lay that at the door of those faithless Vatican II Council Fathers who signed the schema on Religious Liberty. How many souls will burn in Hell because these faithless "shepherds" conspired, yes I feel justified to use that strong term, to expose them to false doctrine, and that without even preparing them for it by warning everyone against it?

Finally, the new modernist "church" has displayed a very different concept of "Magisterium": In the Catholic Church, that which is defined or taught in the magisterium, in papal teaching or otherwise universally taught, is crystalized in truth; it is irrevocable and cannot be changed, once clarified on any point of doctrine. But the Novus Ordo concept is clearly and demonstrably much more fluid and flexible and quite revocable. We have already see above where even their own magisterial, or even "papal" teachings as to what the Mass even is, or what the doctrinal content of its prayers should be, have been changed, Article 7 being changed within a year of release, the Arian preface prayer associated with "eucharistic prayer #4" being softened, and "for all men" being changed to "for all" after 17 years of denying the divinity of the Son and the Holy Ghost, and the gravely doubtfully valid "for all" (in one form or another) left in place for 43 years before being fixed (and yet further things, such as "the mystery of faith" as yet still awaiting repair). In the Novus Ordo, a "pope" could officially teach something, and then, after discovering how wrong it is, then later on go "oops, I was wrong; here is the correct teaching…" Indeed, the whole of the "new magisterium" of the Novus Ordo church of the People of God amounts to claiming that much historically taught by the Church for nearly 2,000 years was all just "oops, the Church was wrong, here is the (new) teaching…"

The Roman Catholic Church could never have done such a thing, yet in its name the ex-Catholic Vatican organization did it. To believe that the Vatican organization is still the Catholic Church is to believe that the Gates of Hell <u>have</u> triumphed over the Catholic Church. They most certainly have triumphed over the Vatican organization! Let us step through the Marks of the Church to see if they still apply to the Vatican organization now that it has become detached from the Catholic Church and has also embraced a New Religion:

ONE: The illusion of unity wears ever thinner and thinner among the massive non-Catholic portions of today's Vatican organization. What they retain is only a loose kind of "societal" unity, much akin to that of the Anglicans with their "high" and "low" and "broad" churches, teaching contrary doctrines, yet all under one roof. The collegiality of bishops means that bishops now vote on issues at "Bishop's Conferences," and it is a well-known fact that wherever groups of people vote on issues, invariably there forms parties or factions among them. Amongst the bishops there exists "Conservative" factions and "Liberal" factions. As shall be seen later this book, one might even find a very small Traditional (read: Catholic) faction, while at the other extreme there are bishops who are still pushing for priestesses despite some rare creditable words from John Paul II against that.

The local leaders of the American and European regions are seriously out of step with the that of other countries in granting marriage annulments to all who ask (the local leaders of many other countries have generally held to more Catholic standards). One bishop in Washington State, in every way but name, had set himself up as an American pope, another in Wisconsin made himself famous for being opposed to nearly every Catholic moral virtue. The new liturgy is said in so many different languages instead of Latin (plus the relative handful approved for the Eastern Rites), that even many individual parishes are divided between groups of different language-speaking peoples. The unity of prayer which is enjoyed by Jews in their Hebrew worship, Catholics in their Latin worship, and Muslims in their Arabic worship, is altogether lost to the Vatican organization.

Novus Ordo "believers" are also not really united to each other since most of them don't believe in being truly subject to the Supreme Pontiff. Some want priestesses while others want liturgical aberrations as yet unimagined by any Fr. Bozo and still others want homosexual marriages recognized, and many, all too many, want contraception, or even abortion. Granted, there are conservatives who still believe in the principle of submission to the Supreme Pontiff, but they are just one more small (and shrinking) faction within the Church of the People of God.

Even the publication of the new "Catholic Catechism" failed to unify in that bishops disagree with each other as to whether to embrace it or not, and even those who do embrace it disagree with each other as to its interpretation (just like with the Vatican II documents on which it is substantially based). What is right? What is wrong? Practically every "priest," bishop, religious, or lay catechism teacher with their one-day training in "Building a Community," has their own idea, even about things long since settled by the Church. One? The pretense wears extremely thin and threadbare. They can't even keep it down to "Several."

HOLY: One good hard look at their sacrilegious new sacraments (to say nothing of the abuses they are powerless to stop), and their atmosphere of mundane, banal triviality, should confirm to any objective observer that the word "holy" does not describe the Vatican organization in any way. More "Catholics" of the Novus Ordo variety *per capita* have abortions and public drunkenness than any other group of people, and that extraordinary holiness of the saints is simply not found at all. There is almost no more reverence for the Blessed Sacrament, even where it may be validly consecrated, since their "extraordinary" Eucharistic ministers often spill the Sacred Blood and trample it without a thought. Many sacraments are of doubtful validity, depriving the people of the power of God to overcome sin and achieve anything for the Kingdom of God. Religious orders and congregations have shut down, and the few remaining are but shadows and mockeries of the great religious orders and congregations of the Catholic Church of history. There are no special areas set apart as holy and only for the priests, making all places equally holy, in other words, equally unholy. Miracles, though at times claimed, do not stand up at all to even the slightest objective scrutiny. Finally, where the Catholic Church was started in the First Century by our Lord Jesus Christ and partakes of that holiness, the present Vatican organization defined itself into existence in 1964 and is therefore no more capable of being holy than its many and faceless founders. Need I say more?

CATHOLIC: The Vatican organization no longer even claims to be for everyone, since in their ecumenical dialogue they are obliged to accept and agree that "Catholicism," even of their strange sort (let alone authentic Catholicism), is not for everyone but only for "Catholics." They also no longer teach the entire council of God, another meaning of "Catholic." One seldom hears about any subject which is unpleasant or unecumenical. Some topics have dropped clean away, such as the merits of the Saints, and even many of the Saints themselves. Even the Saints who remain are seldom heard of, as are Hell, Purgatory, Sin, the value of an individual soul to God (as opposed to some collectivist whole of the community), or the proper use of periodic abstinence. Explicit agreements of "non-proselytizing" have been reached between them and the schismatic East Orthodox, the Jews, the Lutherans, the Chinese "Patriotic Church," and so forth.

The entire character of their services is markedly alien to the Catholic character of the pre-Vatican II Church. Indeed, to apply the name "Catholic" to the Vatican organization today is an abuse of language. They have already taken a new name for themselves, and I believe that it is just to award them their new self-appointed name in order that all may know precisely which Church is being spoken of. The name they have taken is the "People of God." They are the Church of the People of God, which does not mean that they are God's people any more than the Eastern schismatic "Orthodox" are truly orthodox in all of their doctrines (most, yes, but not all), or that "Christian Scientists" are either Christians or scientists. They want to be called the People of God, let them be called such herein; they can no longer be called Catholics. They are a brand-new church which had no existence for well over a millennium and a half after Christ. Another name by which they have very occasionally referred to themselves by, and by which the Traditional Catholic community frequently refers to the Church of the "People of God" is the "Conciliar Church" (a term first coined by Cardinal

Benelli in a private correspondence with Abp. Lefebvre), owing to its having been invented at the Second Vatican Council.

APOSTOLIC: In refusing to even try to convert those of other religions, the Vatican organization has relinquished the apostolic mission of the Catholic Church. Many Jesuit priests had suffered much to bring the gospel to the American Indians, in many cases being tortured to death by fiendish methods which only the very most savage of the American Indian tribes could even conceive of. What has it come to now? I, the author, have personally seen the Snake God of the pre-Catholic Aztecs honored in a Novus Ordo parish in my hometown. The Vatican organization has no apostolic mission and therefore no apostolic authority.

Even the succession of bishops which can be traced clear back to the original twelve Apostles is gravely threatened by the gravely doubtfully valid new sacrament of "Being an apostle like Gandhi, Helder Camara, and Mohammed;" that has replaced the highest degree of Holy Orders which confers the episcopacy (makes a validly consecrated bishop). The new Church seldom ever refers to any document previous to the Second Vatican Council or John XXIII, and then only to defend some general point which most people would agree with anyway. Indeed, the new Church, with its new religion, has absolutely no official existence previous to Vatican II and John XXIII. It is as if some guy who calls himself John XXIII for no particular reason just comes out of nowhere and convenes some Council which he decide to call Vatican II, again for no particular reason, and from that springs into existence an entire Church, the Conciliar Church of the People of God, a Church ready to hold hands with the whole wide world and sing a song of brotherly indifference, and which has absolutely no continuity with the original Twelve Apostles. Apostolic? No.

THE CHURCH IS THE MYSTICAL BODY OF CHRIST: The Vatican organization on the other hand resembles not so much a shepherd with his flock of sheep as a vivisectionist with his laboratory rats. "Let's try this new liturgical experiment, whoops there goes another hundred, oh well, now let's try something else…" The Catholic Church could no more "mature" into the current Vatican organization any more than Christianity could ever "mature" into Devil worship. It relinquishes and disowns any and all claim to being an authoritative representation of God's interests in this world. Remember, the Vatican organization has detached itself from the Catholic Church, and only the latter can now be the Mystical Body of Christ.

THE POPE IS INFALLIBLE: Why, then, was it possible for a translation of a liturgical prayer which teaches that only the Father is God (the Arian heresy) to receive full ecclesiastical approbation (approval of Paul VI) and to stand unchanged clear to his death, and indeed for nearly seventeen years? Why was it possible for him to change God's Divine Revelation as prayed by the Church from the very beginning? One just has to face the fact that whatever occasional infallibility Paul VI might possibly have enjoyed at some other points of his papal (?) career was certainly not present when he did those things.

THE CHURCH IS INDEFECTIBLE: But today's Vatican organization is not. It has defected from the Catholic faith in so many ways that the sheep are bailing in droves, like mice jumping off a sinking ship. They just can't get out fast enough as they trip over each other running out the door. It is fair to ask whether it will exist after another three centuries or so (probably not, unless they repent and return to Catholicism and re-identify themselves with the Roman Catholic Church by abrogating Vatican II in its entirety and applying to the real Church for real legitimacy).

THE GATES OF HELL SHALL NOT PREVAIL AGAINST IT: Clearly, the Gates of Hell have had no difficulty triumphing over the Vatican organization, now that it is no longer identical to the Roman Catholic Church. Such evil fruit is not something extraordinary, but merely the natural result of the legal and canonically established detachment of the Vatican organization from the Roman Catholic Church. The evidence is irrefutable; the verdict is in: <u>The Vatican organization and establishment no longer constitutes the visible unity or structures of the authentic and historic Roman Catholic Church!</u>

Any reader who has not seen any "Catholic" church other than their local Novus Ordo parish and others like it would have to admit that what I have described is what has become of their local parish. Unless you have understood and accepted the fact that your local parish is not Catholic but another Church (properly called the "People of God") you would have to feel total despair regarding Christ's promise to be with His Church always, until the End of Time. Like Jesus dead and buried, your local ex-Catholic parish is nothing but a corpse, spiritually moribund. No wonder so many are leaving. No wonder that those who remain utter the closing "thanks be to God" with the clear meaning, "thank God that waste of time is finally over." Christ's promises would seem to have proven to be nothing but hot air.

Yet despite everything mentioned above, this lengthy chapter is but a cursory overview of the many evils and scandals committed by the church of the People of God. Worst of all is the overall betrayal against the Faith by the Novus Ordo church of the People of God. Mistaking it for the indefectible Catholic Church, many have had their trust in God Himself rankly violated and shattered. How can anyone trust again? But as following chapters will go on to show, God has indeed kept His promises to His Church.

While I have you at this low and depressing pass, allow me to point out some interesting observations. A number of paragraphs ago, I stated that the test of valid theology is scholarship. That is primarily true, but on occasional and very rare instances, we have the luxury of what can only be described as a kind of scientific experiment. Nearly two thousand years ago, certain Jews and Romans conspired to see what would happen if they killed this Man from Nazareth. They found out the hard way, and have been paying for that lesson ever since. He comes back, "to the Jews a stumbling block and to the Greeks foolishness, but to those who are called,…the Power of God." Today, we are in the midst of another such "science experiment."

One of the great claims the Catholic Church has long and often made in its defense is that if ever they were to change the religion from that of Christ to another, there would be an outcry. It would be widespread, nay universal, as the <u>entire</u> Church would have been outraged at any attempt to change their faith on the part of the leadership of the Church, and well documented. Every Christian denomination other than the Roman Catholic Church has had to claim that the leadership of the Church could, out of their admitted personal corruption in certain cases, change the faith from that of Christ to something else, and no one would ever notice, or only some small few who would soon die out. At the heart of <u>every</u> non-Catholic but Christian church there is always the claim that "You Catholics have deviated from the religion of Christ, and we are restoring it."

Catholics have always responded by pointing out the absence of any documented evidence of any outcry which surely would have existed if the Vatican leadership were to have ever introduced any "new" doctrines or deleted or changed any "old" doctrines. Only now do we get to see before our very eyes precisely what would happen if the leadership of the Church, or at least of that which many still mistake for the Church, were actually to attempt such a change to our religion. **If the Church can be so changed now, if a whole new set of sacraments and commandments and doctrines can be invented and successfully imposed on it as has happened since Vatican II, with no permanent survival of what the Church had always been before, then it could have been so changed at any point in the past, and all of Christ's promises to be with the Church until the End of Time would have to have been just so much empty talk.**

Let's also take a look at another claim certain Protestants sometimes make as a way of claiming that the leadership of the Church could just change the religion as suits their purposes and get away with it: It is claimed by these people that in the era of Constantine the Church was persuaded to let in all sorts of pagan customs, so as to expand it and make it acceptable to most Romans. By incorporating such pagan customs as the worship practiced in other religions, as for example was quite vividly practiced at Assisi in 1986 when John Paul II and leaders of almost every other religion each prayed to their respective gods to bring world peace, we see now that doing that does not bring the pagans in, but merely drives the Christians out (along with practically everyone else, except for a few foolish pagans who, by their participation in the Scandal of Assisi, showed themselves every bit as unfaithful to their own ostensible traditions as to the true Gospel).

I do not want to end this chapter on such a low note. Really, you should not view your local People of God parish as just an entombed dead body, but rather as an empty tomb. Do you remember what an angel, camped at another empty tomb nearly two thousand years ago, had to say? "He is not here; He is risen!" He is risen indeed! Up until this point, all that I have done is lay down the why's and wherefore's of the widespread reaction against the new religion, in whatever many forms it has taken and may well take, though that is only one aspect of what many call the Traditional Catholic community, but what I call the Mystical Body of our arisen (again) Savior that is the visible unity of the Roman Catholic Church!

5

OLD CATHOLICS, EAST ORTHODOX, AND THE SPECTRE OF DISSENT

Much as I would love to begin here at last the grand epic of the survival of the Traditional Roman Catholic Church through these strange times by being Herself reduced to a faithful remnant as the Traditional Catholic community, there yet remains one last ghost which needs to be looked directly in the eye so it can then be summarily dismissed. This "ghost" is the principle of dissent which some traditionalists have mistakenly come to identify themselves with, even to the point of taking the likes of Joan of Arc, Athanasius, or Savonarola as their Patron Saints, not for their saintliness but for the controversy that surrounded them.

Dissent of many sorts has occurred throughout the history of the Church, leaving in its burned-over trail a plethora of schismatic and heretical little factions all over the world. There is a world of difference between what these schismatic and heretical groups have done and what Catholics of the Traditional Catholic community are doing, though it may not be at first readily apparent. Once seen, however, it is unmistakable. In order to better understand and appreciate the steps which the Traditional Catholic community has been obliged to resort to, steps that some misrepresent as "dissent" but are in fact the very opposite of dissent, it is necessary to introduce the reader to some of those other groups and their case histories.

As is also to be seen in this chapter, the Novus Ordo Church of the People of God traces its roots to these dissenting groups. In the many places and ways, the Traditional Catholic community differs from these dissenting groups, the Novus Ordo religion shows itself to be merely just another member of this pathetic and sordid lot. This chapter, therefore, is the story of some of those outside the Vatican organization who are also not Catholic.

The fact that the Church of the People of God should be trying to get "ecumenical" with these groups, and for that matter, with utterly non-Christian groups such as Buddhists, Hindus, Moslems, Jews, and so forth does not mean that members of these non-Catholic groups are also members of the Vatican organization, at least in a strict sense. They are still outside the Vatican organization in the sense that they each have their own pseudo-hierarchies which do not in any real way answer to the leader of the Vatican organization, though they are all counted as being salvific by them. Here, I am primarily concerned with those who still think of themselves as "Catholic" or at least "Christian" in some sense, even though they are not. It is the Roman Catholic Church Herself who has formally proclaimed that they are not Catholic, long ago, before the current confusion arose, at least back before Vatican II, in the days that the Vatican organization exactly equaled the Catholic Church.

For one thing, one must make a distinction between dissent and rebellion. A dissenter is one who disagrees with an authority figure, but who still believes in the principle of authority, just not in the exercise of it as performed by a particular authority figure or leadership group from whom they dissent. A rebel is one who does not believe in authority at all (unless it's one's own). Even though dissenting opinions are typically the product of rebellious minds and hearts, the followers in a dissenting religion may often not even think of themselves as dissenting at all, and most certainly are not in any way rebelling against authority. This chapter is not about rebellion but dissent, and the following chapters, as shall soon be seen, are not about either one.

Another word for those who dissent in the sense meant here is "heretic." Again, I remind you that a dissenting heretic is different from a rebel, who would also be properly called an "apostate." Sometimes the Traditional Catholic community is caricatured as being one of dissent, and here we have our problem: "The Church has gone astray; it must be restored!" Thus goes the cry of a great many heretics. How are Traditional Catholics different?

It is interesting to note that <u>all</u> heresies are either one of only two categories, or else a combination of the two categories. Let us start with a short review of the history of some of the heretical claims. In the opening few centuries of Church history, people living had known those who had known those to whom the Divine Revelation had come in the first century. Heresies in those days necessarily had to focus on some new revelation or supposed "secret" doctrines. The Gnostics, who were in some ways the first ecumenists, came up with idea that there was good in all religions, and so attempted to invent a synthesis of Christianity and other religions, both those of the exotic Far East as well as the Greek and Roman mystery religions.

The Gnostic religion, for example, borrowed the Eastern idea that spirit is good and matter is evil, and blended that with the Gospel so as to claim that an evil God created the Universe and a good God sent Jesus who, being entirely spirit (according to Gnostic doctrine), left no footprints. In this case, the "new" revelation was actually just a "new" exposure of certain individuals to other religions unknown to the early Palestinians. The result was a group of documents called the *Nag Hammadi* texts which the Gnostics simply added to their version of the Bible. The Gnostic heretics also used the strategy of the "secret doctrine" by claiming that there were secret doctrines, known only to the high initiate few, but concealed from the general run of Christianity. Indeed, their name itself refers to themselves as the "knowing ones" because they saw themselves as that initiated few. This is exactly the sort of thing condemned by no less than the Apostle John when he spoke of "antichrists many" in his Epistles. These "secrets" not only amounted to heretical beliefs, but even extended to grotesque and sinful liturgical practices and excesses, shocking even to Pagan Rome. It took centuries for society at large to differentiate authentic Christianity from these Phibionite and other deviant sects, and it was from this that came many of the ancient criticisms leveled against the Church.

Next came the Arians who again made no claims to restoring any early Church but only pronouncing on a question which had not really been asked, namely "Is Jesus God?" Up until that point, Christians had simply taken it for granted that Jesus is God, namely God the Son, the Second person of the Trinity. Arius attempted to reason it out with his rational, but human and limited mind, and ended up saying "I simply can't believe that some mere man walking about on the earth could somehow also really be the Eternal God." To Arius, if the Apostles thought that Jesus was God, he could only pat them on the head condescendingly like a good dog and say, "Ah, isn't that cute? How quaint, actually imagining that Jesus might be God. But WE know better. We are enlightened because we have read up on our Greek and Roman (pagan) philosophy and we know that none of the great Gods and Heroes who walked the earth could ever be equated with that ultimate and remote Divine Entity."

Even as late as the seventh century, the only way to create a new heresy was to claim some new revelation or some new insight or other new thing. In this case the new revelation came from an angel of light who called himself the Archangel Gabriel and told the "great prophet" Muhammad all of the things which he wrote down in the *Quran*. This kind of heresy hasn't stopped, but has slowed down somewhat since the appearance of the other kind of heresy which is the one I really want to talk about, but let me finish this out with a few more contemporary examples.

In the early 1800's two new religions came into being, each claiming to be based on some new revelation. In Persia, a new prophet named Baha'u'llah appeared, claiming to advance the faith from that of Muhammad to that of a higher truth, and so founded the Bahai faith. In the United States, another new prophet named Joseph Smith claimed to advance the Christian faith, founded the Mormon Church, and introduced many new claims clearly unknown to Jesus and the Apostles. Later on, we have Mary Baker Eddy and Madam Blavatsky and their new revelations. In the twentieth century, we have Reverend Moon who claims to be the Third Adam, something which the Bible obviously never provided for in any way.

We also have Charles Manson who claimed to be God and Satan reunited, David "Moses" Berg who taught his "flirty fishies" to use promiscuity as an evangelistic tool, Scientology which all comes from the mind of L. Ron Hubbard, Atheism which was invented by several Germans, an "Aquarian Gospel" written by Levi H. Dowling who consulted the Akashic records (apparently, a memory written in the "fabric" of space) as a basis that we all get reincarnated, the Urantia book which mystically reveals things known or believed only by 1930's "pop" science, and numerous messages from the Ascended Masters passed along by Mark L. and Elizabeth Claire Prophet.

Note that in each of these cases, there is some "truth" which is so great and lofty and sublime that even Jesus and the Apostles could not have known and would not have been ready for, or else could only reveal to the initiated few, but which can now be made public. None of these heresies intrinsically involve any claim that the Church has gone astray, although some most recent ones do gratuitously make such a claim as an extra added bonus. The point is that in the earliest centuries of the Church, this was the only kind of heresy available. Christians still remembered what the early Church was like, and the ancient "Church Fathers" were those saintly ones who wrote down (thereby documenting) details of the ordinary beliefs and day-to-day practices and functioning of the Church. As the remembrance of the early Church faded, the age of the ancient Church Fathers also ceased.

It took clear to the eighth century (at which point the age of the ancient Church Fathers was drawing to a close) before another form of heresy first put in an appearance. This started with the Iconoclasts who were the first to put forth the claim that the Church had somehow fallen from the holy purity She had back in the beginning, and that a restoration to that primitive state was in order. These people condemned to destruction many beautiful works of art which adorned the churches around the world and reminded many believers of the Christian saints and biblical events which many, being illiterate, would otherwise forget about, or even not have learned about in the first place.

In that heresy, there were in fact two different heresies at work. One, a local heresy unique to itself, and the other, a general heresy which has been repeated over and over again since then and which is the basis for all heresies apart than those ones I have just mentioned which were based on new or secret revelations. The local heresy in this case was that images were in themselves bad, and the general heresy was that which I would call "restorationism," that false kind of restoration. It is those of this heresy who must proclaim that "The Church has gone astray; it must be restored!"

Before the Iconoclasts came into existence, everyone knew what the Church had been in the beginning and could see that She had not been changed. It would have been utter nonsense in those days to claim that "Well, Jesus and the Apostles taught and believed one thing, but the Church today teaches and believes another, so we need to go back." That is why all of those earliest heresies had to claim instead that "Well, Jesus and the Apostles did not know or reveal everything, but now is the time that things they did not teach can at last be "revealed," or "figured out," or "made public," or "discovered," or "created," or "channeled," or whatever the case may be.

This false "restorationism" is at the heart of a great many heresies from that of the Iconoclasts onward. In the eleventh century, the Churches of the East, while admitting that they had long been subject to the Bishop of Rome, suddenly advanced the claim that it had not been so in the beginning. In the sixteenth century, the Protestants claimed to be restoring the Church to a Biblical form, and one hears that claim in many small Protestant churches to this day, "We're building a New Testament Church, just like Jesus and the Apostles!" The Jansenists, Döllinger, and the Old Roman Catholics all did likewise.

The Iconoclasts blamed the third century Church for introducing images. The schismatic Eastern churches blamed the fifth century Church for giving preeminence to the Bishop of Rome. The Protestants of the sixteenth century have had a field day being able to move the blame around anywhere from Constantine to Saint Thomas Aquinas and back again, as suits their purposes. The Jansenists blamed the tenth century Church for reducing the penances. Döllinger blamed the sixth century Church for forbidding the clergy to marry.

What do all of these heretical claims have in common? All of them without exception state that the change came somewhere "way back when" a very long time ago (centuries at least). Why do they do this? Let us see just what would happen if the opposite were to take place. If, in 1925, some prominent figure were to claim that the Church changed from one thing to another a mere <u>ten</u> years previous (that would be in 1915 A. D.), can you imagine what would have happened? Let me state it for you straight out.

There would have been tens of millions of Catholics worldwide who remember what the Church was like before, during, and after 1915. Not one of them would recall anything happening to their Faith or the Church that year. Pope Benedict XV was on the throne and spending much of his time trying to contain and limit the evil caused by the "Great War" (what we now call World War One), and to encourage nations to try to get along with each other and make peace, nothing all that unusual for the Church. The catechism remained the same; the liturgical rites remained the same. Nothing interesting happened to the Catholic Church that year. Our "prominent figure" would be at once seen, by all who hear him, to be making an absurd and foolish claim, and no one would take him seriously anymore.

But, if he were to claim that the Church changed from one thing to another a thousand years previous (that would be 925 A. D.), who could refute him? Since most people are not familiar with Church history, many of them would be truly surprised to learn that there has never been such a change. But if only they could step through the history of the Church, Council by Council, Pope by Pope, year by year, sure enough no such event would be found to have ever taken place. Unfortunately, the vast amount of scholarship this would require is way beyond the reach of any layman, or even most priests and bishops. As the Church continues down through time, the size of this tremendous amount of data one must study and be familiar with in order to prove that can only grow exponentially with the age of the Church.

If even the greatest lights of the Church today are scarcely able to take in and digest all of this overwhelming amount of information, how much less can the proverbial "man-in-the-street" who not only has a soul to save, but a family to feed? The heretical restorationist takes advantage of this ignorance on the part of most people by claiming that somewhere, way back in the "forgotten" mists of antiquity, the Church went askew, and somehow nobody seems to have cared or even noticed. Maybe the change had been gradual; maybe the concept of a fixed and unchangeable doctrine had not been conceived of as of yet, "but surely, one way or another, some major distortion must have crept in, and it's our job to fix it," goes the blather of false restorationists from every era.

Let's take a short look at why and how he might do this. For some selfish and twisted purpose of his own, he invents some conception of what he wants the Church to be like. He then goes to the early sources, the Bible, the early Church Fathers, and sometimes other documentary miscellany from the period, and looks for quotes that fit his invented conception. No matter what that conception is, it is never hard to find some seeming basis for it, provided that he is careful to reject, or ignore if possible, any evidence from his sources which refutes his claim, especially if he uses only short quotes from basic sources that don't drill into much detail anyway. He then publicly contrasts this fictitious "first century Christian

Church" with the Church as known in his own day and devotes the rest of his life to trying to persuade everyone to "restore" the Church to his twisted vision of the first century Church.

Suddenly, along comes Vatican II and one sees the supposedly "Catholic hierarchy" following the exact same pattern. Practically the entire leadership of the organization which had long been the Catholic Church all agreed and conspired to turn "the Church" into something it had never been. Interestingly enough, one of the claims they made for this transformation is that it would restore the Church to Her first century format! Suspicious indeed is the heavy borrowing of liturgical details and other "incidentals" from other false restorationists, such as Martin Luther, Thomas Cranmer, and certain Jansenists.

The people who pushed for these changes and who finally got them were themselves false restorationists, for the heresies of modernism and liberalism are also false restorationisms. But all of these higher critics who dissect the Bible, deciding for themselves which statements attributed to Jesus were actually said by Him and which were not, are by their very nature and admission modernists. They claimed they wanted to go back to some supposed primitive Church model. And this is what sort of mind set all of the key periti at the Council pretended to have, Karl Rahner, Hans Küng, Edward Shillebeeckx, and all the rest. Of course the ancients themselves would never have even thought of combing through the various written documents of the Church as they knew them to decide for themselves which quotes were real versus which were later interpolations, etc.!

I have frequently mentioned the schismatic East Orthodox and also certain Eastern Catholic Rites, and the history of these bears some short explanation, since the two are closely intertwined:

In the early days of the Church, the original Apostles traveled to many diverse parts of the world, each taking with them their first-hand knowledge of Christ which they had gained in their several years with Him. Paul's missionary journeys are well known since they are written up in the Bible in some considerable detail. Peter seems to have operated in Jerusalem and then Antioch for some period of time before moving over to Rome. Mark (the writer of the Gospel bearing that name) went to Alexandria. Thomas (the famous "doubting Thomas") went to Parthia, and then India, founding a congregation in the Malabar coastal area. James (who wrote the Epistle of James in the Bible) remained in Jerusalem. John first went to Ephesus, and later wound up on the prison island of Patmos. The others went to various other places.

Each of them, filled with the words and actions of Christ, drew from that rich reservoir the particular liturgical elements most suited for the various cultures they evangelized. The liturgies, although substantially the same, contained certain local variations. Some of these apostles were more successful at establishing the Church in their missionary land than others, and the more successful of these arranged to have successors to their respective Sees. These successors are now commonly referred to as Patriarchs. In the early Church, the main patriarchs were established in Rome, Alexandria, Antioch, and Jerusalem. Constantinople came a couple centuries later as the main secular political center moved from Rome to Constantinople making it an important enough center to warrant a Patriarch of their own.

The Patriarch in Rome was also the Pope, and it was to him that all questions went when they could not be settled by lesser authorities. It is in the fifth century that the Church in the East began to call itself "Orthodox" in reference to their rejection of certain fashionable heresies in the East, namely that either Jesus Christ was two persons (God and Man) or else that He had only one Nature (God only, or Man only, or some weird mixture, or even the claim that God and Man are both of the same Nature) which were condemned at the Council of Chalcedon. The correct doctrine, promulgated at Chalcedon and accepted by all orthodox ("right teaching") Christians of the East and West alike is that Jesus Christ is one Person with two Natures. In this matter, even the schismatic "East Orthodox" are in fact still quite orthodox.

It is interesting to note that there has been repeatedly given from Rome commands that the specific Rites as observed by the various patriarchates were not to be contaminated with each other's disciplines. These commands became especially necessary when, due to the success of the Church in the West, and of the European nations which by law embraced Christianity and became therefore "Christian nations," or "Christendom," certain priests and bishops in the East were beginning to Westernize their Rites. Also, certain Eastern Rite Churches at various times tried to force other Eastern Rite Churches to change their liturgy to be like that of the stronger Eastern Rite Church. Such attempts to westernize the Eastern patriarchates or make the Eastern patriarchates uniform among themselves were wrong in precisely the same way that it would be wrong to "fix" the Sermon on the Mount as given in the Gospel of Luke to read exactly like the Sermon on the Mount as given in the Gospel of Matthew. What matters is that all Christians must draw their liturgical "water" from that same well of Christ's "Living Water," not that all should have similar sized and shaped cups to draw it with.

Indeed, one would do well to remember those prohibitions against mixing Eastern Rites with each other, or with the Western Rites the next time some Fr. Bozo decides to introduce some change, such as "communion under both kinds" or "standing during the consecration," or even the use of leavened bread, on the specious basis that such things are supposedly being done in the Eastern Rites.

While it is true that Eastern Rite Catholics use leavened bread and receive the Body and Blood of our Lord under both forms, and also do a lot of standing in their worship, the way communion is handled is quite different from how it is given under both species in the *Novus Ordo* Rite (with or without leavened bread), and the people do also kneel as well as stand, but the one thing they cannot do is sit down for the simple reason that pews are not provided (except a very few for the feeble and elderly and nursing mothers). What one sees in Pope Benedict XIV's prohibition against mixing different Rites is a tremendous respect for each of the Rites of the Church, Eastern and Western alike. None of them is to be compromised or blended in any way, even with other Rites every bit as Christian and legitimate as their own.

Even so, tensions persisted between East and West, particularly since the wise and just legislation which existed to prevent any Rite from growing at the expense of any other Rite frequently went unenforced or even ignored by many priests and bishops of both East and West. It is important to note that the liturgical and disciplinary differences between East and West were not in and of themselves either the cause or the result of dissent. For most of over a thousand years, the Church both East and West was united. Each side more or less respected the distinctive characteristics of the other while both presented to the world a united front.

There were dissenting groups in the East, particularly those who were condemned as heretics at the first eight Ecumenical Councils of the Church. These were the groups which denied that Jesus Christ was God, or that He was Man, or else claimed that he had only one Nature or else two Persons. Also there were those who claimed that Christian art was bad (Iconoclasts) and those who claimed that everyone including the Devil would eventually be saved (many of the Armenians). It turns out that the same weakness which caused these heresies to spring up in the East also eventually caused the separation of those orthodox believers into the Eastern Schism. That weakness was a tendency to go just a little too far in allowing the secular rulers to choose who gets the power positions in the Church, namely the bishops and patriarchs.

Normally, it is the job of the Church to make that choice, not anyone else. While one must admit that the Church is at liberty to show deference to a secular ruler by selecting a candidate for a See who is pleasing to that secular ruler, the Church can by no means be morally obliged to do so. What happened in the East was not quite so much a "dissent" as an "acquiescence," to a nearer and stronger "authority" than the Christian hierarchy way off in Rome. Perhaps that is why it is that even after nearly a thousand years of separation, the schismatic East Orthodox churches deviate from sound doctrine in only a very small number of ways.

Nevertheless, an element of dissent gradually developed in the East, somewhat helped along by the tensions caused by liturgical and disciplinary differences.

There is an extremely short list of areas where the Eastern Church expressed dissent. It is interesting to note just how willing the schismatic Church in the East was to renounce its short list of heresies in order to be reunited with the Catholic Church for a brief period in the fifteenth century. In renouncing their schism and heresy they needed only to affirm the four following tenets of the Church: 1) the supreme primacy of the Pope, 2) the validity of the use of unleavened bread for the Eucharist, 3) the existence of Purgatory, and 4) the procession of the Holy Ghost from both the Father <u>and</u> the Son. In later years, the West would also come to confirm as true and Christian: 1) Thomistic Theology, 2) the Infallibility of the Pope, 3) the Immaculate Conception of Mary, and 4) the Assumption of the Blessed Virgin. None of the eight issues listed here were ever so much repudiated by the schismatic East Orthodox as simply left up to individual opinion. Real Eastern dissent was expressed in only those individuals who ran ahead of their leadership by condemning these eight Catholic doctrines rather than simply feeling that these issues were still merely a matter of personal opinion.

Some today might feel that the schismatic Eastern Orthodox Church might really have been the way to go since they are not undergoing any "spirit of Vatican II" changes right now. They had no Protestant rebellion and no Jansenists either. On the other hand, they have no central leadership (no pope, not even an office for a pope), and have had no Councils for over a thousand years with which to address any of the very many questions which have been raised up over that period of time. No, despite their valid Orders and Sacraments and sound teaching in all areas other than the tiny handful mentioned above, they are not part of the Church. They are schismatic, and in fact merely the Christian expressions of certain Eastern Despots rather than valid successors of the Eastern Apostolic Patriarchs.

Not to be confused with such are those of the Eastern Catholic Rites in union with Rome, who are not schismatic, and who properly are the successors of the Eastern Apostolic Patriarchs. While many in the Eastern Rites broke off and became schismatic, there were a few, most notably the Maronite Rite in Lebanon and Syria, which were always faithful to Rome. Over time, many members of the other Rites, both lay and clergy, renounced their very few heresies and returned to unity with Rome. In all other aspects, these Eastern Catholics kept their disciplines and liturgical practices. Most large and notable of them is the Byzantine Rite.

Some of the early Protestants bear some mention as further examples of dissent. Martin Luther started out merely protesting certain abuses which were taking place, but soon his vitriol began spraying in all directions as he later came to deny huge areas of Catholic doctrine. Since he quickly progressed from dissenter to rebel, he quickly lost all real interest to the history of the Church and certainly would have faded into the woodwork if only a certain secular ruler had not found Luther's ideas able to work to his advantage. Thomas Cranmer likewise would not have got anywhere with his ideas had not Henry VIII wanted a divorce.

All one has to do is look at the many and ever multiplying Protestant sects to see that there is no Church there, only ever so many more and more little churches dotting the countryside. Even regarding such basic issues as whether or not tongues, Rock music, healings, anti-Catholicism, and Christology, they differ so very much from one sect to another. They disagree with each other about almost everything under the sun, and even about issues which have been long since settled by the Church in the days when there was only One Christian Church. Furthermore, the modern false "ecumenical" movement got its start amongst the Protestants who realized that they must band together if they are to continue to be any sort of a political force in the increasingly secular world. The World Council of Churches (long known to be a Communist front) has nearly every major Protestant sect represented as formal members. No, the Protestant "church" is no solution since their entire structure leaves so very much up to the choice of the individual "pastor" or "believer."

The next major sources of dissent were the Jansenists and the "Old Catholics." The Jansenists were merely followers of a man named Cornelius Jansen who, in the name of a false interpretation of Saint Augustine's writings, taught that severe disciplines supposedly practiced in the early Church should be reinstated. These disciplines called for extreme and long penances, and very rarely (if ever) partaking of the Eucharist. Combined with that was a certain "Calvinism," a "predestination" which taught that God chooses to save some while choosing all others to go to Hell.

At length, one of their members by the name of Cornelius Steenhoven finally got what is generally taken to be a valid episcopal consecration in 1723. His consecrator was a missionary bishop, Dominique-Marie Varlet, who was known to be sympathetic to the Jansenist cause and in fact already suspended on account of it. From Steenhoven they made many more bishops and were able to set up a new "hierarchy" of their own in Utrecht, Holland. This shadow society, condemned by Rome as schismatic and heretical, quietly continued to sustain itself for over a hundred years until further events forced them to combine forces with Johann Ignaz Döllinger and others who rejected the first Vatican Council.

It was Döllinger who gave the "Old Catholics" their present direction, insofar as they can be said to have any direction at all. Once Vatican I had proclaimed and defined the doctrine of the Infallibility of the Pope, several apostate priests and others, under the influence of Döllinger's writings organized and held a mock council in Munich in 1871. While Döllinger himself kept a certain distance from the new schism, he was clearly the main intellectual force behind it. It is interesting to note certain parallels between the Councils of Munich and Vatican II. Both advocated ecumenism, a reduced degree of authority for the Pope, and the reduction or elimination of Latin in the liturgy, along with other liberal ideas. Because of these parallels, some even refer to Vatican II as "Munich II."

One characteristic of the Old Catholics which is of particular significance here is their almost casual readiness to make bishops out of anyone who asks. It is this quality, along with the fact that the Old Catholics were not directly affected by Vatican II, which has caused some desperate Catholics to turn to them during the present crisis. The Old Catholics were not directly affected by Vatican II for the same reason the East Orthodox were not directly affected: They were (and are) already in schism.

The present crisis is not the first occasion that disgruntled Catholics have turned to the Old Catholics for episcopal services. Around 1900, a group of Polish immigrants who felt that they were being given unfair treatment withdrew from the Church and soon thereafter turned to the Old Catholics and formed what is often called today the Polish National Church. A smattering of other national groups, such as the Philippine Aglipayan church, or the French and Belgian Petite Église group, have done similar things for similar reasons.

Disgruntled Catholics are not the only ones to avail themselves of the services of the Old Catholic clergy. Other unscrupulous persons have obtained an episcopal consecration from Old Catholic clergy, often at the cost of simony. These "clergymen" would then each set up some little church of their very own with no connection to any other church, even Old Catholic, and lead some small and unfortunate congregation whatever way suited their fancy. Nearly all of these pathetic individuals trace their orders to Jules Ferette, Joseph Vilatte, or Arnold Mathew.

To some Catholics, disgruntled with the changes taking place in their Novus Ordo parishes, the Old Catholics can seem to have quite a viable claim: "Maybe the Church should not have proclaimed papal infallibility back at Vatican I. Then no one would even seem to have the authority to enforce the Novus Ordo Missae. Maybe Vatican I was the false council (somehow) and the Old Catholics are the true Catholic Church." One can see how easy it is to fall into such a pattern of thinking. Alternatively, some Catholics, desperate to obtain some manner of episcopal leadership of some sort, have at times turned to Old Catholics for ordination or ordinands, or priests to fill in here or there. Even certain great figures in the Traditional Catholic community, namely Francis Konrad Schuckardt, Abp.

Pierre Martin Ngô Đình Thục, Fr. Francis LeBlanc, and even Fr. Hector Bolduc have at various times and places and to varying degrees either accidently aided and abetted or else deliberately availed themselves of the services provided by the Old Catholics, however briefly and incidentally in some cases.

Nevertheless, the Old Catholics represent nothing but dissent and are not in any way a protection from the Novus Ordo religion. For one thing, most Old Catholic groups were obliged to sign on to the heretical "Utrecht Declaration" which affirms Jansenistic doctrines as a condition of receiving an episcopal consecration from the Old Catholics. For another, their Munich Council was practically a prototype for Vatican II. If anything it actually goes even further than Vatican II does. Some other problems are the fact that they have rejected the teachings of the indissolubility of marriage, the primacy and infallibility of the pope, and clerical celibacy. While in the beginning, the Old Catholics accepted the Council of Trent as the last valid council of the Church, they have since backed off from acceptance even of that council, owing to their attempts at ecumenical cooperation with Protestant churches, especially the Church of England. Finally, there is some doubt as to even the validity of their episcopal orders since their theology regarding the sacrament of Orders is defective and therefore their intent to consecrate is suspect. Döllinger himself was very much a liberal and a modernist, the same sort as those who gave us Vatican II and invented the Novus Ordo Missae.

In the course of our current Church crisis, one will often hear the words "Jansenism," "Jansenists" or "Jansenistic" used a lot. Properly and strictly, these terms only apply to those Old Catholics who have formally subscribed to the Utrecht Declaration, a document prepared and agreed upon by the founders of the Old Catholic Church. As it is, not even all who count themselves as Old Catholics or whose priests and bishops trace their sacramental orders to the Old Catholic bishops adhere to the Utrecht Declaration, and so cannot be called Jansenists. That term does not apply to anyone in the Traditional Catholic community, nor for that matter, the Novus Ordo Church of the "People of God." When this term is used in reference to anyone other than Old Catholics adhering to the Utrecht Declaration, it cannot be taken as a serious recrimination, but only as an empty epithet one hurls at an opponent when one has run out of more intelligent things to say.

The next dissenting group was a loose association of people who worked within the Bosom of the Church to try and overthrow the Popes and all of their teaching. For lack of any better name, this group may be referred to as the "Liturgical Movement." That is the direction taken by these unscrupulous persons when Pope Pius X began encouraging the lay faithful to have a fuller participation in the Mass, by which he meant that they should take a more active interest in what is going on as the priest performs the mysteries of God in the Mass. These people latched on to that encouragement and used it as a pretext for introducing the "Dialogue Mass," the first step in the direction of the Novus Ordo Missae.

Within this Liturgical Movement, perhaps five to ten percent of their work was legitimate, involving research into the origins of various details and customs of the feasts and other aspects of the liturgy. Their legitimate work served as a kind of "loss-leader" to establish themselves as "great scholars" and acquire reputations and the clout which comes with that. With the remainder of their work, they displayed their true intent which was to foist on the Church a brand new "Mass" of their invention. This was stated first in very cloudy, veiled language which sounded almost totally orthodox, but gradually as they gathered fame, approval, numbers, and just overall momentum, their nefarious plans were stated in more and more detail.

One of the more prominent names in the earliest days of this Liturgical Movement was Virgil Michel who founded and edited a modernist journal called *Orate Fratres* which expounded and advanced the cause of the Liturgical Movement even as far back as the reign of Pope Pius X, and clear until 1938 when Virgil Michel himself died. Taking over the helm from him was Fr. Gerald Ellard who praised his predecessor Virgil Michel as the "Pioneer of the Liturgical Movement in America." Fr. Ellard wrote three books over a period of sixteen years entitled, *Men at Work at Worship*, *The Mass of the Future*, and *The Mass in Transition*. One finds in these horrid tomes (published during the 1940s and 1950s) a startlingly detailed description of the Novus Ordo Missae, along with the claim (similar to the claims of previous dissenters) that this new "Mass" he is proposing is a restoration to some sort of primitive practice.

The dissent practiced by the Liturgical Movement followed a somewhat different course from the dissent practiced by the previous groups described here. Where the previous groups said boldly and honestly "we disagree" with Catholic authority, this group disagreed with Catholic authority in exactly the same way while saying "we are just expressing your truths in a manner which 'Modern Man' can better understand," a claim which any bishop able to read between the lines would have to have recognized as positively insulting. They would then search far and wide for some lame-brained bishop to give their books, magazines, and other articles imprimaturs, even of the bizarre and unprecedented sort which read "Views expressed remain the author's own though the book bears the Imprimatur."

While the Holy Spirit eternally protects the Catholic Church from embracing such heresies, since Vatican II there has been a major non-Catholic portion of the Vatican organization which enjoys no such protection. As a result, they have allowed the dissenting Liturgical Movement faction to take over and run their new Church. The Church of the People of God is therefore just another dissenting group which has withdrawn itself from the Catholic Church and Catholic communion just like all the others. The only difference is that, owing to the clever strategy of the innovators, the number of people this dissenting group has misled is substantially larger than the number of people in each of the other dissenting groups. Theirs is the religion I have already described in the previous chapter. Since I have already explained there why this dissenting group is not an acceptable alternative for Catholics, I will not repeat that information here.

Dissent did not end with the rise of the Church of the People of God, but since the Church was so drastically reduced in size by the mass defection of Catholics to the People of God, the newer dissenting groups have been correspondingly smaller. The first of these dissenting groups came about while the Liturgical Movement was still a minority voice within the Vatican organization, but by which time real Catholic faith had also become a minority voice. A number of Fr. Leonard Feeney's followers, reacting against the increasing liberalism and erosion of the doctrine that outside the Church there is no salvation, began to fight that erosion (an erosion which really did need to be opposed) by denying the Catholic teachings of Baptism of Desire and of Blood. These particular religious errors would only be attractive to those whose Catholicism is still strong enough to see the value of the more basic truth that indeed, outside the Church there is no salvation. But by their overreaction of denying the Catholic teaching regarding Baptism of Desire and of Blood, they have distanced themselves from the authentic teaching of the Church, albeit only to that limited extent. They were also one of the last of the groups created before Vatican II granted a charter to all Catholics functioning outside the Vatican organization.

At about the same time, a number of Chinese bishops and priests broke away from Rome as a result of pressure from the Communist Red Chinese Government. They founded what is called the Patriotic Chinese Church, but are also in schism very much like the East Orthodox. Again, as in the case of the schismatic East Orthodox, it wasn't so much dissent as acquiesence to a nearer "authority" than far off Rome. Needless to say, the Patriotic Chinese Church is not in a position to speak ill of the Chinese Government no matter what it does (even in <u>mandating</u> abortions!) nor warn its parishioners against the heresy of Communism.

Dissent from the Traditional Church continues today in the form of dissenting groups and troublemakers who seem to exist merely to give authentic Catholicism a bad name. A woman by the name of Sinéad O'Conner, after a mildly successful career as a rock star (during which she also made some religious publicity for herself by tearing up a picture of John Paul II on TV), decided to get herself "ordained" as a priestess in 1999 by Bishop Michael Cox (a "black sheep of the family" bishop whose episcopal orders regrettably trace to one of the heroes of this account) specifically with the intention of using her "ordination" to say the traditional (Latin) Mass. Of course this has nothing to do with real Catholicism nor the real Church even as it exists today as the Traditional Catholic community. She deliberately turned to a bishop who might seem to be associated with the real Church today (albeit crudely and mistakenly and only because of his episcopal lineage) instead of the fake Church of the People of God, where she could easily have turned instead. In 2002, a feminist group of seven other women obtained just such an "ordination" from a People of God "bishop," but Sinéad herself had never been interested in this approach.

To understand the historic precedent for such conspicuous evil and scandal being committed in the name of the Church, one must turn to the ancient history of the Church, in the opening centuries of the Church. Perhaps one may recall that the ancient Pagan Romans and others would spread lies about the Church saying that they ate their children or indulged in wicked rites with cannibalism, lewdness, and the like practices taking place. This was done to provide an excuse for the miserable treatment the ancient Roman Emperors were giving to the ancient Christians, and allowed the public to feel "not so bad" about seeing these fine upstanding fellow citizens being fed to lions and so forth. However one thing that helped this false rumor along was certain secretive groups, such as the Phibionites, who, pretending to some "secret" or "inner" Christian initiation, indeed practiced such things. Though the Church had stronger unity then than now, it was quite easy for certain local groups to take advantage of the great geographical distances between themselves and everyone else in order to practice local distortions of Christianity that often took years or even decades for the Church to root out. Today, such similar aberrant persons and groups take advantage of the present chaos and lack of universal leadership to do the same thing.

Another "group" of dissenters, taking a more mild, but still unsatisfactory position is the "home-aloners." This "group" is not really a group at all since they have absolutely no organization or hierarchy at all. They consist of merely a number of scattered individuals who privately say the Rosary and read the prayers of the Mass to themselves and make a "spiritual communion" by reciting a prayer of desire to receive communion. Unfortunately, what we have at work here is not so much bad interpretations of Canon Law (although there is certainly an element of that), but paranoia. The typical home-aloner is isolated from all other members of the Church, even other home-aloners, on account of their paranoid fear that the other person might not really be a Catholic. Such fear is rooted in despair that the Church is really gone and can never be restored. Such despair is a sin, and is furthermore heretical in nature since it denies that Christ is still with His Church.

A milder from of this dissent would be those who seek Catholic Tradition, not as a way to get closer to God and His Church, but as a way to escape accountability to the hierarchy, or to anyone but their own opinions. Such persons might readily avail themselves of the ministrations of Traditional Catholic clergy, but then reject such clergy as being anything but hired hacks, "sacrament vending machines," with no authority, unworthy of obedience or loyalty, to be flitted from one to another as serves convenience.

The last sort of group of dissenters are those groups which attach themselves to false popes. By false popes I do not refer to the doubtful popes John XXIII, Paul VI, and the Vatican succession to follow, but to other "popes" appearing in various places around the world. An attempt to enumerate these popes can be somewhat confusing since some of them take similar names or numbers. There are two (or maybe even three) Gregories XVII, and although there was only one Clement XV, one of the Gregories XVII was named Clemente, and also there was said to be a Hadrian VII. Here are the main false popes and what can be said of them:

- Clement XV of Canada—"self-ordained" to the priesthood, this "pope" claimed to have succeeded John XXIII and taught that there must always be a living pope, so despite historic evidence to the contrary, this "pope" claimed that each pope must have somehow been chosen (and validly accepted the office) before the passing of the previous pope.

- Gregory XVII of Canada—successor of Clement XV, has priestesses, places visions on a level higher than Public Revelation, and continues Clement XV's claim that one pope must choose another before dying, and had no valid Holy Orders in his group until joined by Bp. Richard Bedingfeld (of the Thục line) in 1993.

- Gregory XVII of Spain—Consecrated by Abp. Ngô Đình Thục, personal name was Clemente Domingues Gomez so some have called him "Pope Clemente," teaches that visions have precedence over Public Revelation, has teenage cardinals, claims to have succeeded Paul VI (not an actual Paul VI but one supposedly concealed in a Vatican prison), led a group in the Palmar de Troya region of Spain, deceased March 22, 2005), successors thus far being Manuel Alonso Corral as Peter II, Sergio María Ginés Jesús Hernández y Martinez as Gregory XVIII, and Joseph Odermatt as Peter III.

- Hadrian VII of Washington—the name rumored to have been taken by Francis Schuckardt shortly before his expulsion from Mount Saint Michael's in Spokane, but this papal pretention was not widely circulated and in any case was not pushed to any degree.

- Michael I of Kansas—a relatively young man at the time of his election, who was voted "pope" by lay members of his immediate family, recognized by no one else as pope, and attempting to serve in the capacity of pope while not yet being a bishop.
- Pius XIII of Montana—Fr. Lucien Pulvermacher's (yes, Fr. Carl Pulvermacher's brother) who attempted to ordain another man to the episcopacy who then attempted the same to him in his attempt to become a bishop. Since one cannot give what one does not have, he does not have the power to convey the episcopacy to any other man, and the man he "consecrated" therefore cannot give him the episcopacy either.
- Leo XIV of Argentina—said to have been elected by an unidentified body of traditional bishops, one Oscar Michaelli was claimed to have taken that name, succeeded by another who resigned within only a few months with no successor, but reliable reports indicate this supposed "Pope" was only a rumor created to sully the Catholic cause.
- Cardinal Joseph Siri—the only man among these contenders who might have made something of a fine pope, but even if the vote were to have gone his way during the 1958 or 1963 conclave, the fact that the man never publicly claimed the post constitutes a rejection of any election he might indeed have received. There is some rumor to the effect that he may also have taken the name of Pope Gregory XVII. Technically, those who adhered to this "pope" are not so much "dissenters" as followers of a wildly creative theory about him having been "The Secret Pope." Since Cardinal Siri passed away in 1989 without any successors ever being spoken of, though a Fr. Khoat Tran claims to have been made a cardinal "in pectore," though he has made no clear papal claims, though there is mention of an anonymous Gregory XVIII.
- Linus II of Germany— Víctor Von Pentz who was voted "pope" in 1994 by certain groups operating from Italy and Latin America, refuses to begin ruling as pope or even to reveal his identity (perhaps he hopes to qualify as "The Secret Pope").
- The Secret Pope—not a man but a hypothesis held by some that there could somewhere be in hiding a secret pope who is continuing to rule the Church secretly. Perhaps he might have been appointed secretly by Cd. Siri before his death. Doctrinally impossible or at least extremely difficult to believe since a pope, in order to become pope, must be so recognized by the Church and must shepherd the sheep of God's flock.

In addition to that list there are unconfirmed reports of an "Emanuel I," an "Athanasius I," and several who took the name of "Peter II," and an extreme attempt on the part of the followers of the "Cardinal Siri Theory" to claim that Cardinal Siri, as the Secret Pope and last valid Cardinal of the Church may have left behind somewhere a private, hand-written document which secretly appoints a successor, which (and who) may someday come to light. Of these supposed individuals, absolutely nothing has been discoverable to this author.

Many of these "popes" were elected in a vision rather than by any live persons, or else by laymen, but certainly not by any cardinals of the Church, nor by any known bishops with known communities of Catholics or seminaries. Sometimes it seems as if these "popes" have nothing better to do than excommunicate the hell out of each other, and everybody else. These false popes, to a man, are each so horrible that it is a wonder that anyone takes any of them seriously. If only reliable papal authority resided at the Vatican, there would be no need for these claimants. It is only the deep desire of every Catholic to be in union with "the Pope," as a living authority, and the lack of anyone clearly identifiable as such in the Vatican (or anywhere else), which compels some Catholics to seek a real and Catholic Pope, even to the point of some very few seeing a papacy in one or another of these who have clearly not proven out.

Many of these dissenting groups have very traditional-looking Masses, sometimes even quite beautiful. Some detractors of the Traditional Catholic community have made the claim that since many of these dissenters and the Traditional Catholics both share the Tridentine Mass (or some other traditional form of the Mass, such as an Eastern Rite liturgy), they ought to be regarded in a similar fashion. In answer to that it must be pointed out that these dissenting groups all have embraced various heresies which have already been condemned by the reliable popes, whereas one searches the entire length, breadth, height, width, and depth of the Traditional Catholic community in vain for even the faintest ghost or hint of any deviation from the Authentic Magisterium of the Roman Catholic Church as taught most emphatically by over 260 reliable popes and 20 Ecumenical Councils. As to why so many of these dissenters would happen to use the traditional Mass instead of the Novus Ordo Missae, the reason is quite obvious: If you wanted to counterfeit American currency, you would not use pink, triangular pieces of paper and put Mickey Mouse on the front and Disneyland on the back!

Even so, many of the current liturgical aberrations decreed by the ex-Catholic Vatican organization have their roots in these dissenting groups. It is not only in the Protestant "Masses" of Luther, Calvin, or Cranmer, but also in the Old Catholics as well. For example, many in France who were Jansenists either openly or at least sympathetic to Jansenism, were making many local revisions to their Gallican Rite such as tearing out the high altars and the side altars, reducing the use of altar cloths, candles, and crucifixes, and forbidding silent prayers and private Masses and even such devotions as Benediction. Things got so bad that the Parisian Oratorian Pierre-Francois d'Areres de la Tour had to write, "They do everything to diminish the cult of the Blessed Virgin, to weaken the respect due to the Pope. They pride themselves on using only Scripture in their liturgies, and in declaring themselves followers of Christian Antiquity, they frequently quote the canons of that age, boldly criticize everything, attack the legends, visions and miracles of the saints." See once again the same pattern of false antiquarianism on the part of the dissenters.

In all of these heretical and schismatic groups, one can see that separation from the Church invariably casts one adrift, free to wander off into heresy. Although different groups have wandered off at varying speeds, the East Orthodox quite slowly, the Protestants quite rapidly, and the church of the people of God somewhere in between at a fairly moderate pace, all have wandered away from the truth. It is a truism that in every division of the Church, there are two sides which must gradually grow more and more different in belief and practice.

Msgr. Charles Journet writes in his book, *The Church of the Word Incarnate*, "Continuity is a sure mark of truth; rupture a sure mark of falsity," and again, "at the moment when two Churches separate, each claims to be the true Church of Christ, and each accuses the other of dissidence." How are we to know which is the authentic Church and which is the dissident? In the chapters to follow we will see how the traditionalists of the Traditional Catholic community are holding fast to the Barque of Peter while the People of God are steadily drifting into every sort of error and insanity. A thousand years from now, should the world last so long, the spiritual descendants of the traditionalist priests will be saying precisely the same Mass and teaching the same doctrine as they do today, and by then no doubt in full and living union with the then living pope of the Roman Catholic Church. What the spiritual descendants of Fr. Bozo would be like should any still exist by such a time is something far too horrible to contemplate. It will not require any great amount of insight to see the continuity of the traditionalists and the rupture of the Novus Ordo People of God.

The Church is something meant to be able to exist for all eternity. An organization that embraces change is an organization which will change over time into something utterly alien to what it started out as in the beginning and will eventually disappear. Only an organization which latches on to tradition can last forever as a bastion of stability. That is why tradition can only be built upon, never negated. That is why the church of the People of God cannot last. That is why Catholics cannot be a part of that ex-Catholic Church, the People of God or any other false or dissenting "Christian" Church, large or small.

Many of these dissenting groups have claimed to be trying to "restore" the Church to some ancient primitive form. Their attempts to return to the primitive Church structure have invariably been every bit as inauthentic as an older child's attempt to imitate a baby by shaking a rattle and shouting "goo goo ga ga!" Even if someone should concoct something someday which somehow really resembled the early Church, there does not exist anyone who was alive back then and who can confirm that "yes, this is exactly the Church I remember." All of these dissenters must claim that the Church as it existed back then has been allowed to disappear for a protracted amount of time, many generations at least, in fact always enough for people to forget what those times were really like to those who lived in them.

In claiming that, those dissenters are denying that Christ is really alive in His Mystical Body, the Church. The Church is, after all, not merely an idea or a concept or a system which someone could just read from a book and reconstruct. She is the Mystical Body of Christ, a continually living entity in which there is continuity from age to age. The Catholic Church has a living continuity which will be carried forth without any change from the pre-Vatican II days clear to the most distant future by people who remember one and can even now see the beginning of the other.

The key point of this chapter is that the Traditional Catholic community is not and has never been about dissent, and even less, rebellion. Dissenters and rebels have always followed the Church, as only that which is real gets imitated, mocked, and parodied. The dissent of such a large majority, whether of the Arian heretics of ancient days, or the Vatican organization's church of the "People of God" today, is no less dissent for all its large numbers of people. The Traditional Catholic community is no more capable of being schismatic than a faithful Catholic pope who teaches the entire counsel of God as taught by his reliable predecessors. If, on occasion, certain individual Traditional Catholics have ever turned to any of the dissenting Christian groups described in this chapter (as some few have), it is only because in their desperation and confusion they simply did not know where to turn. The last 55 years or so have been a very confusing time for all concerned, so my recommendation to any future judicial authority in the Church is to "go easy on them." I have spent enough of this book talking about that which is not the Church. In the chapters which follow, I now present to you, dear reader, the Roman Catholic Church!

6

THE BEGINNINGS OF TODAY'S STAND FOR THE FAITH

On Sunday, October 22, 1967, a little girl, attending Mass with her family, looked at the raised host, and saw at once that something was very wrong, "He's not there!" Her parents tried to shush her, but also wondered and asked, "Who's not there, honey?" "Jesus! Jesus is not there. Can't you see that?" It had been on this particular Sunday that their forward-leaning priest, ever anxious to explore all the new liturgical "options" that were coming out, had first received and then promptly put into use the new All-English Canon of the Mass, a "canon" so flawed that validity would seldom if ever be attained. But the girl had known nothing of all this Sacramental Theology; she had simply been gifted to be able to see our Lord in the Blessed Sacrament, and when an unconsecrated host was elevated as though it had been consecrated, she could see His absence. This was also the day she first learned that no one else could see with their eyes the difference that she could see between a consecrated and an unconsecrated host. Her parents wondered whether this was just a child being noisy or an ominous portent of much larger things at stake.

The Traditional Catholic community got its start in 33 A. D. when it was started by our Lord Jesus Christ. So the Church has always been and so She shall always be. Sometimes She enjoys political influence and other times She is the underdog. Vatican II is often seen (even by its proponents) as a decision of the Church to return to "underdog" status. Certainly, that has been the result. The Vatican organization presents itself to the world as being an underdog, but while its influence is definitely on the decline, it is still very much one of the political Powers That Be. The <u>real</u> Catholic Church on the other hand really has been made into the underdog, especially in comparison to the Vatican organization.

While the existence of a Vatican organization distinct from the Church legally began at Vatican II, thus reducing Her to Her present form as the "Traditional

Catholic community," there were a number of perceptive Catholic priests, bishops, and lay writers who sensed that something was up and attempted to sound the alarm even decades in advance. Alas, far too many Catholics slept peacefully through that alarm on their Blessed Assurance that She shall succeed until the End of Time. While it is true that the Church shall indeed succeed either with or without the vigilance of any one of Her members, whoever is not vigilant will not be a part of that success.

In the earliest years of the fall of the Vatican organization, the Catholic Church expressed her will in those who conservatively held fast to the existing traditions. *The Wanderer* was reminding its readers that even though such changes as pulling the altars away from the wall or demolishing statues of the saints had been approved, that did not mean that they had to be pulled away or demolished, or that precious parish funds should ever be devoted to such ridiculous projects. They would merely concede that permission had been granted to do something which would virtually never be a sensible thing to do anyway. Among both clergy and laity there were both liberal and conservative factions, the liberal having already lost their Catholic faith (if indeed they ever possessed it), and the conservatives who <u>were</u> in fact the only truly faithful Catholics. Neither group understood what had happened at Vatican II regarding the detachment of the Vatican organization from the Catholic Church.

Most important is that there wasn't at that time any distinction between "conservative" and "traditional" Catholics. They were both the same thing and saw the increasing chaos in the Church as a temporary aberration rather than any sort of new "direction" the Church could ever have been "meant" to take. In that they were correct, but regrettably, "temporary" has turned out to be considerably longer than expected by either. They lobbied for things to remain as stable in their parishes as possible, and when they found sympathetic priests they united behind them, and when they didn't they transferred out of their parish to another with a more sympathetic priest. Most true lay Catholics who preserved their faith through those early days can tell many stories of having to move from parish to parish and transfer their children from Catholic school to Catholic school as "the changes" swept through, destroying parish after parish. A Catholic father, sitting at an already injured Mass with his wife and children, and having heard the last heresy from the pulpit he was going to tolerate snapped his fingers and the whole family walked out mid-Mass, never to return.

Faithful priests would deliberately drag their feet about buying any new Liturgical books or ordering any architectural changes, often with the claim that they could not afford these things. Some even became excessively generous to the poor in their community by running soup kitchens and food pantries so as to keep themselves "too poor" to implement the changes so they could faithfully serve parishioners who were by then already coming as much as a hundred miles to worship in their truly Catholic Masses. They would change their schools to boarding schools, both again in order to keep themselves "too poor" to implement the changes or buy the new Liturgical books, as well as to cater to families who lived many miles away and who wanted to send their children to a truly Catholic school.

But most of all, they kept the Mass and other sacraments going as they had been ordained to say and do and remained faithful to an oath which all living Catholic priests (with a clear right to that title) have taken against the heresy of Modernism. One such Father was Reverend John J. Keane who had been assigned to Saint Rita's parish in the town of Lowell, Massachusetts back in the days before "the changes" had swept through the Archdiocese of Boston. Although many other priests in that diocese were getting in step with the new religion, Fr. Keane kept on with the traditional (Catholic) Mass and sacraments, and the laypeople, of that diocese increasingly went to St. Rita's.

In December of 1971, the ex-Catholic leaders of that diocese lowered the boom and declared that only the new (Novus Ordo) Rites would be used. Fr. Keane soon found himself without an assignment. By October of 1973, he had established a small chapel in West Roxbury, Massachusetts called Saints Roger and Mary Chapel, where he simply continued doing what he had been ordained to do at St. Rita's. He had started out with a congregation of about 150 faithful Catholics, but within six years he had about 650 parishioners at Sts. Roger and Mary alone, along with another 100 at St. Patrick's Chapel in Scituate, and another 200 at St. John the Baptist Chapel in Lawrence. Over the course of those six years while his ministry grew, negotiations with the diocesan "authorities" continued in an on-again-off-again manner which got nowhere until finally in May of 1980, "Cardinal" Medeiros finally attempted an official suspension of Fr. Keane which only had the effect of drawing yet more Catholics to Fr. Keane's parishes. As Fr. Keane himself put it, "What the Cardinal has suspended is his right of recognition of what I do." Fr. Keane thereafter simply continued to do what he had always done. Fr. Keane's story is only quite typical of the story of many thousands of truly Catholic priests all around the world.

More unusual is the case of a Redemptorist priest in Belgium, the Abbe Paul Schoonbroodt of Steffeshausen. Like all other faithful priests of this period, he continued to use the traditional sacramental forms and teaching in the three parishes he headed while others around got in step with the new religion. What was unusual about his story was how the Belgian bishop over him, Bishop G.M. Van Zuylen, left him undisturbed and unmolested clear up until his own retirement in 1988. It was only when the next bishop took over that Fr. Schoonbroodt came to be under accusation of all of the usual vacuous claims made against faithful priests by the Modernist leadership of the Vatican organization. They promptly "excommunicated" him.

Another faithful priest was Fr. Paul A. Wickens of Livingston, New Jersey. To the end of his life, he continued to function as he always had in his parish, Saint Anthony of Padua Chapel, and also Saint Michael the Archangel Chapel in Somerset. What makes him worth some small additional mention is that he is the author of a small book entitled *Christ Denied* (See Bibliography), which traced the current apostasy to the heretical writings of Teilhard de Chardin and indirectly, to Evolutionism. This book, written very shortly after the election of Karol Wojtyła (John Paul II) had very high praise for the incoming pope in the hope that some of the damage done during the reign of Paul VI might be undone. Alas, that hope turned only to disappointment as the problem turned out to be far more vast than merely that of Paul VI himself. He was also known for his friendly relations with the Society of Saint Pius X (of which more will be said later) and also Fr. Nicholas Gruner who was the main driving force behind the Catholic magazine, *The Fatima Crusader*.

Yet another such priest was Fr. Grommar DePauw, professor of Canon Law and Dean of Admissions at Mount Saint Mary's Seminary in Emmitsburg, Maryland, peritus at Vatican II, and parish priest of Ave Maria Chapel of Westbury, New York, who founded the "Catholic Traditionalist Movement (CTM)" which despite its name is only one more small group within the Traditional Catholic community. He does have the distinction of being one of the very first priests to take action in this crisis, by founding CTM back in 1965. Soon after, Fr. Francis Fenton founded the "Orthodox Roman Catholic Movement (ORCM)" along similar lines. Again, it's the same old story: no recognition from his ex-Catholic diocese, only a suspension which is of no effect or validity or legal or moral force. By mentioning these four faithful priests, I do not mean to slight any of the many other hundreds of such faithful priests all around the world, but only to illustrate with the stories of these four priests the stories of a great many faithful priests during this early period.

Fr. DePauw was not the first to sense the foul winds of heresy blowing through the Church. As early as 1951, Fr. Georges de Nantes of France, later to be known as the Abbé de Nantes, had read Fr. Yves Congar's book, *True and False Reform of the Church* and realized from the high esteem that such a heretical book could be held that disaster loomed close ahead. He warned the French Cardinals about it but they just laughed at him and ignored him. In the 1960s when John XXIII published *Pacem in Terris* and Paul VI published *Ecclesiam Suam*, his fears were confirmed, particularly by the latter document which exalted individual conscience and contained an evident blueprint for destruction concealed as an attempt to "purify" the Church. Before too much longer he was suspended.

Abbé de Nantes then prepared a brief defending his position and the truth against the new Vatican II religion, requesting of his local bishop that it be forwarded to the Holy Office for consideration. That bishop refused to pass it along, but subsequently de Nantes himself sent it directly to that Office which found it free of error. All the same, since those at the Vatican did not like its necessary and inescapable conclusions, they took the unprecedented step of calling the Abbé de Nantes "disqualified," an altogether unknown penalty in all the history of the Church (and the New Vatican as well), which has never been applied to anyone else. He went on to publish several important books of accusation against Paul VI, John Paul II, and the New Catechism.

Also in France, a village curate with a doctorate in Canon Law by the name of Abbe Louis Coache began sensing that problems were ahead as early as the opening months of the Second Vatican Council. He published a newsletter entitled *Letters of a Country Priest* which advocated (and soon brought about) a revival of open-air Corpus Christi processions. By 1969 he was also suspended but the processions, despite their current lack of recognition, continue to this day. Other early pioneers who quickly recognized the approaching danger were a French intellectual actually named Jean Arfel who, using the pen name of "Jean Madiran," began publishing a journal called *Itineraires* which published articles in defense of the Traditional Catholic Faith until its demise in 1996, and Marcel De Corte, a Belgian philosopher who wrote a book entitled *The Great Heresy* which compared the incoming false religion to a "cancerous sickness in which the cells multiply fast in order to destroy what is still healthy in the Mystical Body."

Nor were such insights confined to the clergy. As far back as the election of Roncalli as John XXIII, Dr. Elizabeth Gerstner, a Vatican lay insider, had learned of his connections to the Masonic lodge and his clear unfitness for the Catholic Papacy. With a growing sense of panic she watched close up the serious changes of spiritual atmosphere that took place in the papal circles almost at once. While John XXIII yet reigned, the very first suspicions as to the papal worthiness, and more to the point, papal status of John XXIII were published by a small paper in Texas by one Tom Costello, making him and whatever of his tiny readership at the time as agreed with him the very first of those to raise doubts as to his papacy.

In 1964, a Norwegian psychologist by the name of Dr. Borghild Krane realized that the fall of Latin liturgy was close and began rallying Catholics around its defense. On the 19th of December 1964, with her friend Dr. Eric de Saventhem as the first President, she founded an international foundation called Una Voce,

"With One Voice." By early 1965, chapters had already opened up in France, and within the next few years, in several other countries including the United States. Operating almost entirely within the Vatican organization, the principal aim of Una Voce is to "ensure that the Roman Liturgy, as codified by Pope St. Pius V, is maintained as one of the forms of Eucharistic celebration which are recognized and honored in universal liturgical life." They also promote the study and use of Latin, because that is the official language of the Church, and of Gregorian Chant, because that is the official music of the Church. Una Voce has continued to expand to this day and now has hundreds of chapters in over twenty countries.

In many various places, small groups of laity would form local associations, sometimes called "St. Pius V Associations," to support the loyal priests and organize catechism groups, and who would set up temporary chapels in barns, garages, public meeting halls, and be serviced by priests volunteering in their own spare time to do therein the Mass they loved rather than the Novus Ordo service they mistakenly felt obligated to do in their parish churches.

In 1967 there was another perceptive layman by the name of Walter Matt, who while serving on the editorial staff of *The Wanderer*, came to disagree with his fellow editors (who were also members of his family) about accepting the new "Mass," which the rest had decided to accept, however grudgingly. As a result of his stand for the Catholic Mass, he departed from that paper and founded another publication, a biweekly called *The Remnant* which was intended to cater to the spiritual needs of the faithful Catholic remnant, namely traditionalists and conservatives. In the years to come, his publication would come to work very closely with Una Voce, and other Catholic traditionalists who prefer to work for the rehabilitation of the Vatican organization from the inside.

Yves Dupont is another Catholic layman who sensed that the Second Vatican Council was taking a heretical turn even while the Council was in progress. In 1962, he began publishing a small Australian journal called "*World Trends*," in which there were intensive studies of the prophetic teachings of the accepted private revelations of the Church. In various issues of *World Trends*, he accurately predicted many of the events which would take place in the wake of Vatican II. He also founded Tenet Books, a small Catholic publishing concern which reprinted various Catholic classics and some of his own writings. He authored the widely selling book, *Catholic Prophecy*, and also an English translation of the very rarely published encyclical of Pope Pius X on a subversive movement known as "the Sillon."

There were of course many other tens of thousands of faithful priests who also started out the same as Frs. Keane, Schoonbroodt, Wickens, and DePauw, but as the pressure increased many of them either resigned or retired or knuckled under and finally went along with the new religion, albeit reluctantly. In these categories would be included the saintly "Padre Pio" (albeit only on one isolated occasion, after which he was permitted to retain the traditional Mass of his order), and the great teaching bishop, Fulton Sheen. Many others simply got old and died off, and were replaced in their parishes by youngsters who never knew what the Catholic faith was. Although the new "mass" had been promulgated on April 3, 1969, it was not until October 28, 1974, that Paul VI himself finally lowered the boom worldwide and prohibited all Catholic worship, with only a very few exceptions. By that time, faithful Catholic priests were either forced to retire, get in step with the new religion, or be finally suspended just like Fr. Keane.

Up until that time, faithful Catholics were able to claim that all of the madness and phony sacraments were somehow not really official and that the Pope was with them (the faithful Catholics), even if he was unable to speak in their defense. Any truly Catholic pope would have been with them since to be with the Catholic faith is to be with all true Catholics and therefore separated from all heretics. Neither friend nor foe of Paul VI really realized what they were dealing with. Until October 28, 1974, faithful Catholics who resisted the changes attributed them to false bishops and false cardinals, some of whom, such as Bugnini or Villot, were exposed as Freemasons. "The pope" was always seen as a passive observer either too overburdened with other things to do or otherwise unable to help the Church in Her agony, but never as the lead culprit behind it all. On that day, the mask fell away as Paul VI did the most blatantly anti-papal thing any supposed successor of Peter has ever done. Once that happened, it couldn't have been more obvious than if Paul VI were to have admitted (as if he were capable of such stark honesty) that "It is I who permits the smoke of Satan to enter the Church."

Of course, there were many other laymen keeping the faith and doing heroic things to preserve it in the face of change. Most notable of them is Thomas A. Nelson, who once prayed to his Lord saying, "There must be something that you need done that I can do," and within three weeks founded TAN Books and Publishers, Inc. in Rockford, Illinois on October 13, 1967 so as to publish and distribute great Catholic books at a very inexpensive price. Most of the Catholic books he printed (or more accurately, reprinted) are classics written long before Vatican II and approved by many truly respectable and reliable bishops and popes. As many found Catholic teaching more and more hard to come by in their ex-Catholic parishes, they took solace in their Catholic reading and discovered many deeper truths which previous generations had largely forgotten about. The classics he published include no less than the Catholic Douay-Rheims translation of the Bible, the writings of such great saints as St. Francis de Sales, the Cure of

Ars, St. Catherine of Sienna, St. Margaret Mary, Thomas a Kempis, and so very many more, and also of great contemporary Catholic historians such as Hilaire Belloc and William Thomas Walsh. His publications have been one of the greatest sources of what limited unity Catholics still enjoy during this chaotic period. All true Catholics love the books he published and are strongly encouraged to purchase and read as many of the books he published as possible.

In Brazil, a political group called Tradition, Family, and Property, or TFP for short, was founded by another layman named Plinio Corrêa de Oliveira. In the beginning, TFP had been founded with the purely political goal of protecting the nation of Brazil from the scourge of communism, and also from its religious ally, the so-called "Liberation Theology." In time, the TFP became international, functioning in many nations, and became more religious in nature as it became a kind of lay order for Traditional Catholics who are concerned for the establishment of the reign of Christ. To this day, the TFP publishes literature teaching the true Faith and warning against blasphemies in the public forum, and also organizing protests and fund-raisers etc. for the cause of the traditional Faith.

Another layman also managed to distinguish himself, but in time his distinction became a puzzle to Traditional Catholics, especially those who did not know him well. His name was Francis Konrad Schuckardt. Back in the 1960s, he was very active in the Blue Army, being elected to its International Council in 1963. While the various liturgical and other changes started coming out faster and faster, he began speaking out against them as a member of that International Council. Alas, the Blue Army found itself unable to go along with both the Catholic faith and the Vatican establishment, and forced to choose, they ejected Schuckardt and with him the Catholic faith in 1967. Upon his expulsion from the Blue Army he and a close friend of his by the name of Denis Chicoine founded what they called the "Fatima Crusade" in Coeur d'Alene, Idaho. Under the societal name of the "Congregation of Mary Immaculate Queen," Bp. Treinen of Boise, Idaho granted Schuckardt and Chicoine and their group a canonical erection on August 5, 1967.

Over the next sixteen years or so, these two men went about the lecture circuit advocating the Traditional Catholic faith and gathering true Catholics from all over the countryside to their small but rapidly growing parish in Idaho. During that time, their parish was serviced by heroic priests of the sort already described in this chapter, namely Fathers Lawrence Brey, Burton Fraser, George Kathrein (who later joined Archbishop Marcel Lefebvre and the Society of Saint Pius X of which more will be said later), Joseph Pinneau, and Clement Kubish. That should have been enough, and up until that point one fails to find any fault with the man's career as a prominent spokesman for the Church.

In time, his ministry gained the attention of a married gentleman by the name of Daniel Quilter Brown who was an "Old Catholic" bishop whose orders traced back to the disreputable Arnold Mathew who in turn had been condemned by name by Pope Saint Pius X, a line generally believed to have sacramental validity but being non-Catholic it can have no apostolic mission. Brown had been born and raised a Catholic, but disenchanted with the changes, turned to the Old Catholics seeking from them a sacramentally valid episcopal consecration which he got them to attempt for him in 1969. Before long, Brown befriended Schuckardt and offered to make him a bishop. As no real Catholic bishop had as of yet asserted himself as specifically concerned with preserving the Catholic Faith or even a valid episcopal succession, he was told by Fr. Burton Fraser, S.J. and several other traditional-minded priests that such an extreme resort would be considered permissible given the dire circumstances of the Church. Schuckardt held off accepting the offer owing to the schismatic source of his Holy Orders, and Brown's not having abjured the error of being associated with the Old Catholics in obtaining them. Eventually, Brown did so abjure and the one remaining impediment to consecration by him was removed.

Finally, on October 28, 1971, Brown tonsured Schuckardt and conferred all four minor Orders upon him (Porter, Lector, Exorcist, Acolyte) while about 40 persons (members of Brown's own congregation, plus a few of Schuckardt's closest associates) were present to witness the event. The next day, Brown administered to Schuckardt the first two major degrees of Holy Orders (Subdeacon and Deacon). On October 31 Brown ordained Schuckardt to the priesthood and on the day after that, consecrated Schuckardt to the episcopacy, all with the 40 persons still present. At first, they divided the United States into a pair of dioceses presided over by each of them, but with the pull of family (Brown being a married man with children) Brown largely handed it all over to Schuckardt. Shortly thereafter, they parted company. Brown then returned to his Old Catholics (so far as anyone knows), crawled back into the woodwork, and ceased to be of any further significance to this account, or to the Church in general.

Always one to take the initiative, Schuckardt founded an order of religious sisters, an order of religious brothers, and a seminary. Before long he ordained his friend Denis Chicoine and a small handful of others to the priesthood. Apart from having obtained his Holy Orders from the Old Catholics in the person of Daniel Q. Brown, one finds few if any irregularities in his ministry during this early period. Things would change later on, but not until December 8 of that year would he announce to the rest of his parishioners where he had got his episcopal consecration from. One cannot fault those who being remained with him in any way during these early years, though there were others who left.

Guérard des Lauriers was a Dominican priest who taught philosophy in the Lateran University in Rome, and who had been a confidant and confessor of Pope Pius XII. In the firm tradition of his Dominican order, he was a mighty opponent to the changes and chaos which were coming forth from the Vatican organization. Upon seeing the new "mass" which was promulgated on April 3, 1969, at the urging of Vittoria Guerrini and Emilia Pediconi (two Roman society ladies who had inspired *Una Voce*) and the assistance of Cardinal Ottaviani and Archbishop Lefebvre, he led a group of Roman theologians and experts that included Msgr. Renato Pozzi and Msgr. Guerrino Milani (both from the Sacred Congregation for Studies) and Msgr. Domenico Celada in preparing a short but powerful document which criticized that new "mass" in detail. The next step was to bring this document around to quite a number of cardinals and bishops who were ready to back it with their reputations by signing it, and then it could be presented to Paul VI for consideration. Perhaps it might have been too much to hope that Paul VI would simply withdraw the new "mass," but at least one could have reasonably hoped that it would be mitigated to some degree by being made optional or being revised again to bring it more in line with the Catholic faith.

Unfortunately, the Abbé de Nantes, a priest on his staff of theologians, leaked this document to the press through Jean Madiran's publication *Itineraires*, and as a result, many of the cardinals backed off. One cardinal who had already signed on and who stood by his action was Alfredo Cardinal Ottaviani, then head of the Holy Office, the Vatican tribunal responsible for uprooting heresy and protecting the purity of the Catholic faith. To this short critique he drafted a short cover letter in which he stated that "the *Novus Ordo Missae*...represents, both as a whole and in its details, a striking departure from the Catholic theology of the Mass as it was formulated in Session 22 of the Council of Trent." Because of his firm stand with this document, the document soon came to be known as the Ottaviani Intervention. Because of the leak, only one other cardinal signed his name to this document, namely Antonio Cardinal Bacci, a famous Latinist who was then serving on the Vatican Congregations for Religious, Causes of Saints and Catholic Education.

The Ottaviani Intervention (See Bibliography) definitely bears reading since its criticisms of the new "mass" have the official status of the man who was then in charge of keeping the Church orthodox in doctrine at the time, the one Curial officer that even a pope is expected to have to check with regarding what he intends to promulgate as teaching on Faith or Morals (Joseph Cardinal Ratzinger would subsequently serve in that role) before releasing it to the public. Another writer, Fr. James Wathen, wrote a book about the new "mass" entitled *The great Sacrilege*, in which he picked up on the points brought out in the

Ottaviani Intervention and developed them in some rather considerable detail. Yet another writer by the name of Patrick Henry Omlor actually anticipated many of the same points in his 1968 book *Questioning the Validity of the Masses Using the New, All-English Canon* (See Bibliography). Another important and vocal speaker on that topic in those days was Fr. Lawrence Brey, formerly of Schuckardt's acquaintance.

As for Cardinal Ottaviani himself, his entire career was one marked for its staunch orthodoxy. In 1953, he wrote a short pastoral letter about the relationship between Church and State entitled *Duties of the Catholic State in Regard to Religion* which remains to this day one of the clearest official defenses of the Catholic position on this issue ever committed to print. During the Second Vatican Council, he was one of the only four prelates (Bacci was another, the remaining two very probably included Bishop Sigaud who voted against the Constitution on the Divine Liturgy which had provided the excuse for changing the Mass. Exactly 2,147 other such prelates (including Marcel Lefebvre) had all voted in its favor. While Cardinal Bea was pushing for his heretical schema on Religious Liberty (which was finally signed during 1965 sessions of the Council), Cardinal Ottaviani was pushing for a solidly orthodox schema on Religious Tolerance, which is the correct attitude towards other religions. During that Council some 250 or so of the participating prelates (including Ottaviani) came to oppose practically every facet of the new religion being outlined at Vatican II, and gradually came together (lead by a holy bishop, Dom Geraldo de Proenca Sigaud) as the *Coetus Internationalis Patrum*, the most prominent minority group within that Council, but who were outvoted at practically every turn.

There are some who have put forth the claim that Cardinal Ottaviani later backed down and withdrew his support for the Ottaviani Intervention. That claim is entirely based on a letter which the Cardinal is supposed to have written to some French clergyman by the name of Dom Gérard Lafond which has him saying that he never saw the document which he had written a cover letter to, and that whatever fears he had about the new "mass" had been put to rest. By all evidences, that letter is simply and purely nothing but a cheap and poorly done forgery.

There is one other group which bears mention here, even though they come somewhat later chronologically, and that is the Society of Traditional Roman Catholics (STRC), founded in 1984 by Martin Kupris and James De Piante in order to work for the full restoration of Traditional Catholic rites in North Carolina. They published a small quarterly paper called *The Catholic Voice* which soon caught on and was read by many Traditional Catholics all around the world. In time, Fr. Kevin Vaillancourt would become the editor of this fine paper clear until its demise in 2013 (2010 as a free paper). Unlike many other groups which come along later, this one kept well the clear vision and unity held by all Traditional Catholics during this early period. Most of the long surviving early heroes of the battle for the Faith remained most friendly with this group.

Archbishop Marcel Lefebvre was a bishop of uncommon accomplishment as a missionary in Africa and as the Superior General of the Holy Ghost Fathers, one of the largest missionary orders. He had been born on November 29, 1905, ordained a priest on September 21, 1929 having doctorates in Philosophy and Theology from the Gregorian Pontifical University, and initially installed as a parish priest in the French town of Lille. Shortly afterwards he began his ministry in the African town of Libreville in Gabon and in 1934 became Rector of the Seminary there where he instituted a first-class educational system which boasted at least three bishops and two heads of state among its alumni. He was consecrated bishop on September 18, 1947, and began serving in Dakar, and there as bishop, and later archbishop, he founded many seminaries and twenty-one new dioceses.

In 1962 he was voted Superior General of the Holy Ghost Fathers, a missionary order, and also asked by John XXIII to serve on the preparatory commissions for the upcoming Second Vatican Council. The preparatory commission on which he served prepared seventy-two totally orthodox schemas for consideration at that Council, and the first session (1962) accomplished nothing but the rejection of each and every one of them, one by one. During the Council he joined the *Coetus Internationalis Patrum* and found himself "on the outs" with many of the Modernists. By 1968 he was no longer able to function as Superior General because of the new ways and constitutions which were being forced on that order (along with all others). As a result, he then resigned with no other ambition than to live in a small apartment in Rome on his small pension and quietly devote the rest of his life, as "titular Archbishop of Tulle," to prayer and contemplation. But God had other plans for him.

Throughout his priestly career he had established a reputation for himself as a staunch supporter of the (traditional) Catholic faith. Almost no sooner did he go into retirement he was approached by young men who simply wanted to know where they could get a good priestly formation. For a short while, he simply directed them to the University of Fribourg where an old friend of his, Mgr. Francois Charriere, Bishop of Lausanne, Geneva, and Fribourg, had suggested. Before long, however, it became clear that Fribourg was increasingly infested with the new spirit of Modernism and not a suitable place for his seminarians. What happened next can only be described as an extraordinary act of God.

In 1968, a man by the name of Alphonse Pedroni happened to learn through a chance encounter that a property in the small Swiss hamlet of Ecône which had belonged to a novitiate of the Canons of Saint Bernard was up for sale, and that a communist group was interested in buying it, and in demolishing its chapel to make way for a shopping mall, bar, movie theater, or discotheque. With some money he and four friends of his came up with, he managed to buy the novitiate with the intent to give it to the Church in some way. By October of 1970, he and his four friends had decided to donate the property to the retired Archbishop to be used for his new seminary. Permission from the local regular bishop (Mgr. Charriere) to found a seminary in his diocese had already been granted verbally on June 6 of the previous year, and again in writing on August 18.

On November 1, 1970, Lefebvre gained official recognition in the form of a Decree of Erection issued by Mgr. Charriere, and on February 18, 1971, he obtained a Statute of Pontifical Right from Cardinal Wright, Prefect of the Congregation for the Clergy, who wrote a couple years later that "this Association...has already exceeded the frontiers of Switzerland, and several Ordinaries in different parts of the world praise and approve it. All of this and especially the wisdom of the norms which direct and govern this Association give much reason to hope for its success." For the next several years after that, the seminary progressed nicely with many more seminarians coming for a traditional priestly formation and all blessings and support from Rome, and even the incardination of some of its priests into various dioceses. As an action which would providentially serve as a precedent for incardinating priests into his order, Fr. Urban Snyder and two other priests were incardinated directly into his own Society in 1971 and gaining Vatican approval by indult for these incardinations.

However, since none of his priests were willing to say the new "mass," they soon found considerable opposition, first from the other French bishops who were totally taken in with the heresy of Modernism, and then later with higher and higher "authorities" within the Vatican establishment. In early 1974, the villainous Modernist elements began to circulate the false rumor that Ecône was some sort of "wildcat seminary" even though its official canonical status was a documented fact. In November 1974, Ecône was officially visited by representatives of Paul VI who found no fault with Lefebvre or his seminary, but determined to find some fault with him they said some scandalous things which were offensive to the pious sensibilities of the seminarians and which did not bode well for the future of the practice of the Catholic Faith in Vatican City.

In response to these things, on November 21, Lefebvre gave a short speech of encouragement to his seminarians. Although intended only for their ears, the press soon picked up on it as being the first truly well-stated manifesto of the Traditional Catholic Faith in these times. It has since come to be known as a "Declaration." It read in its entirety:

We cleave, with all our heart and with all our soul, to Catholic Rome, the guardian of the Catholic Faith and of the traditions necessary for the maintenance of that Faith and to Eternal Rome, mistress of wisdom and Truth.

On the other hand we refuse and have always refused to follow the Rome of the neo-Protestant trend clearly manifested throughout Vatican Council II and, later, in all the reforms born of it.

All these reforms have contributed and are still contributing to the destruction of the Church, the ruin of the Priesthood, the abolishing of the Sacrifice of the Mass and of the Sacraments, the disappearance of the religious life, to naturalist and Teilhardian teaching in the universities, seminaries and catechetics, a teaching born of liberalism and Protestantism and often condemned by the solemn magisterium of the Church.

No authority, not even the highest in the hierarchy, can force us to abandon or diminish our Catholic Faith, clearly laid down and professed by the magisterium of the Church for nineteen hundred years. "But," said St. Paul, "though we or an angel from heaven preach any other gospel unto you than that which we have preached unto you, let him be accursed." (Galatians I. 8).

Is not that what the Holy Father is telling us again today? And if there appears to be a certain contradiction between his words and his deeds as in the acts of the dicasteries. We abide by what has always been taught and turn a deaf ear to the Church's destructive innovations.

It is not possible profoundly to modify the "*lex orandi*" without modifying the "*lex credendi*." To the new Mass there corresponds a new catechism, a new priesthood, new seminaries, new universities, the charismatic and pentecostal Church—all opposed to orthodoxy and to the age-old magisterium of the Church.

Born of liberalism and modernism, this reform is poisoned through and through. It begins in heresy even if not all its acts are formally heretical. Hence it is impossible for any informed and loyal Catholic to embrace this Reform or submit himself to it in any way whatsoever.

The only way of salvation for the faithful and the doctrine of the Church is a categorical refusal to accept the Reform.

It is for this cause that with no rebellion, no bitterness, no resentment, we carry on our work of training priests under the star of the timeless magisterium, convinced that we can render no greater service to the Holy Catholic Church, the Sovereign Pontiff and future generations.

t is for this cause that we hold firmly by all that has been believed and practised in the Faith, in morals, in worship, in the teaching of the catechism, the moulding of a priest and the institution of the Church, that eternal Church codified in her books before the modernist influence of the Council made itself felt, awaiting the time when the true light of Tradition shall scatter the darkness clouding the skies of eternal Rome.

In so doing, by the grace of God, the help of the Virgin Mary, of St. Joseph and St. Pius X, we are assured of remaining faithful to the Holy Roman and Catholic Church, to all the successors of Peter, and of remaining "*fideles dispensatores mysteriorum Domini Nostri Jesu Christi in Spiritu Sancto.*" Amen.

One can see in this declaration an embryonic understanding of the distinction between that which is really the Catholic Church (Eternal Rome) on the one hand, and the Vatican organization which has gone over to the false new religion (Modernist Rome) on the other. All the same, Lefebvre would go to his death without ever understanding fully what had happened to the Church at the second Vatican council.

Much as having such a declaration on record caused Ecône to gain prominence among real Catholics, it also occasioned excuses for the ex-Catholic Vatican organization to start trying to shut it down. Paul VI's declaration against the Tridentine Mass had gone on record only few weeks before then. All of this was only the beginning of Lefebvre's troubles. Of greater concern was the retirement of Bishop Charriere which meant the loss of an important ally, and his replacement with Bishop Pierre Mamie who was no particular friend to the Traditional Catholic Faith.

There were other bishops who held the line and remained Catholic despite the pressure from all sides to get in step with the new religion. Most heroic of those bishops was Antonio de Castro Mayer, who along with Bp. Sigaud, who was Bishop of Jacerzinho, and their political ally Plinio Corrêa de Oliveira kept parts of Brazil faithful during the years that the rest of Latin America was falling for the heretical "Liberation Theology."

Bishop de Castro Mayer was born June 20, 1904, obtained his Doctorate in Philosophy from the Gregorian University in Rome and was ordained on October 30, 1927. As a priest in the diocese of Sao Paulo, he served as a professor in a provincial seminary, Canon of a Cathedral, and General Counselor of Catholic action. On May 23, 1948, he was consecrated a bishop and assigned to the diocese of Campos, Brazil, just north of Rio de Janeiro. Even as far back as 1950, this perceptive bishop was writing pastoral letters which warned against the heresy of Modernism which was even then devouring seminaries and other organizations of the Church. Letter after letter was written filled with such high-quality information, yet readable by the common man in his diocese. In 1956 he founded a minor seminary in the small town of Sao Sebastinoa de Varre Sai which expanded in 1967 to become a major seminary. In time, he had instituted a number of Marian devotions and catechism classes throughout his diocese. By the late 1960s his diocese had become the best educated (in their faith) diocese in the entire world, and not a moment too soon.

Elsewhere, most people were spiritually asleep, simply doing whatever they were told by their Catholic but defecting leaders without even thinking. As these leaders ceased to be Catholic, most people simply continued following their ex-Catholic "authorities" into religious error, which is what the revolutionaries were counting on. There is a very good reason that such modernists, liberals, and other such revolutionary heretics must work from the inside. These people are motivated by their own lusts and sensuality and all of the seven deadly sins, including sloth. They lack the willingness or wherewithal to build up anything from scratch as the Church had to. The only way these people ever get into any power is to take over the power positions of organizations which were built up by the labor and sacrifice of others who do not happen to be modernists, liberals, or revolutionaries like themselves. That is why they try to take over the Church instead of merely building some other Church on their own. They want the results but aren't willing to pay the price.

In the well-educated diocese of Campos, on the other hand, all of the people knew their faith well enough that when the false new "mass" was promulgated in 1969, almost nobody went to it. Any priests who wanted to say the new "mass" and propagandize for Liberation Theology found themselves doing it in an empty church. From time to time, such a priest would arrive, try to drum up business for the new religion for a while, and after failing to get any support from the lay faithful they would finally end up just going away to greener pastures. Bishop de Castro Mayer had taught the Catholics of his diocese well. Other than such rare visitations, all Masses said within this diocese remained the Catholic Tridentine Mass. The seminary up at Varre Sai kept all the old ways, teachings, and disciplines, although it soon had to be moved to the City of Campos to prevent its coming under control of the revolutionaries. They wanted to deprive Bishop de Castro Mayer of the seminary by having his diocese cut in two, Varre Sai being in the portion which the Bishop would no longer have under control. Needless to say, once the seminary was moved to the City of Campos, there was never again any talk of splitting up the Diocese of Campos into two dioceses.

Such a state of affairs persisted in Campos clear through the 1970's and clear up until November of 1981 when, forced into retirement, Bishop de Castro Mayer was replaced by Bishop Carlos Alberto Navarro. Throughout those years, Bishop de

Castro Mayer ruled his diocese in a totally Catholic manner and yet also managed to keep peace with the Vatican organization and Paul VI and the John Paul's. Some might wonder how this happened since there really had been no special permission from the Vatican organization for this diocese to remain Catholic.

One must take into account the fact that revolutionaries are, when pushed to the wall, practical people just like everyone else, and they had a very good pragmatic reason for allowing this faithful bishop to keep his diocese Catholic: Money. As the other dioceses in Latin America lost their Catholic faith with the new false liturgies and also by embracing Liberation Theology, they lost the regular attendance of most of their members and most of the financial generosity of what few people were remaining. By the mid-1970s, the typical Latin American diocese could only stay out of the red by selling off an average of about three disused Catholic Church facilities per fiscal quarter. Very few dioceses could sustain themselves financially over an extended period, and only one did so without ever having to sell off anything: Campos. Much as the ex-Catholic bishops hated de Castro Mayer for being such a reproof to them, they needed him. After all, SOMEBODY had to finance all of their horsing around! And there were after all only a limited number of Church buildings left over from the times when Catholics in huge numbers gave and sacrificed freely of their time and money to build them.

Alas, in November of 1981, the peace between the Last Catholic Diocese and the ex-Catholic Vatican organization came to an abrupt end with the installation of bishop Navarro. The man was a complete martinet whose marching orders were clearly to abolish all Catholic faith, worship, and piety from "his" diocese. Over the next five years he went from parish to parish attempting to convert each to the new religion, first with veiled threats, then with open threats and public challenges, and finally with official legal action he would get with the support of corrupt judges. Neither priest nor parishioner would ever have anything to do with the new "mass." No priest would say it even if ordered. If the bishop himself came and started to say it in any Church, the parishioners there would all quietly get up and file out of the Church.

Finally, priest and parish alike would be compelled to give away the keys, a last Mass would be said, and the Blessed Sacrament would be removed from the tabernacle and carried in procession to another place. The bishop would then find himself in charge of a practically empty church building. Often, he would find it cheaper to close it down and sell it off rather than bring in some new "priest" to say the new "mass" for the five to ten casually dressed tourists who might show up instead of the hundreds of regular parishioners who had formerly attended each Tridentine Mass there.

The exiled parish, without a building, would work to get a building and in the meantime meet in some of the most funky places, such as movie theaters and school classrooms. One parish had just completed their Church building only a couple months before they were driven out and had to start building all over again. Other than having to foot the bill for a new facility, priest and parish continued to function exactly as they always had. The systematic removal of Catholic priests and congregations and faith from the Church buildings happened in parish after parish until August 31, 1986, when Padre Fernando Rifan of the City of Campos said his last Mass in his regular church, Nossa Senhora do Rosario do Saco, handed over the keys, and joined the couple dozen or so priests who were similarly exiled. The faithful priests, who had come to call themselves the Society of Saint John Vianney (after the patron saint of parish priests), were now all in exile, along with their bishop de Castro Mayer. Clearly one can see the force of even one faithful bishop. If only many more were like him! One can also see that it is not the priests (and bishop) and parishioners of Campos who changed their religion, but the Vatican "authorities," particularly as seen in the person of bishop Navarro. More details of the events in Campos, Brazil can be read in the book, *The Mouth of the Lion* (See Bibliography).

The Traditional Catholics opt for the Church of living memory; they do not chase after some hypothetical ancient Church, now long lost and forgotten, then rebuilt from scratch, but staunchly stand by the Faith as it was already known to all Catholics of previous generations and who were (and are) still alive to witness to the fact that the Faith as practiced by Traditional Catholics is absolutely identical to the practice of the Catholic Faith as they knew it growing up.

Sometimes one hears the claim, "These parishes with their Tridentine Masses and traditional worship are only an imitation of the Catholic Church as She existed before Vatican II; they are not the real thing but only a copy." As I have already proved, the Vatican organization is no longer identical to the Roman Catholic Church, but merely a secular organization of which some Catholics happen to be members and others are not, and which furthermore has a great many members in it who are not Catholics. As I have also proved, the new "*Novus Ordo*" religion which is believed and practiced by the "People of God" and which is the primary thrust and interest of the Vatican leadership is not at all the Roman Catholic religion, but in fact just some other religion much like Lutheranism or Episcopalianism or Mormonism or whatever. There remains only one kind of Church which even looks like the Roman Catholic Church, and that is all of these traditional parishes in which the teaching is solidly Catholic, uncompromisingly orthodox, where only Catholic morality is taught and practiced, and where all sacraments follow the Tridentine norms.

Now, either these traditional parishes <u>DO</u> in fact themselves together constitute <u>THE</u> One, Holy, Catholic (and Roman), and Apostolic Church, or else they do not but are only some sort of imitation. I offer here proof that these traditional parishes are the Catholic Church. Proof is by contradiction: let us suppose that the Traditional Catholic parishes are merely imitation Catholic Churches and not actually the Catholic Church at all. It is ever the nature of the Church to preach the Gospel to all creation, edifying souls through Her valid sacraments, infallible teaching, and lawful discipline. This cannot be done by a society totally hidden in a forgotten corner of the earth like some "Shangri-La," nor performed by prelates who by their lives and profession give no reason to think they are Catholics. Since there is no society out there which embraces and spreads the Catholic faith (other than the Traditional Catholics ruled out by the assumed proposition), there is no other candidate for anyone's being the Catholic Church. One must therefore conclude that the Catholic Church no longer exists; She has completely vanished off the face of the earth! But Catholic doctrine (*de fide*) teaches that the Church shall endure forever and cannot cease to exist. That is a contradiction!

Either the Church exists or else the Church does not exist, but not both. The organization of the Church's society, the real and actual Mystical Body of Christ, can be injured through widespread confusion, ignorance about an unprecedented circumstance, and an absence of an operative and universal leadership. But the content and character of Her Faith and Morals, both as directly believed and practiced, and expressed (professed) in Her liturgies, can never be corrupted. The claim that these "independent" Traditional Catholic parishes are only imitations of the Catholic Church but not the real thing leads to contradiction, and therefore it is false. This only leaves the alternative proposition, namely that these traditional parishes <u>DO</u> in fact constitute <u>THE</u> One, Holy, Catholic (and Roman), and Apostolic Church. Q. E. D. Therefore, it is heresy to claim that the Traditional Catholic parishes are all merely imitations of the Catholic Church since that would be a denial of the *de fide* Catholic teaching that the Church shall endure until the End of Time.

If there be any imitation "Catholic Church" around, it would be the various dissenters such as the schismatic East Orthodox, certain "High Church" Protestants such as the Episcopalians, the Old Catholics, and the Novus Ordo Church of the People of God. One finds within each of these a divergence from the Catholic Faith, Morals, Worship, and Discipline which is not possible within the Catholic Church, and which is simply not to be found amongst the Traditional Catholic groups and orders. And to be brutally frank, the Novus Ordo "church" of the "People of God" is in precisely the same category. Whether it be imitation tradition, "ancient ways," or even imitation hierarchies, the phoniness of each is easily seen through.

One important observation is the fact that in these early days of Catholic résistance to ex-Catholic Vatican shenanigans, there was no truly great leader. One cannot point to any man and say of him, "He started the Traditional Catholic movement or community; without him it would not exist." Although one of these figures, namely Archbishop Marcel Lefebvre, would later go on to become of rather considerable importance to this historical account, even in some ways the "backbone" of the Traditional Catholic community (and therefore the Church), as this era drew to a close he was still pretty much relatively unknown, just the Rector of a small but truly Catholic seminary located way out in some Swiss boondock. What that observation means is that the Traditional Catholic community is not of man's making but God's making. It was God Himself and not any mere man who was and still is at the helm of His Church and keeping it alive through the actions of all of these holy bishops, priests, religious, and laymen.

A pattern one does see here over and over again in the lives of the faithful priests and bishops is that since in all sincerity they were mistaking the Vatican organization for the Roman Catholic Church, they all did their level best to try to stay at peace with it while continuing to do what they had been ordained to do. It was only when such peace became completely impossible for them to obtain that they found themselves forced to choose between their Catholic Faith and what they imagined to be the Catholic Church. Without even realizing it, those who chose to "lose their life for Christ's sake" by standing fast by their Catholic Faith regardless of what the Vatican organization told them to do or thought of them in fact remained completely within the visible structures of the Roman Catholic Church and thereby found their life. Conversely, those who tried to "find their life" by appeasing the Vatican leadership, even to the extent of sacrificing their Catholic Faith soon found themselves outside the visible structures of the Catholic Church (even while they were as yet within the visible structures of the Vatican organization) and thereby lost their life.

In all fairness to the Vatican leadership, I must make note of the fact that, as promised by the Vatican II documents, part of the Church would continue to "subsist in" the Vatican organization, which means that at least some would be allowed to continue their Catholic faith and worship with the full permission and blessing of the Vatican establishment. Apart from such "foot-dragging" priests such as Schoonbroodt or bishops such as de Castro Mayer (who furthermore referred to the perpetual Indult granted by Pope Pius V to use the Mass he had promulgated in 1570), there were only three places that Catholic worship was officially allowed by the end of 1974: 1) Elderly and retired priests ("dogs too old to learn new tricks") were allowed to say Tridentine Masses by themselves (without a congregation), 2) an "Indult" (special permission) had been granted to Cardinal Heenan (of parts of England and Wales) to be allowed to use the 1967 edition of the Roman Missal on an occasional basis, and 3) the Eastern Catholic Rites which were not in any way yet affected by the promulgation of the Novus Ordo Missae nor any others of the new "sacraments" because they were only meant to replace the sacraments of the Latin (Western) Rite. The 1967 Roman Missal was the last edition promulgated by the Vatican establishment which was still within the confines (however barely) of acceptable discipline and also their last intrinsically valid Mass. Already by late 1967, the new and invalid vernacular "eucharistic canons" were being sent out to parish priests as supplements to the 1967 Missal.

If one wants to make the claim that only such traditional parishes as are approved by the Vatican leadership count as the hierarchy of the Catholic Church, one then ends up believing in a "Catholic" church which only exists in certain places as permitted by some entirely secular establishment (which is what the Vatican organization really is), but not all over the world. Such a church, in lacking universal jurisdiction over the entire earth simply cannot be the entirety of the Catholic Church because it is not Universal, which is what "Catholic" really means. Such a church, as so defined, cannot therefore be the Catholic Church, but at most only a mere portion of it, and that much, at least, I admit that it is.

7

THE ADVANCE OF MARCEL LEFEBVRE AND THE SSPX

One might easily put down the resistance to change during those earliest days to mere human nature, but had that been so it would have soon died out. Both the remaining faithful Roman Catholics and the heretics in the Vatican organization underestimated the situation quite seriously. The Catholics, as yet still making the assumption that the Vatican organization was still simply the Roman Catholic Church, therefore assumed that it must soon return to its senses. Surely the Catholic Church must if She is to be an eternal organization capable of lasting through all of the vicissitudes of time and place, assuming it were capable of such a deviation in the first place. Conservatives and Traditionalists alike were hoping to just "ride it out" or "wait it out" until such time as the madness would be over with. At the worst, maybe it would take clear until Paul VI dies. Then the next pope would be sure to put things back in order. Then again, maybe Paul VI might wake up some day and realize the enormous damage he has been doing and begin trying to undo it.

But as we move into the next period, Conservatives and Traditionalists begin to part company as the madness dragged on and on, and fewer and fewer places of sound doctrine and reliable sacraments could be found. Over time, Conservatives came to be reduced from fighting for the true Mass to fighting for sound doctrine or reverence in worship, or even the use of Latin hymns. Finally their reduction brought them to the pathetic role of fighting to continue the use of incense or bells, or having statues and altar rails left standing in their parish churches, just like a mother having lost her child might cling tenaciously to that child's things. Eventually they got carried off with all of the rest, just a great deal slower. Many of them were quite old and began dying off. By all human standards, the pre-Vatican II Catholic Church should have been nothing more than a vague and confused memory in the back of the minds of a few elderly people. Another couple generations and it would be forgotten entirely.

Just as Conservatives and Traditionalists both underestimated the seriousness and possible duration of the crisis, so it is also true that the revolutionaries and radicals who sought to change the Church into some model of their own making had absolutely no idea just what they were up against. Since they don't really believe in God (except in some vague Masonic sense of some impersonal "Great Architect" who creates the world and then leaves man to his own devices), they think of the Catholic Church as merely a human organization. Since humans have made it (goes the reasoning), humans therefore have the power and right to change it or abolish it as they choose.

As for those of a rebellious disposition (the goats), the revolutionaries did not need to concern themselves on account of them, since they already had them on their side. But the sheep, being the faithful and innocent and naïve and trusting sheep that they are, were expected to blindly follow their leadership into error. Is that not the natural behavior of sheep? If the Pope says, "Today, we must worship the Devil," will not the sheep obediently worship the Devil? The revolutionaries, not believing in God, had no reason to expect that the sheep "will by no means follow a stranger, but will flee from him, for they do not know the voice of strangers."—John 10:5

In a strange and roundabout sort of way, one can even be grateful for their efforts. After all, without the betrayal by Judas, how would Jesus have ever been caught in order to be crucified? In a similar manner, the betrayal by the Vatican leadership of the Church is even now providing an opportunity for the most spectacular demonstration of God's power and glory and existence since the resurrection of Jesus Christ Himself. As it was in the case of Judas, those in the Vatican leadership who have betrayed the Church do not stand to gain any real and eternal advantage for themselves personally from their folly, but God works even the Devil's worst evils into His own divine plan.

As time dragged on without any end in sight to all the madness which was going on in "the Church," Catholics became increasingly concerned about the future of the Church. Although thousands of faithful priests still remained worldwide, most of them were getting up there in years and others were caving in to the new religion. Soon it was not thousands but only maybe a thousand or so faithful priests left. Even worse, what few seminaries as still remained open at all were no longer turning out faithful priests but all sorts of clowns, revolutionaries, and other faithless losers who clearly didn't even know what a sacrament was, let alone a commandment of God. All the new seminarians were learning anymore was holding hands, singing *Kum Ba Yah*, and "building a community," whatever **that** meant. The prospect of "priests" like these becoming "bishops" (or "Pope"?) was unthinkable. Where would the Church get faithful priests and bishops for future generations?

It is with the increasing consciousness of this concern that attention became attracted to the seminary Archbishop Lefebvre had established in Ecône. As the seminary grew in importance to the faithful, so did the attacks upon it. Also, once Lefebvre's declaration became widely known (in early 1975), it immediately became a rallying cry for all true Catholics all around the world. With that, Lefebvre moved from the background into the foreground. This chapter then, is the story of Archbishop Marcel Lefebvre and the Society of Saint Pius X (SSPX) a fraternal association of common life which he founded and led up until his death in 1991; it is the story of the largest portion of the Church which was forced to function outside the Vatican organization, and for that matter the largest single priestly order of the Church which can truthfully be called Catholic. As of this writing the SSPX is still larger (in terms of the number of priests, religious, and lay faithful) than the size of all other Traditional Catholic orders and other clergy and congregations put together.

Bishop Mamie who had replaced Bishop Charriere, the diocesan Bishop of Lausanne, Geneva, and Fribourg, was no particular friend of the Catholic tradition. Although he did not start out as a particular enemy of tradition either, he was not truly committed to the Catholic Faith and only all too easily made to cooperate with the other French bishops in the persecution of Lefebvre and the seminary in Ecône. The first two attacks on the seminary had taken place even before the famous 1974 declaration. The first was an illegal attempt by Bishop Mamie to use his diocesan authority to close down Ecône. That is illegal on the technicality that even though a diocesan bishop has the authority to set up a religious society within his diocese, or to allow an existing religious order to open a house in their diocese, only the bishop who himself personally has done this can revoke the permission so given. In particular, a bishop who succeeds a bishop who establishes a religious order in his diocese must have recourse to Rome to shut it down.

When Bp. Mamie failed to shut down Ecône, the second wave of persecution began. This was the rumor, spread by the entire group of French leaders, that Ecône was some sort of "wildcat seminary." While some foolish or ignorant persons bought that lie, many more seminarians just kept right on coming. Though a smattering of priests had already been ordained at Ecône, beginning in 1971, all of them had been transfer students from other seminaries. Finally, in 1974, the first seminarian, of what would soon come to be hundreds, was ordained who had been entirely trained and formed under the Archbishop's tutelage. It was also in 1974 that the heat began to be turned up. Having failed to close down the "wildcat seminary" either by abuse of diocesan authority or malicious rumors, at length they finally got Rome involved. In March, a group of cardinals met in

Rome to discuss the Ecône problem and by June they had decided to send the two visitors who had so scandalized the seminarians in November.

By January of 1975, the two visitors had prepared their report and delivered it to Rome. Although very few people appear to have ever seen it, by what few indications known it seems to have been for the most part a favorable report. However, there was that famous declaration, and that alone provided the excuse for the legal machinery of Rome (which normally moves along at a glacial pace) to creak into high gear. Abp. Lefebvre had been invited to an informal meeting in Rome in February with Cardinals Wright (the same who had so highly praised Ecône only a couple years previous), Garrone, and Tabera. Upon arrival, it turned out that he had been invited on false pretenses. The meeting was little more than a berating and "chewing out" session in which the Archbishop was roundly criticized for his declaration. It was a tribunal in which he was already judged guilty and granted no recourse.

Exactly the same fiasco was repeated in March. He was accused of "breaking with Rome" and being the leader of the traditionalist community and all sorts of "evil." At the same time Fr. de Nantes was publishing articles in which he was predicting that Lefebvre would break with Rome, and all the better for him if he does! Lefebvre was shocked at this and wrote in response an open letter to Fr. de Nantes saying that "if a Bishop breaks with Rome it will not be me." That episode so clearly demonstrates Abp. Lefebvre's personality, and the reason why he earned so much respect from even many of his opponents. No matter what disagreements he had about the way Paul VI was running "the Church," his respect for the person of Paul VI as duly elected Pope and profound respect for the office knew no limits.

Unfortunately, such respect was entirely lost on Paul VI who frankly resented having his failing plans and programs criticized in the public forum. The three cardinals attempted to shut down Ecône, as if their involvement constituted Bp. Mamie's recourse to Rome. Again, it was of no canonical effect because "recourse to Rome" does not merely mean that some cardinal in Rome says something and there it is. There must be a fair trial by competent authority, and that had been specifically denied Lefebvre. Lefebvre wrote Paul VI asking for a fair trial or at least as much forgiveness as he had shown to several truly dissident theologians and seminary professors such as Hans Küng. In June, Paul VI wrote a letter to Abp. Lefebvre in which he affirmed the decision of the three cardinals and attempted to persuade Lefebvre to discontinue his heroic stand.

The same day that letter was written (and therefore before it had arrived), Abp. Lefebvre ordained three more priests who were incardinated into various dioceses. He had over a hundred seminarians in training and over a dozen professors in Ecône, had already opened houses in Albano, Weissbad, and Armada, and without intending it, had become a sort of *de facto* spokesman for Traditional Catholics around the world. The house in Albano, near Rome, had been officially permitted by Bp. Mamie and also Bp. Maccario of the diocese of Albano. Despite its being quite active in training priests, officially the seminary of Ecône no longer existed. Several professors were lost when their religious superiors summoned them back. Perhaps about a dozen students also left at that time, but there remained hundreds of seminarians who simply continued their training, and there were dozens of applicants willing and even eager to take the place of the few who left. By December of 1975, the rumor mill was in full gear.

The Swiss Bishops' Conference published a "dossier" on Ecône in which they simply repeated numerous calumnies against Lefebvre that Cardinal Villot had made. Worst of all, bishops all around the world were herein advised not to incardinate any priests from Ecône into their dioceses. This bears some discussion since not many people know about the process of incardination. Normally, before a priest gets ordained, he must be incardinated into either a diocese or a religious order. Incardination means that he is expected to serve within that diocese or religious order once he is ordained. Later on, he may transfer if his bishop releases him and if another bishop is willing to accept him into his diocese or religious order (all transfers from other seminaries to Ecône had been lawfully approved in this manner), but some bishop somewhere must approve the ordination in the first place. Typically, that bishop also participates in the ordination ceremony.

The first few Lefebvre priests had already been incardinated into various dioceses and therefore had no problem, but after the promulgation of that infamous "dossier," very few of Lefebvre's priests ever got incardinated directly into any diocese. Even that did not slow up even in the least the steady stream of bright young men eager to place themselves at the service of God and the Catholic Church by entering Ecône and the other seminaries Lefebvre had opened in various parts of the world. Also starting at about this time were both men and women offering themselves to be monks and nuns at the service of the Church through the SSPX.

Over the course of 1976, the wound continued to fester even more as Paul VI, who had been up until that point very much a background figure in this problem and one who furthermore had obviously never been fully informed of the facts, began to take an active participation in the persecution of Lefebvre. Speaking as if he had a better understanding of what "tradition" is than the entire line of previous popes, he said in one speech that "there are those who, under a pretext of a greater fidelity to the Church and the Magisterium, systematically refuse the teaching of the Council itself, its application and the reforms that stem from it," by which he meant to refer to Abp. Lefebvre and all Catholics in sympathy with him.

Finally, again without due process or a fair hearing or trial (and therefore illegally), through the mediation of Abp. Benelli Paul VI threatened to suspend Lefebvre if he did not turn aside from providing for the future of the Church which Benelli disparagingly referred to as a "blind alley." Lefebvre was specifically told not to ordain any more seminarians, in fact, none of the 15 seminarians he was to ordain that year had been incardinated into any diocese, so effective had Cardinal Villot's "dossier" been. As reported in *The Biography of Marcel Lefebvre* by Bp. Bernard Tissier de Mallerais, pages 483-484, operating on the advice of several canon lawyers and his friend Bp. Adam, Lefebvre took the radical step of incardinating those priests directly into his own order, the Society of Saint Pius X. Abp. Lefebvre had based the claim of his ability to incardinate priests into his order on the strength of a letter of praise from Cardinal Wright which was practically tantamount to a "decree of praise," recommending approval for such incardinations, the opinion of Bishop Adam to the effect that "since your society is spread throughout several dioceses, it certainly has the power to incardinate within its own ranks, and the precedent, as granted by indults from the Sacred Congregation for the Religious by which three priests had already been incardinated within the SSPX. The arguments for incardination, though arguably "colored," were therefore nevertheless probable.

By this time, in view of the inability to incardinate priests, other than into his own Society, the priests thus incardinated would be sent into many places around the world, all without regard for the local Novus Ordo leaders, who really and morally had no say in the matter, regardless of any mistaken opinion on their part of the contrary, since, thanks to the legislation of Vatican II and *Lumen Gentium* in particular, "no minister, even an admittedly Catholic one or not, can be excluded from performing [all ministerial] functions within anyone's parish or diocesan territory, even altogether without permission or consent."

Frustrated in his attempts to shut down the Catholic Church, Paul VI raised the stakes much higher by attempting to suspend Abp. Lefebvre on July 1, 1976. Again, no legal recourse was ever offered and no due process of law ever observed. The suspension of Lefebvre on that day goes on record and remains to this day by far the most irregular "suspension" of any clergyman in the entire history of the Vatican. Both Paul VI and the villainous cardinals in league with him in Rome on the one hand, and Lefebvre and his priests and seminarians on the other, fully well knew that this suspension was an illegal and irregular one and therefore of no legal or moral force.

The press (both Vatican and secular) on the other hand published a very different view of these events. Both portrayed Lefebvre as some sort of rebel who had broken with the Roman establishment and was ready, willing, and (in the eyes of the secular press at least) able to take on the entire Vatican establishment. Neither press really understood. The Vatican press presented Lefebvre as some sort of rebel who was being disobedient to the Church. The secular press praised Lefebvre as if he were some sort of twentieth century Luther ready to start a new Church. Both of course were entirely wrong, but the voice of sanity was barely heard at all. Feuds and fights (even where clearly fabricated) always sell more copy than warmth and friendship, and both presses played this card to the hilt.

Lefebvre did not break with Rome; a far more accurate thing to say is rather, that "Rome" broke with Lefebvre. It was the Biblical story of Cain and Abel all over again. God did not accept Cain's fruit offering "which earth has given and human hands have made" and likewise did not accept the increasingly false and invalid new "mass" at which the bread and wine were merely bread and wine fit for a memorial supper served over a nondescript table. But God clearly did accept the flesh and blood offering, the "sacred and unspotted Host," of His loyal servant Abel, and likewise accepted the true, reverent, and unquestionably valid Masses of Abp. Lefebvre and his priests, and even over a consecrated altar, where the Body and Blood and Soul and Divinity of our Lord was still to be found.

The "mass" of Paul VI had already by then driven hundreds of millions out of the Church, but also produced no sanctity and no saints. What very few saintly ones who could still be found (such as Mother Teresa of Calcutta) had long since been formed and blessed by the old ways of the pre-Vatican II Church. On the other hand, Abp. Lefebvre's order, the SSPX, and also his male and female postulants, his seminarians, and the regular lay parishioners grew quite dramatically in numbers and in sanctity. The sinful envy the Vatican leadership had of Lefebvre and the evident blessing God gave his offerings was precisely identical to the sinful envy Cain had of Abel and the evident blessing God gave his offerings.

When, at about this point, a large group of Traditional Catholics made a pilgrimage to Rome on foot, hoping to gain an audience with Paul VI so as to request a return to the traditional (Catholic) Mass, Paul VI demonstrated the profound depth of his apostolic solicitude by refusing even to see them, but instead chose to be entertained by the Belgian Soccer team. Where the suspension and refusal to hear any appeals over the next couple months or so correspond exactly to the murderous blows Cain inflicted on Abel, what happened next can only be described as the voice of Abel's blood crying out from the ground, a voice which sent the twentieth century Cain, Paul VI, into flight. In the view of this writer, history has already made its judgment, and that judgment is clearly in favor of Abp. Lefebvre and just as clearly against Paul VI.

Abel's blood was heard from the little French town of Lille, Marcel Lefebvre's own native region. On August 29, 1976, Lefebvre celebrated Mass there despite the invalid suspension, and over ten thousand persons came. Originally, and even up to merely a week previous, that was not what Lefebvre had intended. What he intended was that he would go home and say Mass in some small chapel which someone had granted to him in the midst of this crisis, and perhaps about 100 local people from Lille would show up and be blessed by the sacrament. Word of this intended Mass electrified the entire Catholic community and so spread like lightning throughout it. People started showing up from all over the world. Entire tour groups from foreign countries who had come for this were turned away, but came back anyway.

At practically the last moment, the auditorium of the International Fair in Lille was rented with seating capacity for 10,000 persons and not only was that filled, but thousands more had come who all had to stand, in many cases, outside that facility for sheer lack of standing room. In his homily which went much longer than he had planned, he reiterated his reasons for training, forming, and ordaining seminarians, explained that this Mass was not a protest demonstration but a manifestation of Catholic belief and loyalty on the part of those attending. He clarified the fact that he was not any sort of "leader" of the traditionalists but merely a simple bishop doing his job. He decried the perverse attempt to unite the Church to the Revolution which only produced bastard rites of doubtful validity, and he advocated a missionary stance.

Some controversy was generated when he praised the government of Argentina for having restored a significant measure of peace and order for that nation because he had not intended to make any sort of political statement at all. He was not endorsing that government in either its leaders nor its structure, but merely pointing out the prosperity, even in a material sense, which logically follows from taking at least a somewhat Catholic stance on the part of that (or any) secular government.

He summed up by pointing out the rather bizarre fact that any other kind of worship was being allowed in the Churches that numerous Catholic saints had built and worshipped in, but the traditional faith was not. Moslems? Come on in! Buddhists? You too! Jews? No problem; we have room for all! But Traditional Catholics? Forget it; you're not wanted here! The one and only mode of worship these great cathedrals and churches were built for had now become the one and only mode of worship now prohibited in these same cathedrals and churches.

After so great an event, an astonishing thing happened. Abp. Lefebvre was suddenly invited to an audience with Paul VI. Having attempted for years to obtain one and after being repeatedly told that it cannot happen until he closes his seminaries and renounces his Catholic faith, suddenly here it was, and that without having renounced a thing! It was in the course of this audience that he learned that Paul VI had not been truly appraised of the facts of his case. Paul VI had evidently been told that Abp. Lefebvre has his seminarians take an Oath against the Pope, a calumny clearly calculated to prevent the meeting between Paul VI who bemoaned the autodemolition of the Church, and the one solitary bishop who knew precisely how to put a stop to it and restore the Church to order. Paul VI was also quite hesitant to meet up with Abp. Lefebvre because deep down he knew he was wrong to disagree with Lefebvre, in whom spoke all the reliable popes of the Church.

The Vatican press subsequently denied it of course, but the facts are clear and indisputable: Someone (presumably Cardinal Villot) had deliberately kept Paul VI in the dark about the Traditional Catholic Faith, its adherents, and the need for its existence. Lefebvre and Paul VI achieved some measure of peace and reconciliation and even closed their time together with some prayers. Alas, after this brief peace, on October 11, 1976, Paul VI wrote Lefebvre saying that he must surrender each and every facility, property, seminarian, religious, and every other asset of every kind to the by now very doubtful Vatican "hierarchy" or else face a suspension from his clerical duties.

Needless to say, the Archbishop never surrendered the assets of the SSPX, and Paul VI never actually carried out his vacuous threat to have him suspended. There the matter rested until January of 1978. In the meantime, several other events of interest took place. In 1977, Abp. Lefebvre ordained 16 new priests, and had come to have over 40. His sister, Mother Mary Gabriel Lefebvre, ran a general house (founded in 1974) which moved that year to St. Michel-en-Brenne. Parishes, mass centers, religious and lay followers continued to grow dramatically in numbers. A group of thirty Catholic University teachers signed a manifesto in which they gave public expression to their solidarity with Abp. Lefebvre and the Catholic Church in all of Her pristine greatness for which he stood. It was in February of 1977 that another miracle happened.

In Paris, a large parish church by the name of St. Nicholas du Chardonnet, which had room for thousands, was quietly taken over by Traditional Catholics. On the last Sunday of that month, Catholics began coming inside in greater and greater numbers. One of the (Novus Ordo) parish clergymen there seemed pleased that so many people would show up that morning. Out of curiosity he turned to one of the laymen who had come in with the rest of the new people and asked, "Who are you? Why are you here? We're overcome with delight." The layman answered him saying, "Let's hope you're still delighted in a few minutes' time."

After that few minutes' time, Msgr. Ducaud-Bourget came in, flanked by a deacon and subdeacon. The funky wooden table along with its tattered purple cloth was summarily folded up and packed away while Msgr. Ducaud-Bourget made his triumphant procession up the center aisle and proceeded to say the Tridentine Mass on the high altar. Without any real violence, the Novus Ordo priest and those parishioners who were not glad to see the faith they ostensibly believed in restored to their parish church were gently driven out of the building. The Traditional Catholics just moved in and took over. When the Novus Ordo clergymen called the police they were informed that in France, the government had taken over all Church facilities, although they have continued to allow the Church to use those facilities for Her worship. The Novus Ordo clergy were simply ignored.

It turned out that the Mayor, the Chief of Police, and practically the entire police force were all of a truly Catholic opinion and would not lift a finger to help the Novus Ordo clergymen. Many of them personally attended the Tridentine Mass as said by Msgr. Ducaud-Bourget. For the next several years, the Novus Ordo establishment tried to get it back, but they never could. Although Msgr. Ducaud-Bourget has since gone to his reward, St. Nicholas du Chardonnet continues as a Traditional Catholic parish to this day. Since that time it has been fixed up, cleaned up, and made to look just as it did when it was new.

As Lefebvre continued his formation and ordination of priests and while St. Nicholas du Chardonnet became one of the largest Catholic parishes in all of Europe, abuse continued to be heaped on him by practically every member of the almost entirely apostate Vatican leaders. Although Lefebvre never quite embraced the opinion that Paul VI had somehow lost the papacy, he demonstrated himself to be very close to such an opinion at various times when he referred to the Vatican establishment as being in schism by having departed from the truths previously established and defined by the Church. In one sermon in 1977 he went so far as to say "The Deposit of Faith does not belong to the Pope. It is the treasure of truth which has been taught during twenty centuries. He must transmit it faithfully and exactly to all those under him who are charged in turn to communicate the truth of the Gospel. He is not free.

"But should it happen because of mysterious circumstances which we cannot understand [the loss of authority back at Vatican II], which baffle our imagination, which go beyond our conception, if it should happen that a pope, he who is seated on the throne of Peter, comes to obscure in some way the truth which it is his duty to transmit or if he does not transmit it faithfully or allows error to darken truth or hide it in any way, then we must pray to God with all our hearts, with all our soul, that light continues to be thrown on that which he is charged to transmit.

"And we cannot follow error, change truth, just because the one who is charged with transmitting it is weak and allows error to spread around him. We don't want the darkness to encroach on us. We want to live in the light of truth. We remain faithful to that which has been taught for two thousand years. That what has been taught for two thousand years, and which is part of eternity, could change is inconceivable.

"We have made our choice. We have chosen to be obedient in the real sense, obedient to what all the Popes have taught for twenty centuries and we cannot imagine that he who sits on Peter's throne does not want to teach these things. Well, if that is the case, then God will judge him. **But we cannot go into error because there is a kind of rupture in the chain of the successors of Peter.** We want to remain faithful to the successors of Peter who transmitted to us the Deposit of the Faith [the reliable popes]. It is in this sense that we are faithful to the Catholic Church, that we remain within it and can never go into schism. Since we are attached to twenty centuries of Faith we cannot make a schism. That is what guarantees for us the past, the present and the future. It is impossible to separate the past from the present and the future. Sustaining ourselves with the past, we are sure of the present and the future."

It was on January 28, 1978 that activity between Lefebvre and the Vatican leadership resumed, this time with a letter from Cardinal Seper. In it, as Prefect of the "Congregation for the Doctrine of the Faith" (what had taken the place of what was once known as the Holy Office), he asked Lefebvre to explain and clarify his position in an official way. Over the next several months, some correspondence and even a personal meeting allowed Lefebvre to answer all questions put to him and to defend his position as he had lawfully asked from the very beginning. Throughout this inquiry, relations between Abp. Lefebvre and Cardinal Seper were quite cordial and for the most part the questions were fair and reasonable, and the answers were clear and complete.

The book *Apologia Pro Marcel Lefebvre, Volume Two* (See Bibliography) describes the proceedings between Cardinal Seper and Archbishop Lefebvre in far more detail. Their goal had been to attempt the lawful suspension of Lefebvre which had never actually taken place. They failed because legally and doctrinally Lefebvre had in fact done no wrong and therefore could not be lawfully suspended or excommunicated. He had been far too circumspect. It was during the time that the Congregation for the Doctrine of the Faith, which had been tying itself in knots trying vainly to extract from Lefebvre's answers to their questions some legal basis to suspend him, was suddenly overtaken by events. As one traditionalist priest so brusquely and pithily put it, "Then on August 6, 1978, Paul VI did something which made a great many people happy. He stopped living."

With him died the great push to suspend, or if possible even excommunicate, Lefebvre. The case against Lefebvre became just another stack of paper in the "IN" basket on the vacant pope's desk.

It might be interesting to take stock of the careers of these two men, Paul VI and Lefebvre up until this point in 1978. Paul VI had come into the highest office any human could attain to, being directly in charge of nearly a quarter of the earth's population. Over the course of his papal (?) career, he managed to eradicate the Catholic faith from all but perhaps half a million believers who were either barely on the fringe of, or utterly outside of, the organization he presided over. Even those who embraced his new religion dropped drastically in numbers as parishes closed, baptisms and marriages and religious vocations fell precipitously while marriage annulments skyrocketed, and many religious orders completely evaporated. He had presided over the most disastrous council in the entire history of the Church, the only one which has promulgated heresy, or if not, then escapes by only the most obscure and complex of technicalities. He has the distinction of being the only "pope" who has ever promulgated sacraments of undeniably human origin to replace the God-given sacraments of the Church.

By contrast, Lefebvre started out retired in 1968 with nothing more than a pension just large enough to provide him with sustenance and covering for the rest of his declining years. In a mere ten years he had formed and ordained over 50 priests and formed a society to which at least as many others had either formally joined themselves or at least expressed solidarity and support. And he was just barely getting started. By this time there were in addition to that perhaps another 100 or so priests worldwide (excluding the Eastern Rites) who were faithful to the Church and Her traditions, but who preferred to keep their heads down and stay out of the "Lefebvre" affair and quietly see to the needs of their parishioners.

A few of these others such as Fr. Georges de Nantes, Fr. Noel Barbara, and even the Dominican Fr. des Lauriers were of the opinion that Lefebvre had gone too far in trying to be diplomatic with the apostate Vatican leadership and for that reason kept their distance. For example, Fr. des Lauriers, who was the main instigator behind the *Ottaviani Intervention*, had served as a professor at Ecône for several years, but had been let go after the 1976–77 academic year because of his position against the papacy of Paul VI. These couple hundred priests and the two bishops Lefebvre and de Castro Mayer (plus Pierre Martin Ngô Đình Thục who was also struggling to continue the Church albeit unsuccessfully thus far; more about him next chapter) constituted the entire visible Western hierarchy of the Church, although none of them realized it at the time.

The cardinals of the Vatican organization held a conclave at which they nominated Albino Luciani to be their leader who took the name of John Paul I. It has been said that at this conclave Abp. Lefebvre received several votes (no doubt from what few cardinals remaining still truly deserved the title), clearly the only time in recent centuries anyone who was not a cardinal ever received so many votes at a papal conclave. Although Lefebvre himself would have adamantly denied it, the authority void at the Vatican had made him the nearest thing to a true leader of the Church to exist in the closing twenty years of his life. Since he refused to be considered the head of the traditional community, then at least he must go down in history as having been its backbone.

John Paul I presided over the Vatican organization just long enough to die under suspicious circumstances which went a long way to strengthen those who held to various crackpot theories as to what it is that had gone wrong at the Vatican. Evidence which is admittedly rather circumstantial nevertheless points rather clearly at Cardinal Villot as having been the one who poisoned John Paul I and cleaned up the evidence. John Paul I was planning to remove Freemasons such as Cardinal Villot from their offices and reform the Vatican bank in some unspecified manner, but he intended no further good. Unfortunately, that (along with a gentle and kindly disposition) seems to have been John Paul I's only real claim to fame. Once again, "the Church" was without a leader and once again the Vatican cardinals held a conclave at which they elected Karol Wojtyła of Poland who took the name of John Paul II.

Meanwhile, none of this stopped Lefebvre from continuing his ministry. In 1978, he ordained 18 more priests, transferred his German seminary from Weissbad to Zaitzkofen but kept Weissbad as a house of preparation, and opened new centers in Madrid, Spain, and Brussels, Belgium. Also, in that year he opened a seminary in Buenos Aires, Argentina, with 12 seminarians. In America, the SSPX acquired a splendid property in St. Mary's, Kansas. This property had once been a Jesuit seminary, but as their numbers declined, they closed it down and sold it off to someone who felt that it should be preserved for historic purposes and therefore used as a Catholic facility. In the hands of the SSPX, this facility soon became a Catholic University (College), the only one to this day in the United States. Even a fire in its chapel later that year barely slowed them down. Another event in America that year was the beginning of their American publication, *The Angelus*, in their South West district, then based in Dickinson, Texas. Their other district in the United States, the North East district, was based in Oyster Bay Cove, New York.

With the arrival of new Vatican leadership, first briefly with John Paul I, and then again with John Paul II, there spread hope throughout the Church and the Vatican organization that normalcy would at last be restored. The fault, they were sure, had been that of Paul VI who would undeniably go down in history as the worst Pope the Church ever had, bar none. As long as the man was blamed, it was easy to believe that a replacement of the man would fix the problem. Furthermore, John Paul II, as the seminarian, priest, bishop, and cardinal Karol Wojtyła, had long been a very conservative leader and seemed quite open to a return to the traditional worship of the Church.

Whatever doubts Lefebvre may have secretly harbored as to the papacy of Paul VI towards the end of that man's peculiar career, and despite his expressed doubts as to the procedure of John Paul II's election (cardinals over the age of eighty were arbitrarily excluded from the voting process, a dangerous precedent and one which could properly raise doubts as to the validity of the election) Lefebvre decided to accept John Paul II's election as a valid election to the papacy. Much of that acceptance was grounded, I believe, in the character of John Paul II himself which undeniably seemed to be a very amiable one. Had such a one been elected back in 1963, one can believe that the modern mess the Vatican organization is in might never have happened. In charge for barely a month, John Paul II received Archbishop Marcel Lefebvre in a private audience on November 18, 1978.

On their meeting, John Paul II seemed quite willing to work with Lefebvre on the problem of the Liturgy. John Paul II wanted Lefebvre to assure him of his acceptance of Vatican II. Lefebvre gave it, on the condition that it would be interpreted "in the light of Tradition." John Paul II casually remarked that "of course it would be interpreted in the light of Tradition," so he therefore felt that the "problem" had been solved then and there. Unfortunately, Lefebvre and John Paul II had been talking at cross purposes. To Lefebvre, "interpreted in the light of Tradition" meant "understanding the documents in a way which does not in any way contradict the teaching of the Church as established through nearly 2000 years of scripture, fathers, doctors, popes and councils." To John Paul II it meant merely that "they need only be understood as I personally see fit to approve," with no reference to the previous teachings of the Church.

Then Cardinal Seper was brought to the discussion and almost no sooner had he arrived he turned to John Paul II and said of Lefebvre and the SSPX that "They are making a banner out of the Mass of Saint Pius V." The interview ended shortly thereafter, with John Paul II leaving it all in the hands of Cardinal Seper to sort out the "Lefebvre" problem. In January of 1979 Cardinal Seper and Archbishop Lefebvre began anew the same sort of examination they had gone through the year before, but this time under the somewhat different guidance of John Paul II.

By June of 1979, Cardinal Seper had once again obtained satisfactory answers to his questions and once again found himself at a loss for any basis to suspend or excommunicate Lefebvre. Over the next several years, a thin trickle of correspondence passed between Ecône and Rome, to no real avail and with no real consequence, to either side. During this time, Lefebvre continued to turn his attention to forming his priests and managing his fraternal association, the Society of Saint Pius X. In 1979, Lefebvre moved his American seminary from Armada (which property he kept as a priory and retreat center) to Ridgefield, Connecticut. Also that year, the North East district started its own publication, *The Roman Catholic*, which initially had a format and content quite similar to that of *The Angelus*. As the society grew, a few difficulties began to show themselves and Lefebvre, in the interest of keeping unity of purpose among his priests and seminarians, had to make some difficult decisions. The next year, a new University opened in France; there are now only three truly Catholic Universities in the entire world, two in France and one in America, all operated by the SSPX. No other Traditional Catholic order is as yet large enough to have opened even one.

It was not very far into the reign of John Paul II that believers all around the world began to worry over the fact that John Paul II seemed quite content to leave the Vatican organization pretty much as it was. At least he would slow things up from getting much worse, but extremely little if anything was being done to restore "the Church" to order. Little by little it became clear that the problem had not merely been Paul VI. That man had been dead now for several years and yet his legacy of grievous damage to the Church just didn't seem able to die.

Conservatives and traditionalists both found themselves in a quandary. Both had figured that the trouble would end with the death of Paul VI, but it didn't. Some conservatives converted to the new religion Paul VI had promulgated while others became traditionalists. Traditionalists themselves began to argue amongst themselves over what had gone wrong. While you, dear reader, have the benefit of the theory and observations from earlier chapters of this book precisely what it is that happened to the Church, namely its legal detachment from the Vatican organization, Catholics of this particularly painful period from 1979 to 1984 didn't have a clue. Theories and explanations began to multiply.

The problem was that different theories often resulted in different solutions as to what to do as Catholics in this crisis. Lefebvre, not having the answer to the problem, could not believe that anyone else had the answer either. Perhaps no one's answer satisfied him. For lack of an adequate answer, all he could do was exert what limited authority he had over the priests in his order as their Superior General to continue the status quo, such as it was. Finally, a horrible scandal forced him to take action.

One priest he had ordained back in 1978 by the name of Juan Fernandez Krohn, had got in his head the idea that John Paul II himself was somehow the problem and if only he weren't around, the Church could start to heal. Being of a somewhat violent temperament, he decided to take matters into his own hands; he took a knife and attempted to stab John Paul II with it during a visit to the shrine in Fátima almost exactly a year after John Paul II had been shot by an Islamic terrorist. This was in May of 1982, several years after his departure from the SSPX (he had left it in the first few months of 1979), but the fact that he had been ordained by Lefebvre and was once a member of the SSPX made for a public black eye on the society.

Eager to distance himself from the possibility of their being any more such dangerous and scandalous priests within his order he set about purging the SSPX of any priests (even those of clearly peaceable intentions) whose opinions of John Paul II's ministry had even the faintest similarity to those of Fr. Krohn. This made for some rather considerable unpleasantness when quite a number of fine priests had to be ejected from the SSPX. These included nine priests many of whom held leading positions within the American North East district (including the District Superior, the District Bursar, the Rector of Thomas Aquinas Seminary in Ridgefield, the Headmaster of the University at St. Mary's, and his assistant). Over the next several years, perhaps ten to fifteen percent of his priests worldwide had been ejected from the SSPX, ending in late 1985 with four Italian priests, Frs. Franco Munari, Curzio Nitoglia, Giuseppe Murro, and Francesco Ricossa. Of these ejected priests (particularly the nine American priests ejected in 1983) more will be said in a later chapter.

Cardinal Ratzinger, who had succeeded Cardinal Seper in 1982 after the latter passed away in late 1981, praised Abp. Lefebvre's efforts to purge those certain priests from his society, saying in a July letter, "the Pope acknowledges the devotion of Archbishop Lefebvre and his fundamental attachment to the Holy See, expressed for instance by the exclusion of members who do not recognize the authority of the Pope." Although that is a slight misreading of the nature of those excluded priests, the gesture of excluding those priests went a long way toward smoothing diplomatic relations with Rome and John Paul II.

Also, in 1982, Lefebvre had finished twelve years as Superior General of the SSPX and it was time for him to step down, even though this stepping down action took place in 1983 shortly after the ejection of the nine American priests. Succeeding him was Fr. Franz Schmidberger, a German who was fond of mathematics and who had been elected to the post of Vicar General of the SSPX during its first general chapter. Later, in November of 1983, Lefebvre and de Castro Mayer jointly published a "Bishops' Manifesto" which read:

Holy Father,

May Your Holiness permit us, with an entire filial openness, to submit to you the following consideration. During the last twenty years the situation in the Church is such that it looks like an occupied city.

Thousands of members of the clergy, and millions of the faithful, are living in a state of anguish and perplexity because of the "self-destruction of the Church." They are being thrown into confusion and disorder by the errors contained in the documents of the Second Vatican Council, the post-conciliar reforms, and especially the liturgical reforms, the false notions diffused by official documents and by the abuse of power perpetrated by the hierarchy.

In these distressing circumstances, many are losing the Faith, charity is becoming cold, and the concept of the true unity of the Church in time and in space is disappearing.

In our capacity as bishops of the Holy Catholic Church, successors of the Apostles, our hearts are overwhelmed at the sights throughout the world, by so many souls who are bewildered yet desirous in continuing in the faith and morals which have been defined by the Magisterium of the Church and taught by Her in a constant and universal manner.

It seems to us that to remain silent in these circumstances would be to become accomplices to these wicked works (*cf.* 2 John 11).

That is why we find ourselves obliged to intervene in public before Your Holiness (considering all the measures we have undertaken in private during these last fifteen years have remained ineffectual) in order to denounce the principal causes of this dramatic situation, and to besiege Your Holiness to use his power as Successor of Peter to "confirm your brothers in the Faith" (Luke 22:32), which has been faithfully handed down to us by Apostolic Tradition.

To that end we attached to this letter an appendix [not included in this book] containing the principal errors which are at the origins of this tragic situation and which, moreover, have already been condemned by your predecessors. The following list outlines these errors, but it is not exhaustive:

1. A latitudinarian and ecumenical notion of the Church, divided in its faith, condemned in particular by the *Syllabus*, No. 18 (Denzinger 2918).

2. A collegial government and a democratic orientation in the Church, condemned in particular by Vatican Council I (Denzinger 3055).

3. A false notion of the natural rights of man which clearly appears in the document on Religious Liberty, condemned in particular by *Quanta cura* (Pius IX) and *Libertas praestantissimum* (Leo XIII).

4. An erroneous notion of the power of the pope (*cf.* Denzinger 3115).

5. A Protestant notion of the Holy Sacrifice of the Mass and the Sacraments, condemned by the Council of Trent, Session XXII.

6. Finally, and in a general manner, the free spreading of heresies, characterized by the suppression of the Holy Office.

The documents containing these errors cause an uneasiness and a disarray, so much the more profound as they come from a source so much the more elevated. The clergy and the faithful most moved by this situation are, moreover, those who are the most attached to the Church, to the authority of the Successor of Peter, and to the traditional Magisterium of the Church.

Most Holy Father, it is urgently necessary that this disarray come to an end because the flock is dispersing and the abandoned sheep are following mercenaries. We beseech you, for the good of the Catholic Faith and for the salvation of souls, to reaffirm the truths, contrary to these errors, truths which have been taught for twenty centuries in the Church.

It is with the sentiments of St. Paul before St. Peter, when he reproached him for having not followed "the truth of the Gospel" (Galatians 2:11–14), that we are addressing you. His aim was none other than to protect the faith of the flock.

St. Robert Bellarmine, expressing on this occasion a general principle, states that one must resist the pontiff whose actions would be prejudicial to the salvation of souls (DE ROM. PON. I.2, c. 29).

Thus it is with the purpose of coming to the aid of Your Holiness that we utter this cry of alarm, rendered all the more urgent by the errors, not to say the heresies, of the new Code of Canon Law and by the ceremonies and addresses on the occasion of the Fifth Centenary of the birth of Luther. Truly, this is the limit!

May God come to your aid, Most Holy Father. We are praying without ceasing for you to the Blessed Virgin Mary.

> Deign to accept the sentiments of our filial devotion, [signed] S. E. Monseigneur Marcel Lefebvre, International Seminary of Saint Pius X, Ecône, Switzerland [and] S. E. Monseigneur Antonio de Castro Mayer, Riachuelo 169, C. P. 255, 28100 Campos, (RJ) Brazil.

At last, in 1984, John Paul II finally gave permission (albeit on a most grudging and miserly basis) for the Tridentine Mass to be said. On the surface it seemed to grant everything Lefebvre was asking for, but it had been carefully worded so as to exclude those for whom the Mass of Saint Pius V is a banner, namely Lefebvre and the SSPX, and even more so, such priests as Lefebvre had ejected from his society. While Lefebvre couldn't help but praise this small step in the right direction, he knew that to throw his society in with this special permission would have been to win a battle but lose the war.

Over the next three years, the thin trickle of correspondence between Ecône and Rome continued, equally unavailing and equally ineffective. A very few priests of the SSPX obtained Indults, but most more or less ignored it as they went about their business seeing to the needs of the souls in their parishes. Then, in late 1986, John Paul II pulled the most outrageous stunt of his entire career. He organized a joint prayer session with the leaders of just about every known religion on the face of the earth. They gathered at a place called Assisi and each in turn prayed to their respective gods for world peace. One can't help but believe that John Paul II had done this entire fiasco "at" Lefebvre, since immediately thereafter he sent a letter to Lefebvre practically daring him to say that he was not the Pope. Lefebvre refused to take that bait but once again in response repeated what he had said in his 1974 declaration about the difference between Eternal Rome to which his loyalty is ever committed, and Modernist Rome which was showing itself to be all the more schismatic and heretical than ever before. And there the matter stood until June of 1987.

Meanwhile, the SSPX continued to grow and expand into more countries, England, Mexico, Columbia, South Africa, Holland, Portugal, Australia, India, Sri Lanka, Gabon, Senegal, New Zealand, New Guinea, Japan, South Korea, Luxembourg, Chile, and Zimbabwe. Seminaries, priories, Carmelite orders, Dominican orders, Benedictine orders, parishes, and mass centers continued to expand in numbers all around the world. Abp. Lefebvre wrote his first book, *An Open Letter To Confused Catholics*, and toured the world, confirming thousands of Catholics everywhere he went; crusades and pilgrimages were led in which thousands participated. And with each passing year, more and more priests got ordained. In 1987, over 200 priests were members of the SSPX, and perhaps another 100 or so were nonmembers who were also faithful to Tradition.

8

THE BISHOPS OF PIERRE MARTIN NGÔ ĐÌNH THỤC

Perhaps the reader might recall that a few chapters ago, I pointed out the doubtful validity of the "sacraments" of the Novus Ordo religion. In particular, I speak now of the new form now being used for the sacrament of Holy Orders, and most specifically that of the highest rank, which confers the Episcopacy (makes a bishop). Given the unreliability of the new man-made ceremonies, one has to wonder how many (if any) of the "bishops" consecrated to the Episcopacy by the new rite were actually validly consecrated bishops, and how many were not, and who (if anyone) among them the true bishops would happen to be. (One usually says of a priest that he is "ordained," but of a bishop that he is "consecrated.") In the church of the People of God, there is no way one can trust that any particular "bishop" in that establishment is really even a bishop at all, nor that any "priest" is really a priest (unless of course they were consecrated or ordained previous to the changes made to those rites made in 1968). The Church cannot afford to have to worry about such things, but now is obliged to. That is one of the two most dangerous aspects of the conciliar and post-conciliar "reforms," the other being the loss of Apostolic authority, the state of being truly "sent" by the Church, the possessing of canonical authority. Such authority cannot be held by heretics nor by self-appointed leaders, but must always exist continually, nevertheless.

As long as we can know that we have validly consecrated bishops, lawfully sent by the Church, all the other problems of the Church can be reasonably put to rights by having the heretics repent or be rejected and put out, and by instructing the ignorant, and the faithful bishops coordinating as necessary to continue the Church. But without validly consecrated bishops, there can be no sacraments except Baptism and Marriage. And without lawfully consecrated bishops, there can be no juridical or evangelical actions of the Church, no absolutions, no dispensations, no missionaries. It takes a priest (or a bishop) to say Mass. It takes a priest (or a bishop) with at least some form of authority to absolve a penitent from sin in the confessional. It takes a priest (or a bishop) to administer the Last Rites (including Extreme Unction) to a soul in danger of death. It normally takes a bishop to Confirm a soul, or bless the Holy Oils, such as are used for Extreme Unction (although in some rare cases, where authority has granted permission (which itself requires a bishop), a "mere" priest will do). It takes a bishop (a priest will not do here) to make another bishop.

One can easily see from the preceding that if ever the Church should run out of validly and lawfully consecrated bishops, an essential component of the Church's fourth mark, Apostolicity (the succession of validly and lawfully consecrated bishops extending continuously from the original twelve Apostles themselves clear to modern times and to the End of the world), would be forever lost, utterly irretrievable. With no more bishops, it would only be a matter of time before the last validly ordained priest would die, and then there could be no more priests, no more Masses, no more absolutions, no more Confirmations, and no more Last Rites. The sacramental priesthood would vanish and could never be recovered. The Church would go from seven sacraments to two. Furthermore, if the Pope must be a bishop, indeed the "Bishop of Rome," how can there ever be a pope in that case?

Beginning in 1951, in response to the Chinese when they were attempting to create a new "hierarchy" of their own, the "Patriotic Chinese Church," Pope Pius XII freshly re-affirmed and emphasized that the appointment of bishops belongs to the Pope, though again on June 19, 1958 he did temper that law by stating in *Ad Apostolorum Principis* that "But if, as happens at times, some persons or groups are permitted to participate in the selection of an episcopal candidate, this is lawful only if the Apostolic See has allowed it in express terms and in each particular case for clearly defined persons or groups, the conditions and circumstances being very plainly determined." This exception, useless for the Chinese schismatics who attacked the unity of the Church by consecrating Communist sympathizers against his express will as Pope, had application in his time primarily to some Patriarchs of Eastern Rites upon whom the role of appointing bishops had been delegated through long accepted custom. For a bishop to consecrate another bishop in the absence of this permission, called a "papal mandate" would ordinarily have to be considered illicit. An illicit bishop, though sacramentally valid, would receive no canonical mission from the Church, no jurisdiction. He would be part of what could be called a material succession (as held by, for example, the East Orthodox schismatics), but not of the formal succession, not "one juridical person with the Apostles." So things seemed, though for several good reasons that cannot have actually been the case.

Who back then could have anticipated that within twenty years, what would appear to be "illicit" episcopal consecrations would become absolutely essential to the continuation of the Church, and that scarcely thirty years later the first of such would take place? Clearly, such a law as that cannot apply in such a case as this where it was done not as an attack against the unity of the Church but in support of it. There is no room to doubt that Pope Pius XII would count the continuation of the Church as being more important than obedience to this bit of legislation. But there is room to doubt that any violation has occurred.

If it hadn't been for the heroic actions of a few faithful bishops, the Novus Ordo religion would have eventually brought about an end of the Apostolic succession! Something had to be done to preserve alive that sacred and irreplaceable treasure of the validly and lawfully consecrated Episcopate. The entire future of the Roman Catholic Church rests on the actions of the very few bishops who actually did this. So far, there have been at least four such bishops, and at least two of them archbishops. But at this juncture of my account of the Traditional Catholic community (June 1987), only one of the bishops had as yet made any other traditional bishops: Archbishop Pierre Martin Ngô Đình Thục, of Huế, Việt Nam (Vietnam).

The peculiar figure of Abp. Thục is probably the most enigmatic of the bishops from whom all reliable episcopal consecrations of the Church in the future are derived. Of course, Abp. Lefebvre could be a somewhat enigmatic figure as well. Could it be that a defining characteristic of archbishops is to be enigmatic characters? In the time when none of the other bishops were even remotely considering the consecration of any bishops according to the traditional rite, which would therefore have been without the permission of the Vatican leadership, Thục had already run way ahead of the others. It was he who blazed all the trails and made all the mistakes. And, for the most part, they were very big mistakes indeed, though at least they were all honest ones. On the other hand, the only alternative to making such mistakes would have been to do nothing at all, and that would have been terminally catastrophic.

Abp. Thục, together with other members of his family, had also found himself close to the center of another major event of the 1960's which tore up American, European, and Southeast Asian society and which contributed in a major way to the rise of the 1960's counterculture movement: the Vietnam War. It is for these reasons that a fairly detailed account of the history of the man and his family up until this point bears telling.

On October 6, 1897, Pierre Martin Ngô Đình Thục was born to Catholic parents in Huế, Việt Nam. His father, Ngô Đình Khả, was by nationality a Mandarin, and by office the Minister of Rites and Grand Chamberlain to Emperor Thành Thái who had reigned from 1889 to 1907. The family had been among the first to convert to Catholicism back in the seventeenth century, and some of them had even been martyrs for their faith.

Although the French had originally introduced Catholicism to Việt Nam, the Ngô family became estranged from the French when the French conspired to depose Emperor Thành Thái in 1907. In the wake of that, their Catholicism became the more solidly traditional faith of the Spanish and Portuguese, both of old as that of King Ferdinand and Queen Isabella, as well as that which would later rally to the cause of Governor Franco, rather than the relatively easygoing and liberal Gallican Catholicism of certain factions in France which were then operative in Việt Nam.

At this point, the family had a mission to establish Roman Catholicism in mostly Buddhist Việt Nam, and without the help of the French, thank-you. Most of his brothers pursued politics, but Thục instead pursued the priesthood. In the course of this he went to Rome and obtained doctorate degrees in canon law, theology, and philosophy. He was ordained a priest on December 20, 1925.

His distinguished career as a seminary professor began in Sorbonne in Paris, but in 1927 he returned to Việt Nam where he held professorships at both the Major Seminary and the College of Divine Providence. So well had he performed his professor duties that on May 4, 1938, he was consecrated bishop and became the Titular Bishop of Sesina. He organized and set up that diocese in the city of Vĩnh Long and also founded the University of Đà Lạt (Dalat).

This university had to be built up practically from scratch, but when one of his brothers, Ngô Đình Diệm (Diem) had risen to power, he gave Thục the right to profit from a forested area which generated the needed funds. Tragedy was already being served up to the Ngô brothers by the communists in that their oldest brother, Ngô Đình Khôi, had been buried alive along with his son by the communists, though some reports have it that he had been shot first.

In 1955, Ngô Đình Diệm ousted Bảo Đại, the ruling Chief of State over Việt Nam at that time (who had previously stepped down from his former post of Emperor in 1945). The United States followed the recommendation of Francis Cardinal Spellman, the Vicar General for the U. S. armed forces and John Foster Dulles, Secretary of State, by backing the Diệm regime. The United States had backed him in order to fight the communists, but unfortunately, he seems to have devoted the bulk of his efforts against the Buddhists instead. This was a grave political blunder, and theologically unsound.

I say "Buddhists" here, following the practice of most chroniclers of the Vietnam War, but actually only about 20% of the population was strictly Buddhist, the remainder being a kind of mixture of Buddhism, Confucianism, Animism, and occasionally other less known local religions and customs. Diệm failed to recognize the fundamental difference between this Buddhism and Communism as non-Christian forces in Việt Nam. On the one hand, Communism was an invading alien intruder, a subtle Red Chinese invasion absolutely no better for Việt Nam than the invasion of the Japanese had been. On the other hand, this sort of Buddhist is what Việt Nam had been for thousands of years—longer than Christianity had even existed. It would have been perfectly reasonable, Catholic, and politically sound for him to have used military force to drive (or keep) the Communists out. Practically all of Việt Nam, Catholic and Buddhist alike, would have cheerfully backed his efforts in this direction. That is also what the United States wanted him to do and why they backed his regime.

But dealing with Buddhism would have required a great deal of time. Where Communism was merely a new idea that some dilettantes found fashionable, Buddhism is what most of the people truly believed. Threat, coercion, and force have never changed anyone's personal belief. The proper thing to do would have been to win the Buddhists to Christ, one soul at a time. That requires a lot of sacrifice, suffering, prayer, and time. Only when the Buddhists can see the superior depth of the Christian Faith would they be able to convert voluntarily; an involuntary conversion is no conversion at all. Diệm thought he could take a short cut around this process, or perhaps he may not have known how he was supposed to respond to the differences between Communism and Buddhism in his country.

In a Catholic nation which is made up mostly of Roman Catholics, one can very easily and properly pass and enforce laws prohibiting the spreading of all false teachings in the public forum, or at least deny them any official recognition. Diệm desired to pass such laws because he imagined that the faith could be easily spread to large numbers if politically mandated. But Việt Nam was about 90% Buddhist and scarcely 10% Christian. The Church does not approve of the use of military or political force to overcome the long-established traditions which a society has received from their forefathers. That is something quite different from dealing in such a manner with Catholic members of a Catholic society who depart from the Faith and encourage others to do the same. In Buddhist Việt Nam, there was no way for Christianity to be thus enforced. Making such an attempt was further made unsound from a political standpoint because he also opposed the French who were the only other force for Christianity in Việt Nam. A truly Catholic and Vietnamese nation was simply not possible in the near term under such conditions.

Seeing this problem and being irrevocably committed to making Việt Nam a Catholic nation, the next plan of the Ngô brothers (including Ngô Đình Nhu who was appointed Political Advisor) was to consolidate the Roman Catholics in South Việt Nam, even though most of them lived in the north. The rumor was sent forth that Jesus and Mary had gone South. In response to that, nearly a million North Vietnamese Catholics relocated to South Việt Nam, placing a very heavy burden on the already overtaxed infrastructure in the South. Hard times were had by all, but this did have the effect of consolidating the Catholics into a smaller area in which they would be a majority, and also of bringing them into the territory which was politically controlled by Diệm's regime and spiritually guided by the Archiepiscopal see of Huế, which Thục was soon to be raised to on November 24, 1960.

Things got particularly ugly in 1963 when Buddhist monks, in protest to the way all Buddhists were being treated, began drenching themselves in gasoline and setting themselves on fire. Mrs. Ngô Đình Nhu (Trần Lệ Xuân) took an extremely dim view of what those Buddhists were doing, and in an attempt to discourage such behavior she responded to it by joking about how they were using imported gasoline to make barbecues of themselves. Her well intentioned attempt totally backfired, resulting in more such suicides and bringing international disgrace on the Diệm regime. With each new problem, Diệm's solution always seems to have been to crack down harder on those who opposed him.

When Roncalli was elected Pope in 1958, the new spirit of ecumenism began to take over at the Vatican. Cardinal Spellman (who floated his boat down the Tiber river causing some to think that by so doing, he would fulfill the prophecy that the next pope would be a Pastor and Mariner) went from being a likely successor to the papacy ("papabile") to practically a cardinal in exile. Roncalli, as Pope John XXIII, took a very soft stance against the Communists, and an even softer stance against other religions such as Buddhism. This was not a result of Catholic leaders becoming more aware of the difference between Communism and Buddhism in Việt Nam, but simply a policy of going soft on everything non-Christian. It was only a matter of time before nominally Catholic President John F. Kennedy sensed the change and began to think about ousting Diệm. When Buddhist monks began burning themselves to death, Kennedy gave Diệm one last chance. Forget the Buddhists and go get the Communists, or else. Diệm responded by merely taking a yet harder line against the Buddhists and so signed his own death warrant. The order was given to the CIA to have Diệm removed but guaranteeing his safe passage out of Việt Nam.

The CIA-directed thugs eagerly agreed to remove Diệm but took matters into their own hands as to what to do about him. On November 2, 1963, Diệm and his brother Nhu had spent part of the night in Saint Francis Xavier Catholic Church, a Chinese-Vietnamese church in Chợ Lớn (Cholon) City, and the remainder of the night in the home of a prominent parishioner, a Chinese businessman, and called for a limousine to take them out of the country. Instead, a military van showed up and they were bundled into the back. While en route to their destination the van stopped at a railroad crossing where the noise of a passing train concealed the sound of the shots by which they were assassinated. Ngô Đình Cẩn, another brother, was captured and had been promised a fair trial by the new government. What he got was only a kangaroo court which found him guilty and had him executed by firing squad on May 10, 1964, after his having spent an entire year in a small cage.

Where was Thục during all of this? In 1962, he had been called to Rome to participate in the Second Vatican Council. During the summer of 1963 between the first and second sessions of the Council he saw his homeland for the last time. On June 5, a massive celebration was held in Huế in his honor. A few days later, the Buddhists wanted to hold a public celebration of their own but were denied permission by the Diệm government. It was at this point that they began dousing themselves with gasoline.

In the fall of 1963, he returned to Rome for the second session, and there he was while his brothers were being killed, and there also he was detained against his will after the Council session was over. One might fairly argue that he had been detained in the interest of his own physical safety, but in 1968, pro-Marxist Philippe Nguyễn Kim Điền was appointed to his post in Việt Nam while he was made Titular Archbishop of Bulla Regis. His role in Việt Nam had ended quite permanently in 1963 when he returned to Rome for the second session of the Council. This was very much in keeping with a tendency which Paul VI displayed on many occasions of attempting to replace faithful bishops, archbishops, cardinals, and even patriarchs, such as Josyf Slipyj, József Mindszenty, or Stefan Wyszyński, with Communist sympathizers. This would typically be done in order to establish "diplomatic relations" with the Communist governments, which was a very thin disguise for a total acquiescence to Communist rule and of no spiritual benefit or consolation to the Catholics trapped in those countries.

John XXIII had made Thục an archbishop on November 24, 1960, going into the Council. At the Council, he seemed to take a number of rather odd positions. The extremism with which he expressed certain views, which could almost have seemed reasonable in the context of where the Council was headed but which were clearly out of step with the long-established teaching of the Church, might in part be traced to the Oriental tendency to sense the direction being taken and jump at once to the logical conclusion of that direction. Early on in the Council, he had complained quite vehemently that the leaders of many other religions hadn't been invited and ought to have been. There were many European white "Christians" represented among the invited observers, but relatively few Asians, which seemed to him to be racist. He only calmed down when he was informed that they had been invited but had not bothered to come. There had been in fact something of a solicitation, a "brainstorming session" as it were, for "new ideas," no matter how crazy or radical, and when it was Thục's turn and feeling obliged to participate, his unique and different background provided him with some unique suggestions that fit right in. But there is no evidence that he would have countenanced the implementation of any of these ideas.

He ventured that the Mass could reflect the culture of those who are celebrating the Mass. After all, some areas had their own particular and distinctive Rites, separate from the Latin Rite though still part of the Church as a whole. Why not give the same to other regions? (The answer is that the existing Eastern and other Alternate Rites go back to Apostolic times; it would be impossible and obviously inauthentic to synthesize new such Rites for those not reached by the original Apostles, for example, the Chinese, Japanese, or Vietnamese.) As this vein continued, he ventured the idea of priestesses. Why not, given the nonsense which was going on! As he saw the discussions progressing, anything should be approved, even the Tridentine Mass! The strangest aspect of all of this is, what kind of Council it was such that one could mention such things and be taken seriously? His seeming advocacy of such ideas were intended only ironically, i. e. "as long as you are doing these crazy things, why not do these other crazy things as well?" Whatever his reason for having spouted such ideas at all, despite their oddness, he voted against each and every document of the Council, practically the only prelate to do so, and was furthermore a member of the archconservative *Coetus Internationalis Patrum*. The Council approved none of the odd ideas he had mentioned and once it was over, he had nothing further to do with those ideas, assuming he had even wanted those ideas in the first place, which is far from clear.

Finally, the Council ended, and being old, not allowed to return to his homeland, and obviously in mourning for his brothers, he had been given very few duties to perform and much time to think. What had particularly galled him was the way the new pope and many of his associated functionaries, by changing their policies in favor of Ecumenism and Communism, had (he felt) practically incited the Americans and Vietnamese to betray and kill his brothers. He was also disgusted by the easy way the new Roman establishment seemed to get along with the new Communist regime which benefitted from that betrayal. His exile also kept him away from everything he had had in Việt Nam, from an active and conventional archiepiscopal office and all that went with it, to any number of material assets, as had been built up through his efforts, forcing him to live in despair as a pauper in a far-off strange land, a shell-shocked but determined man.

Other than some doubtful reports of real estate fraud (namely, should he have accepted that forested land assigned to him by Diệm?) and a suspicion of nepotism (namely, what role might he have had, perhaps through Cardinal Spellman, in helping his brothers come to power?), there is no clear evidence that his behavior as priest, bishop, archbishop, and Council Father had been anything other than exemplary. Vietnamese Catholics who had known him thought very highly of him. The spiritual crisis he went through did much to keep him out of sight while as Titular Archbishop of Bulla Regia he served as a substitute Assistant Pastor in various parishes near Rome. After some years, he then moved to France. While still in Rome, other events forced themselves upon him.

In March of 1968, four young visionaries, Ana García, Josefa Guzmán, Rafaela Gordo, and Ana María Aguilera, saw an apparition of the Blessed Virgin Mary over a mastic tree (called Lentisco) about a mile south of the Spanish city of Palmar de Troya. Before long, other visionaries got in on the act, Rosario Arenillas, Maria Marin, Maria Luisa Vila, Antonio Romero, Jose Navarro, Manuel Fernandez, and others. Clemente Domínguez y Gómez joined their group of seers in the latter half of 1969. So far, things seemed to go along reasonably, claiming that "to restore purity in the world, it is first necessary **to restore the Authentic Sacrifice of the Altar, the Latin Tridentine Sacrifice of St. Pius V**, and not the so-called Banquet of these times...**the New Mass, I have already said, is the antithesis of the Mass; which means that not only does it not atone, but it augments the offence**" [**Bold** in original], something our Lady could quite properly and reasonably said, though of course this is not to imply that we have any particularly good reason to think that she actually did show up and so speak at this time and place. In the early 1970's, the local Novus Ordo leader condemned, out of hand, the visions, though only the four original visionaries, assuming the local Novus Ordo leader to be a Catholic bishop, obeyed him and withdrew from the group. But with the true and original teaching of the Authentic Magisterium of the Church so boldly professed, and already receiving persecution from a self-appointed persecutor of all things Catholic, many of the most serious, devout, and sincere Catholics from the area, and some others drawn from further regions around the world, joined the group.

In time, Clemente Domínguez y Gómez somehow made himself or became their chief seer or visionary, to the decline of all the others, but this was someone our Lady would not have wanted to deal with. Despite some early visions featuring persons in Dominican habits, the group came to be known as the Order of Carmelites of the Holy Face. Such inconsistencies, at first small and insignificant, would in time grow much larger, but only years later. One incipient problem (which no one realized at the time) was that he put his own visions ahead in value of even the Magisterium of the Church. Another problem was their taking Paul VI as a real Catholic Pope (such that their own first "Pope" was the successor of Paul VI, validating everything done by Paul VI). Not even their claim that Paul VI had been replaced by some impostor can help them much since this substitution supposedly took place some years after the close of Vatican II, thus ratifying (for them) the whole of Vatican II and all the instability it introduced, strikingly similar to that of the fallen Vatican organization. They were becoming a breakaway sect, not of the real Catholic Church, but of the new fallen Vatican organization, sort of like the American Episcopalians breaking away from the Anglicans of England. So long as his visions more or less pretty much supported that Magisterium, things went well enough, and so things went, clear to and throughout the time that Abp. Thục was involved.

In those visions he had been told to have himself made a bishop by a traditional bishop, so his representatives first went to Ecône where Lefebvre had his seminary. Lefebvre himself was too busy to look into their case so as to give it the attention it demanded and he certainly wasn't about to consecrate any bishops at that point, but he knew Abp. Thục and felt he could trust him to look into the matter and do the right thing, and that he would have the right and power to do whatever truly proved necessary, whatever that should turn out to be. For that reason, he pointed his visitors to Thục. Lefebvre had come to know Thục as a fellow member of the *Coetus Internationalis Patrum*, and also Thục had spoken at Ecône in some previous years.

These representatives of Clemente, led by a priest who had once taught at Ecône, a canon of Saint Maurice named Father Revas, came to Thục's small apartment in the Italian village of Arpino. They told him that they had a car waiting outside ready to take him to Palmar de Troya where the Virgin Mary was expecting him to perform a great service for her. They also led him to believe that Archbishop Lefebvre had recommended that he go and consecrate Clemente. (Lefebvre had only recommended that he look into the matter, but as far as Fr. Revas was concerned, all he really wanted from Lefebvre was a name and an address and not the content of any message Lefebvre had for Thục.)

In Palmar de Troya, in the last week of 1975, Thục ordained Clemente Domínguez Gomez and four others, Manuel Alonso Corral, Luis Henrique Moulins, Francis Coll, and Paul Gerard Fox, to the priesthood, and on January 11, 1976, he consecrated Domínguez, Corral, Camilo Estevez Puga, Michael Thomas Donnelly, and Francis Bernard Sandler, all without going through the normal channels for approvals, and with very little if any priestly formation. It was this pivotal action which changed his destiny. Normally a great deal of care goes into the selection of a bishop, hence the role a pope, or at least one's fellow bishops, would have in the selection process. In consecrating individuals that he barely even knew and had not been personally recommended by any pope nor by any trustworthy bishops (what very few could be said to exist), the die was cast.

Was he rebellious? Had he lost his mind? These were the kinds of thoughts many had in response to this consecration. And unfortunately, Thục himself has been granted extremely little in the way of any platform to explain himself or answer any questions people had. And as will be seen, most Catholics, even Traditional Catholics, were just not ready for all that Thục knew. Indeed, had he set out to explain himself before taking that step, he rightly feared that he might not live long enough to answer all possible objections (many of them insincere) and opposition he might face before being able to perform any consecrations. At the time it seemed best and most prudent to do it immediately and then explain himself at leisure later on, after the fact, should God grant him the years to do so, or trust in Divine Providence that professional theologians of the Church would at least eventually figure out what he did and why, and see the lawfulness and validity of it. Regrettably, his opportunities to explain himself proved few and minimal, as even his friends and supporters did not fully understand him, and much that he suspected he himself could not yet be certain of.

However, in hindsight, a great many reasons (at least subjective in some cases) that he could have done it, and should have, indeed have been found. For one thing, he had been granted, by Pope Pius XI on March 15, 1938, to be made a Papal Legate, and there was granted to him "all the necessary powers," every papal prerogative communicable to others (which would include even the power to nominate and appoint bishops, as might a Patriarch to some Eastern Rite). For another, he had been given the assurance that Abp. Lefebvre (and presumably those of their fellow *Coetus Internationalis Patrum* confreres with whom Lefebvre had continued correspondence) agreed that he really "should" do these consecrations, that they all supported him, though they could not be present. For yet another, our Lady and even Pope Paul VI (the real one held incommunicado in some dungeon, not the fake public persona substituted in his place) had so called him to perform consecrations, or so he had been quite persuasively informed, a reasonable development not at all surprising to him.

For yet another he knew, having been a professor of Canon Law that a law ceases to have any legal or moral power if it becomes impossible to comply with as written, and especially in the face of dire necessity. He knew that episcopal consecrations performed according to the new Novus Ordo ceremony were at least virtually certain to be invalid, but that of course the Church must have real bishops in order to continue Her very existence. He knew that something was up with Paul VI. Perhaps he was not Pope (Sede Vacante), or perhaps he was trapped by "handlers" and others who rendered it impossible for him to function, or else maliciously hated the Church and was bent on abolishing the sacramental episcopacy, or even perhaps, per the belief of the Palmerians, that the "Pope Paul VI" seen and known by all at that time had been some impostor foisted upon the Church, the "real" Paul VI presumably being held a prisoner, incommunicado, and in any case clearly unwilling or unable to grant any papal mandates publicly through conventional channels for any real Catholic bishops to be made. Ergo, the law requiring a papal mandate simply could not be insisted upon, and he therefore has every right and duty to consecrate without one.

Finally, he knew at least something of what Vatican II had attempted (changed?), namely, to include schismatic East Orthodox (and other illicit but valid episcopal successions) as part of the Church, as not only valid but licit as the ecclesial means for the salvation of souls, effectively awarding them all a juridical place in the Church. If even schismatic and heretical churches that possess valid episcopal sacraments (all without ordinary papal mandates, of course) could be considered as legitimate, then how much more so are episcopal consecrations of actual Catholics despite the circumstantial impossibility of a papal mandate?

And what were the subjective aspects of these reasons he had? One was that Abp. Lefebvre (and those others he knew) supported the consecrations when in fact all they recommended was that he look into the matter. Fr. Revas had misrepresented Abp. Lefebvre to Abp. Thục. The other was the supernatural aspect. While one can safely believe that God (and our Lady and any real Pope) would have fully intended that Thục should consecrate the first of the Church's bishops since the arrival of the current circumstances (just as we can safely believe that God had intended that Abraham would be the father of the Nation of Israel), that just as it was not God's intention that the Nation would arise from Ishmael the son of the servant girl, neither was it God's (or our Lady's) intentions that these vital consecrations commence with the consecration of these fake mystics who lied to him. Their deception did not deprive Domínguez and his associates of the sacramental validity of the episcopal consecrations they received, but despite the legislation of Vatican II it did deprive them of the legitimate canonical mission which Thục could, and fully intended to, have conveyed to them.

The Modernists in Rome excommunicated him and Domínguez by September 17, 1976, at which point Thục repented from consecrating the Palmar de Troya bishops and obtained absolution from Paul VI for his mistake. The mistake (at least as he knew within his own mind) had not been the fact of consecrating bishops, but in his choice of consecrands. It was not yet clear to Thục whether Paul VI was actually Pope or not or what was going on with him, but he could think of nowhere else to go. Not long after Thục returned to Italy, their true intentions and heretical beliefs became evident, in particular their willingness to follow Domínguez into whatever beliefs, and even a distorted liturgy, he would teach, regardless of what Magisterial teachings he contradicted. When, upon the death of Paul VI, Bp. Domínguez had himself declared Pope Gregory XVII in a mystic vision, Thục even more came to regret having consecrated him, and publicly condemned them as not being real and legitimate bishops of the Church. Some misrepresented that condemnation of his as being a denial that he had validly consecrated them (or, simulated the sacrament – a grave sin), but in fact it was their claimed legitimacy and jurisdiction or canonical mission which Thục denounced as invalid, owing to the crass and criminal manner in which they had deceived and exploited him. If, per the rumors, he were to have subsequently claimed to have only simulated his consecrations of the Palmar de Troya leaders, it would not be because of any actual withholding of intent at the time, but to discourage as extremely as possible any attempt by Palmar de Troya to continue their schismatic and now heretical succession, and to discourage all others from having recourse to them.

For several years, Thục was content to lie low living in poverty in Arpino, and then later with a Buddhist Vietnamese family in Toulon, France. While regretting his decision to consecrate Clemente Domínguez y Gómez and his cronies, he was clearly very angry with the Roman leaders for having not even tried to reestablish him in "his" See in Việt Nam, nor obtain any reparation for what had been allowed to happen to his brothers. But ever anxious to continue the Church he continued to seek prospective bishops for the Church. In that, many others would approach him for Episcopal Orders, most notably Catholic-minded clergy who had obtained their sacramental orders from the Old Catholic succession, and who had doubts about their Orders they had received from other Old Catholics, nearly all of whom he turned away, but a very few did truly seem to qualify. This would again create activities on his part which Catholics long had difficulty comprehending.

Opponents of the Church heaped spread lies about him, claiming that he had performed these consecrations for the various Old Catholic and other sects these men had obtained previous Orders from. But had he really become an Old Catholic? Was he willing to consecrate just anyone who approached him? Indeed, once a person gets known for something it's amazing just how many will try to get the same thing, though it does not appear that he had understood that particular aspect of human nature, and even less the peculiar custom among Old Catholics to collect episcopal consecrations like scalps on one's belt. To him it was mysterious: "Why does everyone wish to be a bishop?" Before long, he would enlist the help of a few professional laymen to filter out all the bum candidates approaching him and help him find legitimate candidates. But before that several had already been consecrated by him, to the perplexity of Catholics.

As Bp. Louis Vezelis once wrote, "Once a lie has done its work, truth must labor hard and long to dislodge it. Lies fly as if on wings; and truth comes slowly limping far behind." Many, far from the facts and with only misinformation and even disinformation to work with, felt Thục's activities to be gravely scandalous, as though he had lost either his faith or his mind. What is not well-known was that he did not simply consecrate them upon being approached, but had spent considerable time, several months at least, during which he had sounded out the men, investigated them to what extent his limited means enabled him, considered each case quite carefully, rejected who knows how many, until finding a few who reasonably seemed to pass muster as men who, despite whatever unfortunate past, truly seemed fit to be hierarchical leaders within the Church. For example, one of the most "scandalous" of these choices was Jean Laborie, who had gathered two "episcopal consecrations" from the Old Catholics and was also alleged to have been a homosexual and to have involved himself in Satanic practices. But this ignores the fact that Jean Laborie had once been held in high esteem by Cardinal Alfredo Ottaviani (in charge of the Holy Office of the Roman Curia), and had actually enjoyed a distinguished career before being devastated by "the changes," left without moorings in his life, and seemingly with all legitimacy only all too grateful to Thục for an opportunity to work to restore the Church he had thought to be lost.

The first such had happened even before the Roman excommunication and absolution, namely on July 10, 1976, he consecrated another man to be bishop, a certain P. E. M. Comte de Labat d'Arnoux by name, one of these would-be bishops who drew his orders from an Old Catholic succession. For whatever reason, the issue of what had become of this bishop just never came up during his September approach to Paul VI.

Then came Jean Laborie, who had been consecrated by an Old Catholic bishop in 1966, and again in 1968 (in case the first wasn't valid), and possibly other times as well, approached Thục, and was conditionally consecrated by him on February 8, 1977. The next month, on March 19, 1977, Thục consecrated another Old Catholic-line bishop Claude Nanta de Torrini, and the year after that on October 19, 1978, he consecrated Roger Kozik and Michel Fernandez.

After the Palmar de Troya fiasco, Thục continued to seek prospective consecrands who would honor the Church, use this extraordinary opportunity as a great grace to be of real importance and value to the Church. For it was not (or not merely) a valid consecration these men were ostensibly seeking, but regularization into the hierarchy of the Catholic Church. They approached Thục, persuaded him of their repentance of their former schism in which they may have lived, or in any case derived their episcopacy from, and that they shared Thục's vision to rebuild the Church. The Church needed real bishops, and these men, one after the other formally pledged and promised that they would faithfully serve Holy Mother Church as faithful bishops to whom Her care could be reliably entrusted. Unhappily, none of these Old Catholic-line bishops are known to have functioned as intended and Thục had been woefully exploited. Thục's own goals were simple enough; as he expressed it, "I have nothing to lose. There is only one thing: to continue the Catholic Church," and "When I ordain priests and consecrate bishops, perhaps, the Catholic Church will have a chance to continue." Unable to bring anyone else along with him, he ran way ahead of everyone else, into taking actions that few could understand at the time.

Up until this point however, the bishops he had consecrated were all unmitigated disasters for the Church. Domínguez and his group have since gone on to consecrate hundreds of bishops and even appoint many "cardinals" at the service of "Pope" Gregory XVII (of Spain). Some of these "cardinals" are (or were at the time of their selection as "cardinals") teenagers and totally untrained and unqualified. It was one of the bishops, breaking from this group that attempted an "ordination" of Sinéad O'Connor. The entire false Church of Palmar de Troya places private revelation above the defined truths of the Church and deviates quite widely from the universal and historical Magisterium of the Church. Roger Kozik and Michel Fernandez were actually prosecuted for fraud in the secular courts and even served jail terms, and the others appear to have done nothing or at most set up some "little church" of their own which was of no real or lasting value or interest. The problem was Thục had no help, no assistants to do his fact checking, and no means to hold anyone accountable to their promise to function as best they could as authentic Catholic bishops. Some help then arrived, in the form of three men, Eberhard Heller, Kurt Hiller, and Reinhard Lauth (Ph.D. degreed college professors, not medical doctors, and though obviously pious and educated lay Catholics, their formal education regrettably does not appear to have included the divine sciences) who would vet future candidates for consecration, and protect him from all the would-be consecrands who would only disappoint him.

So Thục pressed on with more bishops, the first three of whom were vetted by his lay doctor friends, though Dr. Lauth left shortly after only the first, then followed by his final bishop who would again be of an Old Catholic line. The one semi-success in this endeavor would not come until September 25, 1982, when he consecrated his last bishop, Christian Marie Datessen, who had been ordained by the Old Catholics, but though failing to demonstrate a clear profession the Catholic Faith in the years subsequent to his consecration, did at least have a succession which would include such faithful and Catholic-minded bishops as Peter Hillebrand and José Ramon Lopez-Gaston.

It is only the first three of the final four bishops, those vetted by the doctors, he consecrated who would come to be of the most direct and obvious interest to the future of the Church, being in themselves truly Catholic and Apostolic priests fit to be consecrated as bishops. The first and most important of these three bishops was the one-time philosophy professor of Lateran University in Rome, the Dominican priest, Fr. Guérard des Lauriers, of Ottaviani Intervention fame, and one-time professor at Ecône. During his professorship at the faculty of Ecône, Fr. des Lauriers and several priest friends had spent a great deal of time attempting to come to an understanding as to what had become of the Church and eventually their research centered on the great question of whether or not Paul VI was really even a pope at all.

After all, a pope is supposed to be a source and center of Catholic orthodoxy, and clearly Paul VI had failed in that capacity, and it would soon begin to look as if John Paul II was not about to do much better. Maybe somehow (thought Fr. des Lauriers), they are not really popes. But then he had to grapple with the question of where the visible leadership of the Church resided if not in the Vatican Electee. There were those who characterized Paul VI (and then John Paul II) as extraordinarily "bad" popes, but popes nevertheless, and others who took the opposite position that they could not have been popes at all, despite appearances, a position called sedevacantism. This term comes from the Latin "sede," seat and "vacant," vacant. The Chair of Peter is vacant, according to this understanding of events. There is nothing unusual about the Chair of Peter being vacant. That occurs upon the death of each pope and lasts until the election of another pope. What set apart the position of Catholic "sedevacantists" from any conventional occurrence of a "Sede Vacante" status of the Church was the claim that the elections of certain modern popes (Paul VI and the John Paul's and those coming later, at least, with some suspicion hovering over John XXIII as well) were either invalid, or else if their elections were valid, then they must have subsequently lost their papal office somehow, most likely through heresy. Both positions were already known, and des Lauriers, desirous of rising above such a dispute developed a thesis called "Sedeprivationism." By this notion, Paul VI (then John Paul II) would be "pope" in one sense ("materially") but not pope in another sense ("formally"). Of this distinction, more will be said next chapter.

These considerations were clearly not attempts to break with "the pope" as much as they were to continue holding the office of the papacy in the esteem that so great an office is worthy of, while also accounting for the abject failure of the men themselves to function at all as if they held such a capacity. Since the teaching and actions of the new Vatican leadership were so seriously out of step with the teaching and practice of the reliable popes, one way to preserve the integrity of the Church is to conclude that the newer "popes" are not popes at all, either entirely, or at least in a formal sense! After the 1976–77 academic year, Lefebvre dismissed Fr. des Lauriers from the faculty of his seminary at Ecône. Whatever validity the theory may have had as an explanation for the crisis in the Church, Lefebvre felt it was too scandalous and that having such a professor on his staff might impede any diplomatic gestures he was making toward the Vatican leaders.

On their own with only their small congregations to support them, Fr. des Lauriers and a few of his closest priest friends began publishing a series of studies which presented the case for the sedeprivationist position he had taken on the pope issue, a position also known as the Cassiciacum Thesis, a term derived from a work titled Les Cahiers de Cassiciacum (The Cassiciacum Notebooks). Independent of these notebooks, the lay professors associated with Thục were turning their attention to des Lauriers as a possible consecrand, and he for his part was beginning to consider accepting the episcopacy, so he went to Toulon, France, where Thục had relocated. At first, Thục and the three laymen doctors were opposed to his consecration; none of them accepted Fr. des Laurier's sedeprivationist theory, and only after some considerable discussion and negotiation during which Dr. Lauth persuaded them that des Laurier's attachment to his sedeprivationist theory was only purely academic, did the other two doctors and Thục agree to consecrate him to be a bishop on May 7, 1981. Once that happened, his priest friends abandoned him; his congregation shrank. He had to endure a considerable amount of abuse and public infamy for this heroic act, but in the end, it has increasingly become clear that this action was the beginning of the continuation of the Church. Thục, and Drs. Hiller and Heller, were again disappointed when des Lauriers proved to be far more attached to his theory than he and Dr. Lauth had let on, and Dr. Lauth lost everyone's trust, hence his departure.

Abp. Thục and the two remaining doctors next turned to Mexico, where they found two priests, Adolfo Zamora Hernandez of Mexico City and Moises Carmona y Rivera of Acapulco who had, on their own, embraced the position of sedevacantism and also came to accept the invitation to be made bishops in order to help rebuild the Church. Fr. Carmona had been named Irremovable Pastor of the Divine Providence parish in Acapulco by Bp. Raphael Bello Ruiz (the regular diocesan bishop), but on May 5, 1977, Bp. Ruiz attempted to excommunicate Carmona for keeping the faith. Since Mexican Law is somewhat like French Law in this regard, Bp. Ruiz found it just as impossible to remove Fr. Carmona as the apostate French leaders had found it to remove the Catholics from St. Nicholas du Chardonnet. So they went to France and on October 17, 1981, Thục consecrated them to be bishops.

These three new bishops were in a stronger position than all of the other bishops consecrated by Thục put together, as they had all purely been Catholics, never associated with any schismatic group, and of all the rest only Datessen appears that he might have repented of his schismatic past, or at least bearing

Catholic fruit. Of all the rest, they had deceived Thục into consecrating them, and as such proved certainly incapable of receiving a valid canonical mission, even though their Orders were sacramentally valid. He has also expressed sorrow for having done all the unsuccessful consecrations he had performed. However, Thục has never wavered from his stand that these first three men of his last four chosen were truly and validly and lawfully consecrated as bishops.

There is therefore every reason to believe that the three episcopal consecrations of Catholics are valid and lawful and ought to be so recognized by the Church. And one might also hope the same for his final consecration of a man converting from schism, though this is less certain. Even so, many had cast doubt upon Abp. Thục and his consecrations, even those of the three Catholics. Many did not understand him. A big part of the problem was that he had no platform with which to explain himself or even answer any questions people might have had of him. As priest, bishop, and Archbishop of Huế he enjoyed the position of one backed by an immense ecclesial infrastructure of priests and fellow bishops, of a cathedral, of a chancery office fully staffed with trained and competent officers, and a blood brother who was a Head of State. Now exiled and impoverished, bereft of all his relations and his reputation, he had only himself and those things the Church had providentially given him, his episcopacy, his canonical place in the Church (which included being a papal legate), in short, God and the Church's authority on his side but practically no one and nothing else, he had no one to help him set up a seminary, to do background checks and vetting of the candidates that came to him, no publications with which to publish his own account, announce his consecrations, and also announce his excommunications of those who, having vowed, pledged, promised, and swore by everything they supposedly held dear, to serve as true bishops of the Catholic Church, then reneged on all that to return to their little churches or whatever else suited their fancy.

And very few had any real idea just how serious Thục could see that things were. The Vatican-led society (despite whatever few and fewer local exceptions) was defecting in all the ways the Catholic Church could never defect. What little remained, as those few and fewer local exceptions, would have to provide for the entire future of the Church. One does not have to formulate an opinion about the Pope question to realize the far grander scope of the problem, and hence what would be needed as a remedy. Rebuild practically the whole hierarchy from what tiny seed he, seemingly alone, carried and preserved alive and untainted? Who in his own lifetime, even of what paltry few friends he had, would have been ready to accept such an extreme view of things? And to those who did not, the thought-progression was "I don't understand what he is doing; it looks weird; therefore, I am scandalized." With such a paucity of real information available about him, it was easy for those who didn't comprehend him to believe all manner of malicious stories and gossip as would be invented in an attempt to discredit him and deface and defile his memory. Some claimed that he was either a criminal or demented, neither one of which fit the truth. But evidence would only prove to be of worth to those who don't reject the truth of Abp. Thục.

It is the case of Fr. des Lauriers which merits close scrutiny. Fr. des Lauriers, having been the main intellectual force behind the Ottaviani intervention, clearly knew that the validity of Church sacraments was being threatened by the new rites. He came to Abp. Thục, secure in the knowledge that Thục's consecration would be valid. Certainly, Thục himself would be able to follow the correct book, and Fr. des Lauriers would have noticed if the wrong book had been used, or if anything else improper had been done.

While he did notice some small omissions in the ceremony, in particular the reading of the papal mandate, and perhaps one or two other optional parts, as one who was sufficiently educated to have been a professor at Lateran University, he knew enough theology to know that those omissions in no way threatened the validity of the consecration he received. As Bishop de Castro Mayer has acknowledged, "If it's valid for Guérard (des Lauriers), it's valid for me." On another occasion, the Papal Nuncio to the United States, Pio Laghi, also acknowledged their validity. As for the case of the two Mexican priests there is no reason to believe that Abp. Thục had done anything different with them than with Fr. des Lauriers. The three Catholic priests can be quite safely regarded as validly and lawfully consecrated Catholic bishops.

Now, three (or four) out of fourteen bishops does not sound like a very good record. However, for the sake of putting all of this into perspective, let us compare his record to the record of those bishops chosen by the Vatican leadership over the same time span (1976–82) during which Thục consecrated his fourteen bishops (ignoring, for the moment, the question of whether the men chosen to be bishops by the Vatican establishment were validly made bishops or not).

Out of hundreds of "bishops" made by the Vatican establishment over that same period, NOT ONE has ever clearly and unambiguously taken a stand for the Traditional Catholic Faith. While some small percentage (maybe ten percent if one wants to be optimistic) allow or have grudgingly allowed Traditional Catholic worship on an "Indult" or "extraordinary form" basis (more about those in later chapters) and a much smaller number may even have shown support for some Catholic virtue, even these have allowed the heretical Novus Ordo religion to be the main thrust of their labors and efforts. Others have positively hijacked "their" dioceses on the claim that they are Corporations Sole and have demonstrated complete disregard even for their ostensible pope, let alone real Catholic teaching and worship. If only the Vatican leadership could have chosen their "bishops" anywhere near so well as Abp. Thục did!

There was yet more to what Abp. Thục did which many did not understand. The two Mexican priests, unlike Fr. des Lauriers, were faithful to him and finally began to operate as Thục had intended for all of his bishops. But there is more to operating as a faithful bishop than many have come to expect. These priests were simple country priests, who would to on to consecrate some yet further simple country priests very much like themselves, honest and sincere, who were also willing to be guided by Abp. Thục.

The first man consecrated by Carmona (with Zamora assisting as a "co-consecrator") was a Texan by the name of George Musey who was also just a simple country priest, on April 1, 1982. Only a couple months later on June 18, 1982, Carmona (this time assisted by Zamora and Musey as co-consecrators) consecrated two more Mexican priests, Benigno Bravo y Valades and Jose de Jesus Roberto Martinez y Gutierrez. While these two Mexicans appear to have functioned reasonably (Bp. Bravo passed away in 1985, but Bp. Martinez served faithfully in Mexico for many years until his death), George Musey together with Carmona and Zamora consecrated Louis Vezelis on August 24, 1982. That co-consecration of Bp. Vezelis marks the last known episcopal consecration on the part of Bp. Zamora. Bp. Zamora became inactive soon thereafter and passed away on May 3, 1987.

The story of Bishops Musey and Vezelis bears some telling since it demonstrates the nature of the episcopal actions taken by some of the other such early Thục bishops from Carmona and Zamora. Unlike George Musey who was just a simple country priest, Louis Vezelis was a somewhat more educated Franciscan who became increasingly disturbed about the changes being made even to his order as it went over to the Novus Ordo religion, to the point that on April 19, 1978, he left that Franciscan community and became an independent Franciscan. He started his own Franciscan order on the claim that their bishop was the "Bishop of Rome," somewhat misleading in that, though all Traditional Catholics are in union with the Pope, the "Bishop of Rome," or at least the office even if vacant, that did not necessarily imply a union with John Paul II (to be explained next chapter), and in fact didn't in his case. In 1980 he began publishing a small newsletter called *The Seraph* which came out (and continues to come out) irregularly.

At about the time he learned of the Thục consecrations, and especially the consecration of George Musey, he abandoned that approach which had only spread confusion and injured his reputation, and openly adopted a theological opinion similar to George's, befriended him, and obtained an episcopal consecration from him, together with Bishops Carmona and Zamora. Once consecrated, Vezelis together with Musey were themselves set up as the only Catholic hierarchy in the United States, with Vezelis taking everything east of the Mississippi except Florida and Musey taking everything west plus Florida. Having Abp. Thục to guide them into the appropriate episcopal behavior as prescribed in Canon Law and aware of the power and influence their valid episcopal consecrations brought them, they possessed full and regular jurisdiction over the United States, now carved up into only two "mega-dioceses." Many found this step incomprehensible and regarded it as presumptuous or worse. Some of the priests aligned with Abp. Lefebvre were already in the country and serving all without any diocesan blessing, certainly not from the local Vatican representatives, and not about to accept the diocesan leadership of Bps. Musey or Vezelis. Let us first discuss why it is that what they did was truly right and proper, after which we will look at how and why it became impractical and ultimately unworkable.

In bringing the Gospel to all the Church all over the world, it is reasonable that the Pope (who heads the whole Church) would be assisted by local leaders, representatives, bishops in fact, who together with each other and the Pope comprise the hierarchy (divinely established and maintained ecclesiastical leadership of the Church). From almost the beginning it has been the practice of the Church to divvy up the world into geographical regions, often corresponding (albeit somewhat loosely) to secular divisions among nations and provinces, etc. as well as physical boundaries such as mountains, rivers, and oceans. That is not the only way the world's population could be divvied up among the bishops of the Church, but by far (under most circumstances) the most practical. It is a setup which exists due to ecclesiastical law. And what ecclesiastical law has set up, it can also modify, adjust, reinterpret, or even dismantle if need be. When a region of multiple Catholic dioceses suffers a significant lost of Catholic population, one practice that thereby becomes acceptable is to combine the dioceses into fewer dioceses, or even just one. The "mega-dioceses" of Bps. Musey and Vezelis were each such a combination of the many former dioceses that had previously comprised the Catholic dioceses of the United States. They had agreed with each other as to which would have what parts of the United States, and with the rest of the bishops that they would have the United States and the others would divvy up the rest of the world. Even their not having been appointed to any diocese, particular or "mega," by the Pope need not be a problem. During the 1200's, there was a three-year period during which the Church had no Pope. Yet during that time some at least 21 bishops were appointed, consecrated, and even set over dioceses lacking a bishop for the previous bishop having died.

It falls to the remaining Catholic hierarchy, whoever that comes to consist of, to interpret Her laws and apply them to a given situation. Most of all, the Church must be preserved: the doctrines, the morals, the liturgy, the insistence of unity based on subjection to the Pope (which translates into a desire for a Pope to be elected when there is no Pope), and the overall hierarchical structure. This last area is the only place where some real flexibility can be permitted. Ordinarily they try to keep the historic dioceses all going in a more or less conventional manner, but sometimes they can't, and if the Church should also happen to be between Popes, especially for a protracted time, then if there are not enough bishops then those remaining must make their own choices either as to who to consecrate, or who to set over multiple dioceses (effectively combining them in practice), or what combination of those two strategies to use. Is this really what they are supposed to do?

Let's use an illustration. Suppose a plague kills off 99.95% of the human population, randomly throughout the whole world, all in a matter of weeks or less. Only the remaining 0.05% are blessed with some rare genome which renders them immune to the plague. Obviously, this would seriously reduce not only the world's population and the Church's population, but also the number of religious and clerical, nuns, monks, priests, and bishops. Instead of thousands of bishops, only a dozen or two. Instead of hundred of thousands of priests, only a few thousand or less. Chances are microscopic that the Pope would survive, so let's assume he doesn't. And while there are some several dozen or so Cardinals, the chances of even one of them surviving is also significantly less than 50% so let's assume none of them survived, either. Out of the entire Curial staff, perhaps one person survives, and that is a secretary whose job there is to type up memos as dictated, answer the telephone, and make sure the dignitaries around her all have coffee. While humanity and civilization in general must struggle to pick up and continue, the Church must do the same. Is it not obvious what the officers of the Church, the few surviving bishops there are, must do?

First, they would have to see to the immediate needs of the Faithful all around the world. They would find out who among all the bishops is still alive, then each look to see if any of the priests they know might be fit and ready to be consecrated as bishops, but obviously that will not be anywhere near enough bishops to replace all those deceased. Many of the surviving priests will not be ready or willing to serve as bishops. Even some historically schismatic bishops might be accepted in as Catholic bishops, providing of course that they disavow all their former connection to any schism and heresy. Each would negotiate with their neighboring bishops as to who would take which groups of dioceses, and once some more bishops can be made, how those arrangements might be adjusted to accommodate the newly consecrated bishops. Many churches would be closed but a scattered few would be kept open, to be staffed by the few remaining clerics, and word would be spread among the Faithful where the remaining open churches are. Religious Orders of nuns and monks would also have to be consolidated, with remaining members of smaller orders of which very few persons have survived being consolidated together into new "ad hoc" Religious Orders or Congregations. Then, with some stability established for all members of the Church, that all the Faithful will have their sacraments, teaching, and guidance of the Church, the bishops would next take on the more long-term needs of the Church, beginning with the restoration of the papacy. Theologians have already discussed such a scenario: the bishops would meet in Council (an "imperfect" Council, at least in its convening), decide upon who shall comprise the electors, the electors shall elect a Pope, and then the Pope once elected and accepting the office truly is Pope and all submit to him, unified in his person under his infallible leadership. From there, the Pope coordinates all further restoration of the Church.

Now imagine some miscreant who comes along, thinking he can interpret the Church's laws better than the officials of the Church. He sees these bishops, each one presiding over hundreds of former dioceses, of which they might have been given one, if even that in some cases, by the Pope, and some of these bishops consecrated without any Pope on hand to approve them, and others having come up through the ranks of the East Orthodox or other schismatic lines but now claiming (to the satisfaction of the rest) to be Catholics, and so he accuses these bishops of being a "false and schismatic hierarchy," "presumptuous," "self-appointed," and therefore "a new church, set up as a rival to the original Church, false, scandalous, and utterly disreputable." Is not the insanity and unreasonableness of such an accusation obvious? For what I have described is not some "brand new fake hierarchy starting up from scratch" but in fact all that remains of the original apostolic hierarchy, taking extreme steps in response to an extreme circumstance. This is how Abp. Thục and the bishops he consecrated and some to follow further in the succession saw themselves and their action. As late as April 1992 Bp. McKenna listed some nine bishops consecrated by Thục, the three Catholics, plus six of those coming from the Old Catholics but regularized by Thục (including Bp. Datessen) and assumed to be functioning correctly as a "providential assurance indeed of the Apostolic Succession."

So, what broke down? The principal reasons were psychological, on the part of many persons, the general run of the Church, not only laity but even many considering (and pursuing) the priesthood. The biggest real problem was this: in my illustration I pictured a loss of the vast majority of mankind to a physical plague, leading to physical death, something anyone can trivially recognize when it has happened to someone. But what if the death were not physical, but spiritual: schism and heresy unto apostasy? From the purely canonical standpoint of who has the right to hold a place of authority and jurisdiction in the Church, either kind of death equally removes a person. But when death is only spiritual and physical life remains, determining that spiritual death becomes something of a challenge. Such a huge apostasy has happened, and yet things all just seem so "normal." How can that be?

Far too many simply took the conventional diocese holders, now corrupted out of the Faith, as if they were still real Catholic bishops, so of course the real traditional hierarchy must have seemed like some new and rival and "parallel" hierarchy, difficult to take. And it didn't help that many of the few faithful Catholic bishops had only been educated as simple country priests. While there is no divinely set educational standard required for Catholic bishops, the custom has long been that the bishops would be selected from among the most educated priests (or monks, in the case of the Eastern Rites who draw most of their bishops from the monasteries rather than the parish priests). It must have seemed strange, bishops making diocesan claims at the same time others were doing the same, and by all superficial appearances, holding all the historic offices.

And the fact is that almost anything a person does can be made to seem criminal or sordid or scandalous as suits fancy, especially if the person has no platform to explain themselves or answer questions. And then add rumors, lies, half-truths, and omit all the important details. For example, it was said that Bp. Vezelis had even set up an Old Catholic priest as a regular parish priest in one of the churches in his diocese, in Ohio. What is not revealed is that he, as bishop of the Church, had regularized said priest into the Faith, and as his bishop monitored his behavior carefully, as befits Catholic bishops. Some new clergy, seeing how real diocesan Catholic bishops were being ill-spoken of as though they had appointed themselves (not actually true, but an easy accusation to make in the absence of real information), instead felt they would avoid the stress and responsibility by making no conventional territorial (diocesan) claims.

Most of all, the vast majority, even of Traditional Catholics, were just not quite ready to see the necessity that the Church take the extreme steps as taken by Abp. Thục and his bishops. Surely, common sense, or a more Catholic spirit, was sure to return to the Vatican and Vatican-led figures, if only we could just be patient, keep our own faith, pray, and write a lot of letters encouraging, exhorting, admonishing, and begging them to please return to the Faith. The Lefebvre priests of the SSPX had taken this tack, as did also a "second wave" of bishops and priests stemming from the Thục succession that began to make its appearance coming into the 1990's. Remember that the "overall hierarchical structure," being determined purely by ecclesiastical law (as interpreted by the collective leadership of the Church), is the one thing wherein a real and significant degree of flexibility is possible. We live in times in which conventional dioceses do not make anywhere near as much sense as they did in former ages, with people now ever on the go and on the move. Catholics were also halting on two opinions, whether to accept the traditional hierarchy or the Novus Ordo leaders as being in at least some sense the "office holders."

In many ways what we have resembles how it was in sixteenth and seventeenth century England. England had priests and bishops (in the beginning of their schism often the same men as had been appointed by Catholic authorities), now belonging to an English schismatic church (the Anglicans), occupying the same parish churches and cathedrals, while priests still united to the Pope in Rome had to go underground. Those priests had to hide, for example in a "priest's hole," in some Catholic's house, or other places, for fear of being caught and killed by slow torture. For such real Catholic priests, it was meaningless to attempt to attach them to conventional parishes or even diocesan territories, as they might be often forced to flee for great distances and would do what they could to help Catholics wherever they ended up. These priests, a few faithful locals, but mostly trained overseas (primarily at the English language seminary of Douay-Rheims in France), were given faculties to function as such in any part of England, wherever their travels and adventures would lead them. How much easier for the general run of English society to assume the Anglican priests and bishops to be the "legitimate" holders of parishes and dioceses, as nearly everyone did, and the priests faithful to the Rome as mere "acephalous" and "wandering" clerics of no apparent spiritual jurisdiction, and certainly of no conventional territorial parish or diocesan jurisdiction. Their jurisdiction was real, but aterritorial in nature, so like nearly all traditional clerics today.

And as it turned out, though few (if any) traditional clerics would even dream of citing *Lumen Gentium* as the basis for any detail of their manner of ministering, there remains self-same principle which made the local Novus Ordo leaders merely "honorary" rather than with real authority (in addition to their departure from the Faith), namely that "no minister, even an admittedly Catholic one or not, can be excluded from performing [all ministerial] functions within anyone's parish or diocesan territory, even altogether without permission or consent." In short, while Bp. Musey remained the rightful Catholic bishop of the "mega-" diocese of the Western United States, he had not the right or power to exclude either the Lefebvre priests or priests of, for example, Bp. Pivarunas, from setting up shop or ministering within his diocese and without his permission.

The Church must play the hand She has been dealt, instead of the hand She would wish for. There are times and places, such as England back then, or the whole world now, in which insistence upon such a conventional diocesan structure is just not practicable. The Lefebvre priests already so operated, already undermining the diocesan authority of the early Thục bishops; later waves of priests would find it seemingly "more respectable" to so function as well. There were many opinions and explanations to be had as to "what had happened" and Catholics could not agree even among themselves upon such basics as where the Church is, or whether the Papal chair is occupied or vacant. How much more complicated still to decide between rival claimants to diocesan sees. And how unjust to insist upon having an uninformed populace, untrained in theological and canonical matters, be obliged to decide between them. How far easier and more practical for priests and bishops to function as such wherever they can and leave it all to some future Pope to resolve all the diocesan questions.

While it was more correct than they knew to claim that, as Catholic bishops of that part of the Church subsisting outside the Vatican organization, they enjoy true jurisdiction over the Catholic faithful, the early Thục bishops ended up injuring their own reputations by claiming that their jurisdiction was fully that of regular diocesan bishops, clear to the point of having the authority to rule out all other bishops from functioning in "their" territories and that all American priests must submit to their authority, either to one or to the other, depending on where the priest was located.

Such a claim, despite its soundness in the theological abstract, merely scandalized many to the point that very few other Traditional Catholic priests ever had anything to do with them. The other bishops in Mexico and other parts of the world had done similar things at various points and in various groupings, and that too soon broke down for the same reasons. However, now that all of the original such have passed away, the position is now an extreme minority one; only Bp. Giles Butler (ordained and consecrated by Bp. Vezelis) still attempts to function as the ordinary bishop of the eastern portion of the United States. Even such bishops of that first wave as survived well into the second wave, such as Bp. Roberto Martinez of Mexico, eventually shifted from the one position to the other. Now, Traditional Catholics tend to be attached to a congregation or society, usually headed by one or more bishops, which societies operate canonically as "ecclesiastical districts tantamount to dioceses." This seems to be at this time the most practical structure for the Church's hierarchy to assume.

Bishops tracing their episcopal line to Abp. Thục have proven many, as Thục would have wished (and doubtless does wish); the following deceased bishops have served the Church loyally:

Michel Louis Guérard des Lauriers, consecrated by Thục, May 7, 1981, deceased February 25, 1988.

Moisés Carmona-Rivera, consecrated by Thục, October 17, 1981, deceased November 1, 1991.

Adolfo Zamora Hernandez, consecrated by Thục, October 17, 1981, deceased May 3, 1987.

George J. Musey, consecrated by Carmona and Zamora, April 1, 1982, deceased March 29, 1992.

Benigno Bravo Valades, consecrated by Carmona and Zamora June 18, 1982, deceased 1985.

José de Jesus Roberto Martínez y Gutiérrez, consecrated by Carmona and Zamora June 18, 1982, deceased 2008.

Louis Vezelis, consecrated by Musey, Carmona, and Zamora, August 24, 1982, deceased January 1, 2013.

Gunther Storck, consecrated by des Lauriers April 30, 1984, deceased April 23, 1993.

Conrad Altenbach, consecrated by Musey May 24, 1984, deceased October 19, 1985.

Ralph Siebert, consecrated by Musey May 24, 1984, deceased 1986.

Robert Fidelis McKenna, consecrated by des Lauriers August 22, 1986, deceased December 16, 2016.

J. Elmer Vida, consecrated by Mckenna July 2, 1987, deceased March 30, 1993.

Oliver Oravec, consecrated by Mckenna October 21, 1988, deceased July 9, 2014.

John E. Hesson, consecrated by Oravec June 12, 1991, deceased August 26, 2017.

José Ramon Lopez-Gaston, consecrated by Olivares June 29, 1992, deceased May 5, 2009.

> *Note: Olivares traces his Thục lineage through Bp. Datessen, of whom it is said that he had partially trained at Ecône and that, despite coming through some seriously questionable connections, "Those who knew him spoke of the solid Traditional Catholic Faith that he professed, and this was the reason why Bishop Thục agreed to consecrate him conditionally on September 25, 1982." Some questions remain as to his loyalty to the Church, and in any case, he consecrated Pierre Sallé on June 27, 1983 (now deceased), who in turn consecrated Guy Jean Tau Johannes de Mamistra Olivares on March 28, 1987. Since I have been unable to verify the loyalty of these three figures and the orthodoxy of the latter two, while the validity of Orders is not in doubt, a clear conveyance of the Canonical Mission from Abp. Thục to Bp. Lopez-Gaston cannot be established by this author. Nevertheless Bp. Lopez-Gaston and others of this succession seem to have served creditably and loyally and with apparent orthodoxy, perhaps worthy of being seriously considered for acceptance by those bishops who have a clear Canonical Mission as had been Bp. Yurchyk*

> *of the Russian Orthodox. It is also unknown to me what manner and depth of education and training and formation these bishops stemming through Bp. Datessen and also those consecrated by Bp. Slupski have received, and what flocks, if any, they have been set over and accepted by.*

Francis Slupski, consecrated by Mckenna October 12, 1999, deceased May 14, 2018.

Paul Petco, consecrated by Slupski March 11, 2011, deceased October 20, 2018 (some questions regarding his personal morals had been raised).

The following Thục bishops appear to be serving the Church credibly, and are still alive (as best known 2019):

Mark Anthony Pivarunas, consecrated by Carmona September 24, 1991.
Julio Aonzo, consecrated by Lopez-Gaston December 27, 1992.
Gary Gonzalo Alarcon Zegada, consecrated by Lopez-Gaston August 1, 1993.
Daniel Lytle Dolan, consecrated by Pivarunas November 30, 1993.
Emmanuel Korab, consecrated by Lopez-Gaston June 26, 1994, conditionally in November, 1999 by Hnilica.

> *Note: The conditional consecration by Hnilica, assuming it occurred as generally believed, would have provided Bp. Korab with the Canonical Mission of the Church.*

José Franklin Urbina Aznar, consecrated by Lopez-Gaston June 26, 1994.
Juan José Squetino Schattenhofer, consecrated by Urbina February 11, 1999.
Martin Davila Gandara, consecrated by Pivarunas May 11, 1999.
Geert Stuyver, consecrated by Mckenna June 16, 2002.
Donald J. Sanborn, consecrated by Mckenna June 20, 2002.
Markus Ramolla, consecrated by Slupski May 23, 2004.
Michael French, consecrated by Korab October 16, 2004.
Robert Neville, consecrated by McKenna April 28, 2005.
Giles Butler, consecrated by Vezelis August 24, 2005.
Andres Morello Peralta, consecrated by Neville November 30, 2006.
Luis Armando Argueta Rosal, consecrated by Urbina April 10, 2007.
Luis Alberto Madrigal, consecrated by Vezelis December 12, 2007.
Robert Dymek, consecrated by Slupski December 7, 2011.
Bonaventure Strandt, consecrated by Vezelis August 15, 2012.
William Greene, consecrated by Slupski in 2012.
Joseph Selway, consecrated by Sanborn February 22, 2018.
Simon Scharf, consecrated by Korab August 4, 2018.
Merardo Loya, consecrated by Squetino January 12, 2019.

The following bishops, though probably alive in most cases, appear to have no active ministry, though in the past they did serve:

Peter Hillebrand, consecrated by Sallé July 27, 1984, conditionally July 17, 1991 by Carmona, believed to be retired.

Franco Munari, consecrated by des Lauriers November 25, 1987, now inactive having truly left the ministry.

Michel F. Main, consecrated by Musey December 8, 1987, presumed dead.

Richard F. Beddingfeld, consecrated by McKenna December 17, 1987, defected to fake claimant in Canada (St. Jovite) in 1993.

Guido Alarcon Zegada, consecrated by Lopez-Gaston August 1, 1993, presumed inactive.

Commonly heard in traditionalist circles is the expression "Thục-line bishop" which usually means a bishop who traces his episcopal Orders to the three Catholic priests he consecrated, though the more respectable following from Bp. Datessen as listed here might also count. For the time being, the above list pretty much comprises the main figures; there are others less well known, for example Anton Thái Trinh (consecrated by Bp. Slupski, date unknown). Beyond that, there are some writers who still use the phrase "Thục-line bishop" to refer as well to the handful of Old Catholics he regularized and the Palmar de Troya bishops which now number in the hundreds, generally disreputable and many doubtful, and perhaps a Bp. Neal Webster, a follower of Fr. Feeney's error and the first to claim episcopal status, supposedly consecrated by Bp. Slupski but date unknown. Happily, most of the more recent Thục-line bishops from des Lauriers, Carmona, and Zamora, and even some from Datessen, have been quite admirably able to defend their reputations and demonstrate their good character.

On February 25, 1982, Abp. Thục declared that "As a bishop of the Roman Catholic Church I declare the See of Rome being vacant and it is my duty, to do everything to assure the preservation of the Roman Catholic Church for the eternal salvation of souls." Over a year later, on May 26, 1983, Thục, together with the Mexican bishops he consecrated enlarged on this point by making the following public statement:

> The Roman Catholic Bishops, united with His Excellency Archbishop Ngô Đình Thục, declare:
> That we support him in his valiant public declaration made regarding the vacancy of the Apostolic See and the invalidity and illicitness of the New Mass. We hold with him that the Apostolic See has been vacant since the death of Pope Pius XII by virtue of the fact that those who were elected to succeed him did not possess the canonical qualifications necessary to be legitimate candidates for the Papacy.

> ...Based upon the Bull Cum Ex Apostolatus Officio of His Holiness Pope Paul IV, we hold that Angelo Roncalli was never a legitimate Pope and that his acts are completely null and void.
>
> We declare that the New "Mass" is invalid...We declare that the introduction of this New "Mass" also signals the promulgation of a new humanistic religion in which Almighty God is no longer worshipped as he desires to be worshipped...Those who have accepted this New 'Mass' have, in reality and without taking notice of it, apostatized from the true faith; they have separated themselves from the true Church and are in danger of losing their souls, because outside the Church founded by Jesus Christ no one can be saved. For this reason, we invite the faithful to return to their Faith from which they have strayed.
>
> We reject the heretical Decree on Religious Freedom which places the divinely revealed religion on an equality with false religions. This decree is a clear and evident sign of the denial of our holy traditions by the apostate and schismatic hierarchy.
>
> We declare that no one can oblige us to separate ourselves from the true Church, from that Church instituted by Christ Himself and which is destined to last until the consummation of the world just as He instituted it...We give thanks to God for the integrity of our Faith and we beseech His grace that we may be able to persevere in it. We pray for those who have lost this Faith by accepting the heretical changes that have given rise to a new Church and to a new religion.

For Thục and his bishops, it had become clear that such a great apostasy, coming from even the highest levels in the Vatican, represented a clear departure from the Papacy on the part of their leadership. So, had he made mistakes? Yes, but those mistakes were all honest ones, coming from a totally good place. He meant well and did well, and where some things he did failed to turn out well he cannot be justly faulted in the least. In the next chapter we will explore what that would mean, and the various theories as to how that could happen.

Shortly thereafter while living in New York with Bp. Vezelis, Thục was kidnapped and taken to Carthage, Missouri and held there incommunicado by Vietnamese priests in union with Modernist Rome until his death on December 13, 1984. After his death, when he was no longer in a position to protest what was being said in his name, the following was published by them:

> I, undersigned, Peter Martin Ngô Đình Thục, Titular Archbishop of Bulla Regia, and Archbishop Emeritus of Hue, wish to publicly retract all my previous errors concerning my illegitimately ordaining to the Episcopate, in 1981, several priests, namely Revs. M. L. Guérard des Lauriers, O. P., Moses Carmona, and Adolpho Zamora, as well as

> my denial of the Second Vatican Council, the new 'Ordo Missae', especially the dignity of His Holiness, Pope John Paul II, as actually legitimate successor of St. Peter, published in Munich in 1982.
>
> I wish to sincerely ask you all to forgive me, praying for me, and redressing all scandal caused by such regrettable actions and declaration of mine.
>
> I would also like to exhort the above mentioned priests who had illegitimately been ordained to the Episcopate by me in 1981, and all others whom they have in turn ordained bishops and priests, as well as their followers, to retract their error, leaving their actually false status, and reconciling themselves with the Church and the Holy Father, Pope John Paul II.

Clearly, that statement was either an outright forgery or else signed by him under the greatest duress, in view of the lifelong commitment he had made to continuing the Church and opposing the apostasy that had come to be so widely accepted. Had this come from Thục it would have mentioned all of his episcopal consecrations, including Palmar de Troya and the "Old Catholic"-ordained clergy including Bp. Datessen, and also his 1983 declaration, and would have provided reasons for his supposed change of mind. Even the turn of phrase used is not that of Abp. Thục, clearly indicating its spuriousness. Like Abp. Thục himself, the Catholic Thục-line bishops know why God has allowed them to be bishops, and that is to do their share in preserving a valid apostolic succession on into the future. Also, despite the shaky start some of them got, nearly all of the better-known Catholic Thục-line bishops now function quite well as responsible and capable leaders of the Church.

9

THE ADVANCE OF THE SEDEVACANTISTS

For a moment, imagine yourself not knowing anything of what you have read in the earlier chapters of this book, particularly regarding the Second Vatican Council, namely how the Vatican organization was legally detached from the Roman Catholic Church. Perhaps you were in such a state before you started reading this book. Picture yourself facing a most painful mystery. You see "the Church," once solidly founded on the Rock of St. Peter now cast adrift with nothing but shifting sand to grab on to. Desperate to reconcile such phrases as "the Mystical Body of Christ," or "the Church, One, Holy, Catholic, and Apostolic" with what you saw happening in your local parish, how could you help but be willing to consider almost any theory, no matter how crazy or far out, which could possibly explain what had happened to the Catholic Church?

With such a large number of once beautiful, devout, Catholic parishes practically all at once going to seed, losing members, financial support, sanctity, and even that Catholic atmosphere or ethos or mind-set or character, hundreds of millions of Catholics all around the world, like yourself in such a case, are at a loss to explain or understand how all of that can be. Clearly, something has gone gravely wrong, and with so many people being personally confronted with things that are going wrong, you will find a tremendous diversity of theories to explain it. Not understanding what happened to Catholic authority back in Vatican II, the following list is just a sampling of some of the main theories you might have encountered for consideration.

There is the "prisoner in the Vatican" theory which claims that "the Pope" everyone sees is actually an impostor while the real Pope is held captive in some Vatican dungeon. The impostor, being not really the Pope, is therefore not infallible, but quite able to foist his heresies on the naïve and unknowing Church. That is one of the more exotic and far out theories which would otherwise be quite difficult to take seriously and would not have been taken seriously by anyone had only everyone understood what happened at Vatican II. That theory does not in and of itself require a non-Catholic mind set. After all, granting that such a thing were to have happened, wouldn't the things we have seen happen become entirely possible? But many of those who have embraced it had nowhere to turn but the Palmar de Troya group and/or discreditable Bayside apparitions, or else went off on their own in comparable directions.

Another exotic and far out notion is sometimes called the "Cardinal Siri Theory," a claim that Cardinal Siri was actually the one elected to be Pope in either of the 1958 or 1963 conclaves, and that he has secretly functioned as Pope Gregory XVII until his death in 1989. Unfortunately, Cardinal Siri never did show any clear support for the traditional Faith or Mass, beyond a vaguely conservative leaning, let alone exercise any Papal prerogatives, and now that he is dead, who are his successors supposed to have been?

There is the "End of the World" theory which claims that we have entered some special period of time known as the "End Times," or the "Three Days of Darkness." There are many prophecies both in Scripture as well as in the Private Revelations of the saints regarding a major apostasy just before the End of the World. One therefore explains the fall of the Church in terms of that last great apostasy. Again, this theory need not apply now that a far more realistic and likely cause of the disaster has been located, and while the End of the World is always a possibility it need not be the case this time around, which might disappoint some. The result of embracing this theory is often that one ends up hunkering down in a bunker or hiding in the mountains, trusting no one, and thereby being deprived of the sacraments and Catholic fellowship.

While one can readily see that the first three theories are fallacious, and even (particularly in the case of the first two) rather odd, offbeat, and unlikely, at least they each permit one to keep a Catholic frame of mind. There need not be any guilt or doctrinal error on the part of those who have resorted to those theories, providing they have avoided the Palmar de Troya group, the Bayside (or other false) apparitions, and not become cut oneself off from all Catholic sacraments and fellowship, becoming what is called a "home-aloner."

There is another theory which poses somewhat more of a threat to one's faith, but which can be rendered at least doctrinally harmless, if coupled with some other theory. This is the conspiracy theory, namely that the Church "hierarchy" has been taken over by subversives of various kinds, whether Freemasons, Masonic Jews, Communists, Illuminati, or other typical archvillains that the conspiracy theory enthusiasts often go on about. One must grant that some sort of conspiracy has been at work bringing about the Second Vatican Council and all of the damage it has brought to the Church. However, without the benefit of some other theory to explain why God has allowed this conspiracy to succeed, where all other previous conspiracies throughout Church history have failed, those who embrace this theory are in danger of falling into the heresy of believing that some extra bit of cleverness on the part of Satan would have the power to repeal God's promises regarding the indefectibility of the Church.

Still another similarly serious theory, but less controversial, is the thought that sociological and/or historical changes, particularly in the United States and Europe, have caused many clerics to become soft and to have enough time on their hands to concoct the novelties of Vatican II. Indeed, the abuses have been worst in these areas, and it was their easy financial prosperity that gave them unwarranted clout at Vatican II, but again, to put the doctrinal rectitude of the Church at the mercy of such mindless and random sociological or historical forces is to deny the role of God in protecting His Church. Whether by planned conspiracy or by unplanned sociological/historical pressures, this theory requires another theory to supplement it in order to explain things without denying God's promises to His Church. The next three theories are much more serious because they intrinsically involve actual damage to one's faith.

There is the "maybe Vatican I was wrong" theory which leads one to suppose that maybe the Old Catholics or even the East Orthodox or some other dissenters are the "real" Catholics. Aside from the evident loss of faith such a position entails, it even causes a person to identify with dissent which is where the fall of the Vatican organization actually came from in the first place. Keep in mind that the subversive Liturgical Movement is made up of the same sort of dissenters with the same heresies as the Old Catholics or the East Orthodox which this theory leads one to join. It is as if one jumps out of the frying pan into the fire.

Then, there is the "maybe Catholicism is wrong" theory which is the position which lies at the heart of all of those who left the Church when "the changes" came along, regardless of whether they became Protestant, Jewish, New Age, or just lost all apparent interest in religion. Seeing something, which could never be changed, suddenly get changed quite understandably destroyed the faith of many in the Catholic Church. Were it really possible that the Catholic Church could ever be changed the way the Vatican organization has changed during and since Vatican II, then that would have conclusively proved that all of Christianity was a hoax and a fraud from the very start. It was the separation between the Catholic Church and the Vatican organization which allowed the latter to fall, and it is the recognition of that separation (even on an intuitive and instinctive but inarticulate level) which has made and will make it possible for the Church to continue today and tomorrow as the Traditional Catholic community, or "movement."

Finally, there is the "Truth has changed" theory. This is by far the most widespread theory. Every active member of the Novus Ordo Church of the People of God has on at least some level (even if only an unconscious one) embraced this theory. In this case one follows the leadership of the Vatican organization no matter what they teach, and no matter what they contradict. Maybe religious liberty and indifferentism used to be sinful but God has since changed His mind and so they are no longer sinful. Perhaps next year God will change His mind again and decide that 2 + 2 = 5, or that the speed of light is 55 miles per hour, or that one must commit adultery to get to Heaven, or that tabernacles must be moved (again) to birdhouses across the street, or that all crucifixes should be torn down and replaced with Satanic pentagrams, or that Paul VI was a saint, or that not even a Pope has a right to judge the sin of homosexuality! (Wait, that last has already happened, unless one wants to construe that as an admission on the man's part that he is in fact no Pope!) Who knows what will be "The Truth" tomorrow?

A very unusual and clever theory which never got much headway despite its interesting features and initial attractiveness is the "Western Patriarch" theory. This theory operates on the basis that Paul VI only changed the Western (Latin) Rite, and so therefore was doing so not as Pope but merely as Western Patriarch. One can at least see in this theory a small germ of the truth because the Novus Ordo religion does, by even its own definition, lack universal jurisdiction. Alas, the one great weakness of that theory as presented here is that it lacks any explanation for the Novus Ordo teachings on ecumenism and religious liberty or indifferentism. And now that the other Rites are also being adversely affected this too is invalidated. Unlike the other theories, there are no groups or priests who support this theory. It is the following theories which are of the most serious merit and therefore of real interest to all true Catholics:

There is the "authority has been gravely abused" theory which is that the Pope and the bishops are abusing their authority. Many who take this position believe that such abuses ought not be followed. This is the approach taken by most members of the SSPX. It is based on the belief that the areas in which the Vatican hierarchy have erred are all in the realm of discipline, which has never been guaranteed to be infallible.

Many others who believe the "abuse of authority" theory are not clearly aware of the contradiction between the new disciplines on the one hand and Faith and Morals on the other. All they know is that they don't like the new disciplines and so they request to be excused from them. Sometimes their request is granted, and they receive some rare and special permission from the Modernists to "keep the old ways," at least in their public form of worship. Originally this was done under the terms of what was called an "indult" (a special privilege), but later on broadened out a bit with a significant document in 2007. Only then do these people get to keep their Catholic Faith and Worship. Certainly, in the case where they are granted their special permission, they do keep their Catholic Faith and Worship. But what will they do if that permission is revoked?

Since many of the unfortunate new disciplines contradict Faith or Morals, they ought not be obeyed. Discipline is by its very nature inferior to Faith and Morals (and Revelation) and so no discipline which contradicts Faith or Morals can ever have any moral force of Law. The priests and members of the SSPX, and those Catholics practicing their faith with permission from the Modernists, are as yet unaware of the legal detachment of the Vatican organization from the Roman Catholic Church and so therefore find themselves unwilling to admit that their leaders have lost at least something of their former authority, hence their opinion that rightful authority is being wrongfully abused by the persons who hold it, whether as a result of ignorance, fear, or malice. At least such Catholics as either have that permission or else align themselves with the SSPX or comparable clergy have kept their Catholic Faith.

In some ways, the "abuse of authority" might be more properly considered a refusal to admit to any explanation, particularly on the part of the SSPX. Their policy is to avoid any real attempt to explain what has happened to the Church. After all, any explanation capable of holding any water would be almost certain to destroy their diplomatic position and render them unable to assist in the potential rehabilitation of the Vatican organization, for that (aside from aiding the souls in their care) is their true purpose and goal. Admitting that they lack the ecclesiastical competence to pronounce that the Conciliar and post-Conciliar popes are not popes at all, the SSPX treats them as much like popes as they can and try to encourage them to behave more like Catholics.

And finally, there is the "the ostensible hierarchy has lost authority" theory (whether partially or totally), now commonly referred to as sedevacantism. We got a brief glimpse of that in the last chapter. Let us now take a closer and more detailed look at what sedevacantism actually means. Although the term "sedevacantism" and the discussions regarding this theory often focus on the person of the pope or the papacy, it just as properly applies also to that fallen chain-of-command which the man (whatever he may properly be called) is ostensibly leading.

Sedevacantism faces opposition from two directions. The Vatican leaders themselves are quite terrified at the prospect of this theory gaining widespread attention. Such a fear is quite understandable, perhaps even justified: For every "illicitly" consecrated bishop who takes a sedevacantist position owing to the fact that Paul VI or the John Paul's or Benedict XVI is not Catholic enough for him, there are hundreds of Novus Ordo local functionaries, "bishops," who would have been quite happy to adopt a "sedevacantist" position simply because those leaders were still <u>too</u> Catholic for them. With Francis I of course, this later side ceases to be a concern since even most of his own cronies don't find him "Catholic" enough. I can only shudder to contemplate the heinously fallen spiritual condition of anyone for whom Francis I is yet still "too Catholic" for them! Yet, in reality it has been those non-Catholic Novus Ordo "bishops" who most directly stand accused of having fallen and thereby vacated their Sees, even more so than John XXIII through Benedict XVI themselves had fallen.

The other source of opposition has been a theological one. The basic argument of the case for sedevacantism goes like this: If a man who is pope decides to embrace some heresy (thereby making himself a heretic) and to teach heresy, he no longer has the Faith, is therefore automatically excommunicated, and since he is no longer a Catholic, he can no longer be regarded as being the leader of the Catholic Church. He would therefore lose his office and his chair is empty (hence "sedevacantism"). There does exist a strong consensus of Canonists and Doctors of the Church such as Matthaeus Conte a Coronata, Wernz-Vidal, or St. Robert Bellarmine, that WERE a pope to become a heretic, THEN he could no longer be pope and it would therefore be time for another papal conclave. That much has been established beyond all doubts, so far so good.

The theological difficulty enters in on two accounts. For one, virtually all of these same Canonists and Doctors opine that such a thing is not possible since the Holy Spirit protects any living Successor of Peter from becoming a heretic even if he chooses to become one. Granted that is only their expert opinion and as such they themselves admit that it could be wrong, but the other theological difficulty has been by far the weakest link in the sedevacantist's argument:

No one but God can ever rightly judge the Pope. Even in the above described example as discussed by the Canonists and Doctors, the Pope is not excommunicated by anyone else, since he lacks any lawful superior in the earthly realm but incurs an automatic excommunication. The role of the bishops and cardinals in this case is merely to ascertain that the Pope has excommunicated himself (and so they elect his successor), not to excommunicate the Pope. How much less then, is it our role as ordinary private Catholics to excommunicate the Pope. For any of us to decide for ourselves that any of these Vatican leaders is not a pope certainly seems very much like a private judgment, and as devout and informed Catholics we all know how little <u>that</u> is worth!

Fortunately, it is not quite so bad as all that. For a person to say, "I think I know better than the Pope," is clearly an untenable position, and anyone who takes that position is simply not a Catholic. That is categorically **<u>NOT</u>** what the Catholic sedevacantists I am writing about here have done. It is the reliable popes of the Church, who have taught her doctrines in the most forceful language they could muster, who have already condemned the false beliefs promulgated as a result of Vatican II. It is they who said that if anyone teaches such-and-such then that person is a heretic. Catholic sedevacantists are merely those who relay the teaching of the reliable popes when they say of a Conciliar or post-Conciliar pope that he is a heretic because he teaches such-and-such, the same exact "such-and-such" the reliable popes have condemned as heresy.

Just as it is not Catholic for someone to decide that the leader of the Vatican organization is not a pope on their own private authority, or because that leader should happen to disagree with one's personal pet theory or political agenda, it is also not Catholic for one to claim on their own authority that everyone MUST believe that the current leader of the Vatican organization is not the pope. Some of the sedevacantist persuasion have made the mistake of insisting upon that, or of rejecting as non-Catholics those who have not arrived at the same conclusion.

As a Catholic, one has always been free to publish and publicly quote at length the teachings of the reliable popes, to show what the great Canonists and Doctors of the Church have written regarding what is to happen if a pope becomes a heretic, or even juxtapose the true teachings of the reliable popes and councils with the contrary teachings resulting from Vatican II. Having said that much they can only leave the conclusions to their audiences, or speak of their own private opinions or conclusions as such. Until reliable authority (in the person of the next reliable pope) should so state, the Church and everyone in it can go no further.

The big problem with that is that many sedevacantist felt forced to judge individuals within the leadership of the Vatican organization. Such judgments are by their very nature extremely subjective and private and therefore quite fallible. We Catholics have long had the circumstantial evidence that something had gone wrong. For all we knew it might have been a loss of Catholic authority, but there were many other theories which seemed at least as plausible to many of the faithful. Could such a loss of Catholic authority have happened secretly? Without knowing precisely what, when, or how that authority had been lost, sedevacantists felt forced to opine that it could, even though it is a universal consensus of the Church's theologians that it cannot. Even worse, resorting to such speculations carries with it the temptation of committing character assassination, i. e. thinking or teaching that each of these fallen Vatican leaders must be an evil man, deliberately and with criminal intent resisting the guiding influence of the Holy Spirit, or even that he is the Antichrist of Biblical prophecy. Protestants had once said similar things about reliable Popes.

Now that the true legal and juridical impact of Vatican II is known, the subjective and circumstantial evidence (which had for so long hinted at the loss of authority) has now become merely corroborative evidence of the partial loss of jurisdiction decreed back at Vatican II. As far as this writer is concerned, that big problem has now been solved. Those brave souls who managed to march straight ahead, as if it <u>would</u> one day be solved, are at last proven to have behaved correctly in this crisis! Moreover, it is no longer necessary to claim that John Paul II (for example) had been some sort of evil man or Antichrist, a claim which is impossible to sustain in the presence of the man's evident benignity. The man is almost certainly unaware of the fact that by affirming Vatican II he is affirming his own lack of both universal jurisdiction and absolute authority. It is a *de fide* Catholic teaching that such infallibility is exclusively reserved to the bishop who has universal jurisdiction and absolute authority, or those bishops in total union with him during the time that they serve as a College of bishops in an Ecumenical Council.

Even though the Canonists and Doctors opined that the Holy Spirit would almost certainly protect a pope from becoming a heretic and thereby losing his office, they nevertheless have discussed in their writings what the Church should do in the event their opinion is wrong and some pope does some day vanish into heresy. Classically, the Fathers and Doctors (most notably St. Robert Bellarmine) who discussed this question narrowed it down to five basic alternatives:

1. God will never permit that the Pope will fall into heresy.—regarded as most likely but not confirmed, so therefore the other positions need to be explored.

2. Falling into heresy, even though merely internally, the Pope loses "ipso facto" the pontificate.—completely abandoned as a possibility.

3. Even though he falls into notorious heresy, the Pope never loses the pontificate.—completely abandoned as a possibility.

4. The Pope heretic only effectively loses the pontificate upon the intervention of an act declaratory of his heresy.—minority opinion, opposed by St. Bellarmine in favor of the last alternative, but not ruled out.

5. Falling into manifest heresy, the Pope loses the pontificate "ipso facto."—majority consensus, accepted by most and by Bellarmine as the true opinion, providing of course that the first isn't correct.

Unfortunately, many of their discussions focus on the scenario of a pope all by himself somehow vanishing into heresy while a basically sound bishopric or cardinalate still exists capable of discerning that fact and of taking the appropriate action. Many, particularly those within the SSPX and like groups, seem to favor the fourth alternative, but that is the position most adversely affected by the fact that what we now have is almost the other way around. Almost the entire Vatican-led leadership of local leaders and other prominent prelates is so far off base out in left field as to make some of their less questionable leaders such as John Paul II or Benedict XVI actually look good, even if only by comparison. This writer could never trust them to depose the Vatican leader validly since they would only do it so as to install someone more to their liking and I shudder to think what sort that person would be. There exists a growing current of those within the Vatican organization who are pushing for their leader's resignation. Some have done this because they want someone "less Catholic" than John Paul II or Benedict XVI, and others because they want someone "more Catholic" than Francis I (but who would settle for another John Paul II).

Among the ranks of the SSPX, it has always been policy not to discuss such things, or at the very least to restrict such discussions to the level of very private conversations among priests. Indeed, their "position," privately anyway, has been to avoid taking any position even on such basic questions as "Has the Vatican organization lost authority?" At first, they were fairly open with these inner doubts, but over time, many of the priests and seminarians at Ecône and other SSPX seminaries had come to be exposed to a number of theories about the current Church crisis. Owing to differences of temperament and background and the particular teachers and associates to whom they were personally exposed, a few different camps emerged. Publicly, they would increasingly come to deny these inner doubts, even while such doubts continued to be privately entertained.

The main camps, based on schools of thought, which emerged were the hardliners, the softliners, and those who, for lack of a better phrase, have been sometimes called the Lefebvre-liners. Everyone in the SSPX was (and is) obliged to proceed as much as morally possible on the premise that the Vatican leader is a pope until proven otherwise by the formal declaration of some reliable Pope or Council yet to come. A hardliner is one who believes that at least most or all since John XXIII will one day be deleted from the list of successors of St. Peter while a softliner is one who believes that most or all will always be kept on the list of successors of St. Peter. A Lefebvre-liner is one who refuses to entertain any opinion on the question, one way or the other, even in the most private corners of his mind, but who goes with whatever Abp. Lefebvre (or those who succeeded him since) direct.

As long as everyone agreed to do the same thing, regardless of their private opinions, things went along smoothly enough within the SSPX. In the United States, there existed a separation of the North-East and South-West districts, as though they were a couple of dioceses. As it happened, those priests who took a hardliner approach tended to gravitate to the North-East district while those priests who took a softliner approach tended to gravitate to the South-West district. The Lefebvre-liners were more or less equally found in both the North-East and South-West districts. Many of the priests of the North-East district had been with Lefebvre at Ecône during the wretched confrontations between Abp. Lefebvre and Paul VI. Understandably, many of those who had a ringside seat to those ugly confrontations would have trouble believing that Paul VI was a pope at all. The American seminary was placed under the care of the North-East district and a certain professor there, a medical doctor, ventured the theory that the Vatican leadership had somehow lost their authority.

Over the course of the 1970's, Dr. Rama Coomaraswamy, a specialist in Cardiovascular and Thoracic Surgery, and Psychiatry, trained at New York University College of Medicine and who had served in a medical capacity with Mother Teresa in Calcutta, had some rather considerable correspondence with her regarding certain issues he had come across in his private studies of Church history and Canon Law. In the beginning, she had insisted that he agree to have all of this correspondence published, but as she increasingly found herself out of her depth and unable to defend the Novus Ordo religion, she changed her mind and asked him not to publish their correspondence. Out of respect for her wishes while she lived, he did not publish her letters to him. However, he did gather up copies of his own letters to her (minus any references clearly addressing her) into a book entitled *The Destruction of The Christian Tradition* which he published in 1981.

By that time, he had already signed on as a professor at St. Thomas Aquinas Seminary at Ridgefield, the SSPX seminary in Connecticut under the auspices of the North-East district, teaching (rather predictably) Ecclesiastical History. Under his tutelage, the priests of the North-East district acquired a very hardline position against the papacy of the Post-Conciliar popes. Even the good Doctor's book subtly advocated the sedevacantist position, with such phrases as "Pope Leo XIII" on the one hand, but "'Pope' Paul VI" on the other. Being very untrusting of the Conciliar and post-Conciliar Vatican leadership, they strictly adhered to the Missal as used during the reign of the last reliable Pope, Pius XII. They refused to have anything to do with any later liturgies, even the relatively small changes made during the reign of John XXIII. They also refused to accept Novus Ordo "annulments" based on "psychological immaturity" as being valid annulments, which Lefebvre was insisting upon based on his diplomatic outreach to the Vatican leaders, they discouraged recourse to doubtfully ordained priests (?) who had come over to the SSPX without a reliable ordination, and without any willingness to accept a conditional ordination (though clearly one was called for), again tolerated by Lefebvre out of his diplomatic efforts, and raising several other complaints against the arbitrary manner in which the SSPX was being run, such as the expulsion or reassignment of priests without due process.

Lefebvre, suspicious of the growing rift between the SSPX and its own North-East district, sent in (then) Fr. Richard Williamson to investigate the matter. Another event somewhat relevant to this was the promulgation of a new Code of Canon Law by John Paul II in January of 1983. In spite of his criticisms of the new Code, Abp. Lefebvre decided to recognize this new Code as being the Law of the Church. The priests of the North-East district, however, did not. With that, they revealed their true appraisal of John Paul II's status.

Tensions continued to mount between the North-East and the South-West districts as the hardliner position of the North-East district and the softliner position of the South-West district became more and more manifest. Each of their publications, *The Roman Catholic*, and *The Angelus* respectively, published substantially different quotes from Abp. Lefebvre, presenting two very different portrayals of his position. Eager to take the moral high ground, (then) Fr. Kelly, who was the District Superior of the North-East district, took Fr. Bolduc, who was the District Superior of the South-West district, to task for having allowed two Old Catholic priests to serve in a few SSPX chapels temporarily until better priests could be obtained, and for various other liturgical and pastoral concerns. (In all fairness to Fr. Bolduc, the two Old Catholic priests <u>had</u> abjured their error of Old Catholicism.) The two Old Catholic priests were promptly removed by Lefebvre, but they simply went to Florida where they continued claiming to be SSPX priests for a time even though their chapels were never listed in the official SSPX directory.

On account of that, Fr. Bolduc's career as District Superior did not last very much longer, despite his otherwise adequate career as District Superior during which he (together with the saintly Franciscan Fr. Carl Pulvermacher who did most of the actual manual labor of printing and assembling the literature) founded Angelus Press, led the entire South-West district (including the University in St. Mary's, Kansas), and took care of his own parish in Dickinson, Texas.

Finally, in April 1983, matters came to a head. Abp. Lefebvre, greatly concerned over the news sent to him by Fr. Williamson, came over to Oyster Bay Cove in order to deal with this matter personally. He asked Fr. Thomas Zapp, one of three young priests he had ordained for the North-East district only the previous November, to say the mass from the 1962 Missal. Fr. Zapp refused. Eight other priests and twelve seminarians stood in solidarity with Fr. Zapp. Those eight priests were Fathers Donald Sanborn who was then rector of St. Thomas Aquinas Seminary, Clarence Kelly the District Superior, Anthony Cekada the District Bursar, Joseph Collins who was the Headmaster of the University in St.

Mary's, Eugene Berry who was his assistant, Martin Skierka who was another of the three priests ordained by Lefebvre the previous November, Daniel Dolan, and William Jenkins.

On the 27th of April, Abp. Lefebvre ejected all nine of these priests from his society along with some seminarians who were sympathetic to them. Almost immediately, they nine formed the Society of Saint Pius V (SSPV). Fr. Williamson replaced Fr. Kelly as the superior of the North-East district, and also took over control of the seminary at Ridgefield from Fr. Sanborn. Dr. Coomaraswamy also resigned from his teaching post at St. Thomas Aquinas Seminary.

A number of legal battles over the properties started shortly thereafter. It is a fair legal question however. Should the properties go to the side who had formed, trained, and ordained those priests, and under whose auspices they worked and collected funds, or to the side of those priests themselves who had done the real work by laboring in the fields bringing the sacraments and sound Catholic teaching to the hungry souls, along with the majority of those laity as well? The nine priests of the SSPV presented a united front throughout those property disputes. In time that would change somewhat, but not until the property disputes were all settled and certain other events transpired.

The priests of the SSPV lost ownership of the seminary, and some of the church properties, but retained ownership of other church properties (including the one in Oyster Bay Cove) which they had previously taken the precaution of transferring to trusts controlled by them, just in case something of this sort should happen. They also retained control of their journal, "*The Roman Catholic*" which continues to be published to this day. In May, 1984, Lefebvre came to his newly reorganized North-East district and ordained four more priests. Three of them promptly took their valid ordination and placed themselves at the service of the SSPV, namely Frs. Daniel Ahern, Thomas Mroczka, and Denis McMahon. After that, Lefebvre was a great deal more cautious as to whom he ordained.

The (now twelve) priests of the SSPV gathered on June 7, 1984, at their chapel in Oyster Bay Cove and prepared a statement of their principles by which they operated as a religious society:

The Church

> **1.** The changes following the Second Vatican Council have proven so damaging to the Roman Catholic Religion and so detrimental to the sanctification of souls that one can easily discern that "an enemy has done this." This Council marked the culmination of the first phase of a liberal and modernist intrusion into the Roman Catholic Church, which intrusion had already begun in the nineteenth century and to which

St. Pius X alerted the Church in 1907. In his Encyclical "Pascendi" he states: "The partisans of error are to be sought not only among the Church's open enemies; but, what is to be most dreaded and deplored, in her very bosom, and are...thoroughly imbued with the poisonous doctrines taught by the enemies of the Church, and lost to all sense of modesty, put themselves forward as reformers of the Church." This intrusion was made possible because men influenced by modernist ideas gained positions of authority, thereby permitting confirmed heretics and enemies of the Church to overtake our Catholic institutions.

2. The aforesaid intruders have embraced and promoted the modernist and liberal program of the reform of the Church, condemned by the Roman Pontiffs, particularly by Pius VI, Gregory XVI, Pius IX, Leo XIII, St. Pius X, Pius XI, and Pius XII.

3. These intruders have attempted to promulgate, in the name of the Roman Catholic Church, abominable novelties in every aspect of her life, i. e., in the areas of doctrine, morals, liturgy, canon law, pastoral practices, seminary education and religious life.

4. The intrusion of the liberals and Modernists into positions of control has caused the widespread destruction of Catholic Faith, morals and worship and the creation of a new religion—the so-called conciliar religion which is not the Catholic Religion. It should be apparent to all that this new religion is not the Catholic Religion because since its introduction into our Catholic institutions, these institutions no longer manifest the four marks of the true Church, the marks of unity, holiness, catholicity, apostolicity. Thus, those who promote the doctrines and reforms of the conciliar religion do not represent the Roman Catholic Church, which is absolutely and exclusively identified with the Mystical Body of Christ and which is known by its four marks.

5. The Catholic Church was established by Our Lord Jesus Christ for the purpose of teaching, ruling and sanctifying the faithful in His name. The members of its hierarchy are true successors of the Apostles, and the Pope, who as the head of the Catholic hierarchy, is the successor of Saint Peter and the Vicar of Christ on earth. A Roman Pontiff consequently has universal and immediate jurisdiction over all the faithful.

6. To this Catholic hierarchy throughout the ages have been addressed the words of Christ to the Apostles: "As the Father hath sent me, I also send you" (John 20:21). By virtue of its divine institution, therefore, the hierarchy, by its very nature, exercises an authority over the faithful which is the very authority of Christ.

7. To exercise authority over the Church one must externally be a member of the Church. To be a member of the Church one must profess the Catholic Faith. Public abandonment of the Faith severs one from the Church and causes one to lose any position of authority one may have had. For this reason, theologians of all time have held and taught, and Canon Law confirms in canon 1325, no. 2, that anyone who publicly and notoriously defects from the Faith by obstinately denying or doubting any article of Divine and Catholic Faith is a heretic. It is evident that such a person could not possibly rule the faithful, for by analogy to a physical body, it would be impossible to be the head of a body of which one is not even a member.

8. Thus Canon Law equally provides for the tacit resignation from positions of authority of those who defect publicly from the Catholic Faith (Canon 188, no. 4).

9. But those who presently are thought to be occupying hierarchical positions in the Catholic Church are acting, for the most part, as if they do not have the Faith, according to all human means of judging.

10. Among Catholics who are presently adhering to tradition, bishops, priests, and laity alike, we observe a marked difference of opinion concerning the legitimacy of the present hierarchy. We hold that there is certain and sufficient evidence to assert, as a legitimate theological opinion, that anyone who publicly professes the conciliar religion does not legitimately hold any position of authority in the Catholic Church for the reasons stated in paragraph seven. While we do not claim the authority to settle this question definitively, we believe that the legitimacy of this theological opinion is dictated by logic and a correct application of Catholic theological principles. We recognize that the definitive and authoritative resolution to such theological questions rests ultimately with the magisterium of the Church. We thus deplore the attempt of some to settle this question by acting as though they had the authority to bind the consciences of the faithful in matters which have not been definitively settled by the Church.

11. The secondary object of the infallibility of the Church is her rites and disciplines. Because of this secondary infallibility, it is impossible for her to prescribe for the universal Church a law which is harmful or evil. But the Modernists have promulgated, purportedly in the name of the Church, rites and disciplines which are poisonous, evil and harmful to souls. It is therefore certain that these rites and disciplines do not come from the Roman Catholic Church.

The Sacraments

12. Since the Second Vatican Council, the sacraments of the Catholic Church have been radically altered by the Modernists. These alterations contain substantial changes with regard to the ceremonies of the sacraments. In addition, they have effected changes in the very matter and form of the sacraments, thus rendering some of them doubtful or invalid.

13. In any case, therefore, in which the form or matter of the sacraments has been altered, we hold them thereby to be invalid if the change is substantial, or doubtfully valid where the matter or form is not certain, depending on the nature of the alteration effected. A clear example of such an alteration is the approval of grape juice as the matter for the sacrament of the Holy Eucharist by the Modernists operating for the Congregation for Divine Worship.

14. Sacraments in the new religion are further rendered doubtful or invalid (1) by a defect of intention on the part of the minister in certain cases and (2) by the deviations, undertaken by the ministers, in individual cases which corrupt form and/or matter.

15. In the practical order, in the course of our pastoral activity, the Church obliges us to require the reiteration according to the traditional rites, either conditionally or absolutely, as the case may be, of any sacrament conferred in a doubtful or invalid manner. We refer the final determination of the validity or invalidity of the doubtful sacraments to the judgment of the Church when a normal state of affairs shall be restored.

The Sacred Liturgy

16. The Modernists have destroyed the sacred liturgy of the Roman Catholic Church in nearly all of her holy places. The process which brought about this destruction was begun well before the Second Vatican Council and achieved its ultimate expression in the impious New Order of the Mass promulgated by Paul VI in 1969. This destruction was effected by applying to the liturgy the principle of conforming the Church to the modern world. The end result was the New Mass and the many liturgical aberrations produced by it, thereby changing the liturgy from a treasury of Catholic doctrine and piety into a cesspool of protestantism, modernism, ecumenism, pantheism, and virtually every error condemned by the Roman Catholic Church.

17. We consequently reject this New Mass as an evil ceremony, since it is a purveyor of sacrilege, error, and heresy rather than the beacon of Catholic light and truth. We equally reject all the sacramental rites and ceremonies reformed in accordance with modernist principles. In the light of the foregoing, we must conclude that it is objectively a mortal sin to take part in the New Mass.

18. Since the very authors of the New Mass admit themselves that their destructive activity began before the Second Vatican Council, we logically reject the first steps before the Council which led to the general reform of Vatican II, particularly those produced by Annibale Bugnini in his work as Secretary of the Commission for Liturgical Reform. We do not presume to bind others to this rejection of all the pre-conciliar reforms, but we believe it is both right and expedient for the good of the Church to adhere to the Missal of Saint Pius V, reformed by Clement VIII, Urban VIII, and Saint Pius X. While it is possible that there could be differences of opinion concerning the acceptability of the pre-Conciliar reforms, the principle remains the same: that we should follow a determined set of rules used by the Church at some time before the Council.

The New Code of Canon Law

19. We utterly reject and condemn the New Code of Canon Law for the sole reason that it is a legal expression of the modernist distortion of the Roman Catholic Church. Its non-Catholic nature is recognizable by the blasphemous, sacrilegious, and impious practices which it condones and mandates concerning the Holy Eucharist, whereby it sanctions the giving of the Body and Blood of Christ to heretics and schismatics, and the receiving of Communion from heretical and schismatic sects.

Annulments

20. Since the Second Vatican Council, the Modernists have been granting, purportedly in the name of the Church, annulments to married couples for reasons which have no foundation either in the traditional Canon Law of the Church or in the Roman Catholic doctrine concerning matrimony.

21. We consequently deplore this contempt for the Holy Sacrament of Matrimony commonly found among the Modernists operating the diocesan marriage tribunals and the Rota itself. In the practical order, therefore, we refuse to recognize any annulments coming forth from the aforesaid courts unless it can be demonstrated beyond any reasonable doubt that the marriage bond of the annulled marriage did not exist in the first place. For, according to canon 1014 of the Code of Canon Law: "Marriage enjoys the favor of the law, consequently, in doubt, the validity of the marriage must be maintained until the contrary is proved."

Conclusion

22. In the light of the forgoing, we see no other practical course to follow than (1) to adhere with the certitude of the Faith to all of the doctrinal and moral teaching of the Roman Catholic Church; (2) to continue the work of the Church for the salvation of souls, and fulfill our duties as priests by providing the Catholic faithful with integral Catholic doctrine and unquestionably valid sacraments, using the faculties which the Church provides for such critical situations, for "jurisdiction is not granted a man for his own benefit, but for the good of the people and for the glory of God." (St. Thomas Aquinas, *Summa Theologica*, Supplement, Q.8, A.5) Therefore, "since necessity knows no law, in cases of necessity the ordinance of the Church does not hinder." (*ibid.* Q.8, A.6); (3) to reject the destructive modernist alteration of the Catholic liturgy and discipline; (4) to condemn, reprove and reject the poisonous errors of the Modernists, refusing the Catholic name to their tenets, worship, and discipline and thereby rejecting ecclesial communion with them. Mindful of the words of Saint Ephraem, Doctor of the Church, bidding us "not to sit with heretics nor associate with apostates," and that "it would be better to teach demons than try to convince heretics," we deplore every initiative that would seek to make compatible, in one Church, Roman Catholicism and Modernism.

23. These things we declare, mindful of St. Paul's injunction to the Ephesians to "have not fellowship with the unfruitful works of darkness" and in fulfillment of his command to "reprove them." (5:11) These things we do in the firm certitude of adhering to the indestructible and supernatural unity of the Roman Catholic Church, which extends, unaltered and pure, from her foundation by Our Lord Jesus Christ to His Second Coming, from one end of the earth to the other, from the Church Triumphant in Heaven, to the Church Militant on earth, to the Church Suffering in Purgatory, as one unadulterated Church and Faith.

Rev. Clarence Kelly	Rev. Donald J. Sanborn
Rev. Daniel L. Dolan	Rev. Anthony Cekada
Rev. William Jenkins	Rev. Joseph Collins
Rev. Eugene R. Berry	Rev. Thomas Zapp
Rev. Martin Skierka	Rev. Thomas Mroczka
Rev. Daniel Ahern	Rev. Denis McMahon

Octave of the Ascension June 7, 1984

I have included this rather lengthy statement in full in order that the reader may see the eminently reasonable and proper form which the SSPV has given to the position of sedevacantism. On this very sound and solid foundation, they were able to march forward in total union with the Church while having nothing to do with the Modernists who labored to ruin the Church.

While they clearly maintain that the sedevacantist position is a reasonable one, and one which they personally hold, they also denounce those who dogmatically attempt to impose that position on the consciences of others. One also sees in their words the rationale for using the liturgy of the Mass as it existed prior to the reign of John XXIII. At the same time, they are not forbidding the use of the John XXIII liturgy on the part of others (such as the SSPX or the Indult for whom that is their official version of the Mass).

Over the next five years, things went along smoothly enough for the SSPV while they finished off their property disputes against the SSPX and saw to the needs of their parishioners. Lacking a bishop, they were unable to ordain any priests, however a few elderly and saintly priests, such as Fr. Roy Randolph, did align themselves with the SSPV. They also founded a congregation of female religious called the Daughters of Mary, and several Catholic schools. The spiritual and academic standards of these schools were (and are) so high that on several occasions, pupils of those schools have written brilliant, prize-winning essays against the evils of abortion.

If it was the duty of the priests of the SSPV to lend credibility and theological stature to the sedevacantist theory, it has been every bit as much the duty of the other sedevacantist groups to provide the sheer numbers of lay parishioners, religious vocations, and even the preservation of numerous local customs. While there were many sedevacantist groups around the world including the Mexican groups headed by Bps. Carmona and Zamora and the French groups headed by Bp. des Lauriers and Fr. Noel Barbara, this writer prefers to focus on the story of the group founded by Francis Konrad Schuckardt, otherwise known as the Congregatio Mariae Reginae Immaculatae (which means Congregation of Mary Immaculate Queen), or CMRI for short.

While the SSPV had a surprisingly clean and straightforward origin, the CMRI weathered some considerable internal agony. Under Francis Schuckardt the CMRI had grown, but then endured more difficulty than any other sedevacantist group. Where the SSPV ran fairly smoothly as a sedevacantist group, the story of the CMRI is a case where a healthy Catholic sentiment on the part of a large number of people was gradually turned in upon itself, gradually coming to believe that no true and valid Catholic bishop could exist save that of Bp. Schuckardt himself. He had narrowed his opinions and ability to trust so extremely that no bishop could be found to support him. Yet despite this and other brewing problems, because of their being part of the Roman Catholic Church such a state of affairs would not last, and the recovery they have made towards a proper Catholic functioning borders on the miraculous. All other sedevacantist groups all around the world fall somewhere between these two extremes.

After an early period of upset and some degree of instability owing to their not knowing if there were any other Catholic groups than themselves, and not knowing who to trust, by 1987 nearly all sedevacantist groups were emerging as reputable Catholic societies. Such would be the story of the CMRI, but not without first going through several more painful shifts. If you recall, we left off the story of this group after Schuckardt had been consecrated a bishop whose consecration came through the Old Catholic line, namely Daniel Brown, and had soon after that parted company with Brown, who then presumably returned to the Old Catholics.

For a short time, Schuckardt had remained quiet about where he got his Episcopal Orders, perhaps not sure how to reveal his association with Brown, who besides having received his Orders from the Old Catholics, was also married with two children. Under pressure, he was forced to reveal his association with Brown and all of who Brown was. He did, however, successfully evade any substantive charge of being an Old Catholic himself in any way. Brown had turned to the Old Catholics not to join their religion but to obtain valid episcopal orders. It is not clear whether he noticed that though the Old Catholics might be able to give him valid orders they could convey to him no valid apostolic mission. Even so (perhaps on the strength of his own Society's canonical erection in 1967, Schuckardt, having witnessed Brown's abjuration of the Old Catholic source of his Orders had allowed himself to be ordained and consecrated by him. His followers generally accepted his obtaining of his orders from the Old Catholic line on the basis that times had become just that desperate, Abps. Lefebvre and Thục being at that time unknown to them.

Bp. Schuckardt must have been doing something right, since by 1977, the group had grown so large and prosperous that they were able to purchase an impressive facility just north of Spokane, Washington which had previously belonged to the Jesuits as their North-West seminary. Being physically located on a mountaintop with an impressive view of Spokane and having been dedicated to Saint Michael the Archangel by the Jesuits, the facility came to be called Mount Saint Michael, or MSM for short.

They have used this facility as a parish church, grade school and high school, seminary, monastery, convent, retreat center, print shop and bookstore, priory, and international headquarters of the Mount Saint Michael community and also the Fatima Cell community. Many of their nuns, with their distinctive blue habits, have gone on tour as "The Singing Nuns" and have had several commercial recordings released. Their official publications were *Salve Regina* and *The Reign of Mary*, which latter continues to be published to this day. For the next seven years, the community continued to thrive and grow, sending priests to various parts of the world, especially Canada, Mexico, Australia, and New Zealand.

The CMRI had also come to have a K–12th grade boys and girls school, A number of various smaller societies: the Knights of the Eucharist, the Knights of the Altar, the Knights of St. Karl the Great, the Altar and Rosary Society, Our Lady of Cana Cell (for couples interested in marriage), the Holy Name Ushers, several Third Orders, numerous Fatima Cells (small groups that would meet in towns all around the world to organize and coordinate local evangelical activities), also the St. Anne's Home for the elderly and infirm, the Little Daughters of the Immaculate Conception Convent for the mentally impaired, the Singing Nuns, the Kevelaer School for the neurologically impaired, the St. Joseph the Worker Guild, the Mater Dolorous Guild (to aid the dying), plus the spiritual needs of a large number of attached lay Faithful, and considerable assets and properties to manage. As Schuckardt's condition worsened and his need for medications increased, the operation of all these things languished.

There were, however, certain problems brewing. Schuckardt, for all of his perception, initiative, insight, and eloquence, and staunch stand for the traditional Faith, had a number of medical conditions, such as gangrenous intestine, requiring muscular injections of strong painkillers, notably Meperidine, and this may have disrupted his mental stability, and would also inveigle him into some scandalous situations. Being in his own mind and that of his followers "the last Catholic bishop that we know of," the power he wielded over these people went to his head and posed a temptation for him.

As Mount Saint Michael gathered momentum and personnel, it also began to acquire some aspects of a personality cult in the day-to-day operations of its leadership over its people. Francis Schuckardt tried to run everything himself directly, attempting to micromanage more details than his failing health enabled, to the point that much needed actions and decisions, even the mere depositing or cashing of checks from donors, fell so far behind that the checks began to expire. Seeing how liberalism had spread in the world at large, destroying faith and promoting religious indifference, he was ever on the lookout for it within his religious congregation and even among the attached lay Faithful. A number of strict and even arbitrary rules were imposed, such as forbidding television and maintaining a strict control over their reading material, the effect of which was to isolate its followers from society in general, and particularly anyone who might be critical of MSM. To some degree, the philosophy had been to bring forth older rules and disciplines of the Church rather than those which immediately preceded

Vatican II, on the premise that the somewhat relaxed rules may have contributed towards there being a Vatican II. But not every discipline he imposed hailed from any former Catholic practice.

None of the weird practices imposed could be in any way categorized as against the faith, but many were definitely fanatical. For example, the dress code for women was unusually strict, even by Traditional Catholic standards. For example, women are expected to have their heads covered while in Church, but he would expect this not only in church, but at all times. While it is reasonable that consecrated religious would be strongly encouraged to wear a rosary around their neck, over their clothes where everyone could see it, such a policy had never otherwise been imposed upon the laity.

For another example, while the practice of having the men sit on one side of the aisle and women on the other in church is known to have existed at various times and places, and even enshrined in the 1917 Code of Canon Law as a recommendation (Canon 1262), there is no known precedent for them being expected to walk backwards out of church. The rationale for this last was that our Lord, present in the sacrament, is a Royal Personage, entitled to be treated with honor as other royal personages, to whom no one ever turned their backs while in their presence, a rule never before imposed by any Catholic parish or religious order. Discipline of the children was very strict, and sometimes bordered on abuse, such as by shaving their heads. There were accusations of locking them in attics with no access to water, or stuffing jalapeño peppers down their throats. There were also accusations of "marriage wrecking," but this was merely the imposition of Catholic standards, which would, for example, forbid a married person from marrying another, despite a bogus Novus Ordo "annulment."

Coming into the 1980's and only getting worse with each passing year, Schuckardt had strange way of being both demanding and yet inaccessible, remote, distant. His presence at CMRI functions became scarce and he would often show up hours late, and towards the end, even days late, if at all, while everyone waited for him. And once completed with all the many demands of the event, tired and exhausted, he would retreat to the Priory and might not be seen again by the general run of the Faithful for weeks. While learning to wait patiently is a virtue, many of their attached lay Faithful had jobs and other commitments which such waiting interfered with.

Schuckardt also had to grapple with homosexual tendencies, and while his struggles for purity were mostly successful, there were some failures which involved some of the seminarians and monks, causing them to be torn between obedience to their superior and the risk that obedience to some commands might be or lead to inappropriate behaviors. For example, the Meperidine would have to be injected into a muscle, either of an arm or a leg. To avoid knotting, the injected muscle would have to be massaged, and a brother would be called into do it. Suffice to say that there were some occasions on which these massages went astray, and I must leave it at that. These troubles remained private to those living in the Priory house with him, and at first, his long-time friend Father Denis Chicoine stood by him, even publicly denying all charges of homosexual behaviors even to the point of saying that Father Clement Kubish lied because Fr. Kubish had first gone public with accusations of Schuckardt's behaviors.

There were other problems too, perhaps less clear or obvious to the following of the CMRI, but of increasing concern to the doctrinaire leadership of his fellow clerics. Between Bp. Schuckardt's immoral conduct and frequent and extreme tardiness, the CMRI's ability to continue its evangelical outreaches ("preaching the Gospel to all Creation") began to suffer. The energy to do these things was ebbing seriously as his antics became an emotional drain to everyone. Also, circumstances had changed. In 1971 when Schuckardt was consecrated bishop, while there were still some number of traditional-leaning diocesan bishops, none of them seemed strong enough or willing to take the stark stance needed for the Church to see and show clearly her identity as something truly becoming separated from the Novus Ordo schismatics. But by the 1980's the traditional Faith now had many highly visible representatives, including several books and periodicals and Archbishops. Thục and Lefebvre. Schuckardt's claims of being "the last Catholic bishop that we know of" were beginning to go to his head even as other credible bishops were beginning to emerge, making such a description doubly untenable. He began entertaining thoughts that he was a Pope, partially based on a private vision, and partially by default. Where before he would have gladly worked with and submitted to any credible traditional bishop, by this late stage he had narrowed his specifications for a "credible traditional bishop" to something so stratospheric as to be incapable of existing. He also bought into every vicious rumor against every other bishop as anyone knew of and dismissed consideration of any of them on that basis. This too was another factor in the CMRI beginning to become something of a personality cult.

Eventually it got so bad that even Father Chicoine had to denounce Schuckardt publicly on May 27, 1984. Unfortunately, he seemed to feel that the bulk of the congregation might have difficulty appreciating these finer theological points. Many knew nothing of his papal delusions, but those who did seemed to think nothing of it. Perhaps if only he had been the only clearly orthodox Catholic also possessing the episcopal power of Orders, he might well have been able to assume the office and be universally accepted as such, attaining it by universal acclamation. But that was no longer the case as there were now other bishops, truly sent as such by the Church (as he himself had only been "sent" by Bp. Brown), each with superior claims to authority and just possibly even a superior claim of validity of Orders, and clearly he was never qualified to be their Pope. So rather, Fr. Chicoine chose to focus on the concerns more obvious to the people, his immorality (ignoring the fact that immorality does not of itself remove anyone from any ecclesiastical office) and his incompetence (by which he meant his chronic tardiness and neglect of all the details he intended to micromanage which had irked everyone). The immorality could be exaggerated by surprising everyone with what things he had previously held private, and by mentioning things in that context that could have otherwise been accepted as legitimate, such as "knee inspections." The implication (lost on no one) of that last was that he supposedly gained some puerile or unchaste enjoyment from seeing the knees of young boys, though in fact such a thing could be legitimately explained as an attempt to discern those with a serious prayer life as might be evidenced by calloused knees.

Early next month, Schuckardt and the small percentage of those of the CMRI who stuck by him fled to Greenville, California. With his small handful of followers, Schuckardt carried along, growing only most slowly, being carried on only most quietly, nearly isolated and rather ingrown. There had been a number of civil lawsuits, going nowhere and finally deferring to a final lawsuit in Washington State in 1993. The Fr. Chicoine side had quite oddly attempted to argue that the CMRI was congregational in nature but being hierarchical in nature (being part of the hierarchical Catholic Church), the judge refused to be fooled about that. In Washington State, disputes over religious properties are handled differently depending upon whether the religion is congregational (State takes an active role in divvying up properties in some fair manner) or hierarchical (State simply defers to the chief hierarch involved). Acting not with any canonical authority (by which the judge should simply have put it to Bp. Schuckardt to decide who gets what) he awarded the lion's share of the assets to the side which had retained the lion's share of the people. The deposition of Bp. Schuckardt, whether as "pope" or as "the last [and only] bishop we know of" (and with no living superior), took a form not all that different from the deposition of several actual popes during the Saeculum obscurum, a historical period during the first two-thirds of the 10th century in which popes would be elected, even while other popes were living, and accepted by the Romans, and then deposed, forced into resignation. The bishop himself ceased all attempts to rule the majority of the CMRI and the attached lay community up at the Mount, and withdrew, thereafter mostly staying out of sight. He passed away in 2006, but his significance to this account had ceased in 1993.

The reversal of Father Denis Chicoine with regards to bishop Schuckardt has also puzzled some. For some number of years, when Fr. Chicoine would learn of some unsavory action of his friend Schuckardt, Schuckardt would apologize and promise not to do it again and ask Chicoine to please not talk of it to others. Chicoine, good friend that he was, would accept the apology, really believe that Schuckardt would try to do better next time, and cover his faults with silence. But as the other problems gradually attained a kind of critical mass, Father Chicoine turned his back on Schuckardt on behalf of the people in MSM. He had known Schuckardt since the earliest days of the Fatima Crusade, and had been crucial in contributing towards the growth of that Traditional Catholic community. He had been ordained by Schuckardt, but even that seemed to carry with it some slight shadow of doubt, so just in case of any possible lack of any validity of the Old Catholic orders or lack of priestly faculties, he was subsequently re-ordained conditionally by Bishop Musey. After many years of traveling and building up the Church, including several years serving in New Zealand, he died on August 10, 1995.

Meanwhile, what became of MSM and the CMRI now that Schuckardt was no longer in charge? After a year without a bishop, His Excellency George Musey, the Thục-line bishop from Texas who claimed jurisdiction over the Western half of the United States (plus Florida) accepted the post. Bishop Musey spent over a year with them trying to pick up the pieces. They still had a tremendous amount of suspicion taught to them by Schuckardt about other traditionalist groups and bishops, and some of the more tame practices instituted by Schuckardt were still habitual, but Musey did his best to try to bring them in line with the other traditionalist communities and the more "normal" Catholicism as existed universally only a few decades earlier.

The MSM parishioners and clergy were suspicious of Musey but needing the sacraments and episcopal leadership and jurisdiction which only a bishop could provide, they put up with him and the constant pressure he put on them to conform with the norms of Traditional Catholicism. Unfortunately, Musey, a priest whose training was by his own admission hardly adequate, found himself completely out of his depth in dealing with MSM.

While they had made considerable progress under his tutelage, by January 11, 1987, Bp. Musey found himself forced to leave and return to Texas. Musey's opinion of them is best expressed in his own words, "My mistake was thinking I was a good enough surgeon to handle it. I didn't realize the patient was going to bleed to death when I started operating." Father Chicoine's opinion of Musey had come to be equally low by that time as he said of Bishop Musey, "His credibility is zilch."

However much "the patient" may have bled, it was certainly not to the death. The next bishops to help them were Robert McKenna, Joseph Vida Elmer, and Oliver Oravec, Thục-line bishops who had much more of a "hands off" approach to running the MSM. They pretty much would just come, administer the sacraments, and leave. This allowed the Lord to work with MSM directly, bringing their hearts around to the truth. During this time, they made considerable progress in their spiritual position as all the strict and somewhat extreme ways instituted by Schuckardt faded away and a more common or familiar manner of Catholicism, which the people never lost any affection for, made a comeback. The Jesuit facility they had purchased also came with the extensive library the Jesuits had kept there, and an extensive study of these books by the entire remaining MSM leadership went a long way to guide their theological development.

From the beginning, the fathers of the SSPV had nothing to do with any of the Thục bishops, but even more so, nothing to do with the CMRI and MSM which they regarded as terminally disreputable. In the days of Schuckardt, and especially the last six or seven years of his presence there, their reputation had suffered much. The SSPV did not consider MSM's involvement with Musey much of an improvement, even though it really was.

The real trouble with Musey, why he wasn't able to do as much good for MSM as he wished was that having regular jurisdiction, the exclusive authority of an ordinary diocesan bishop over his "diocese," He just assumed that the authority of the office carried all the weight he should have needed. If the priests, religious, or parishioners of MSM got a little recalcitrant, it was too easy for Bp. Musey to resort to saying "I am your lawful superior; you must obey me!" whenever things got too difficult for such a simple country priest to explain, even where he was right. This brings up the issue of jurisdiction, which for a time seemed to pose a problem for Traditional Catholics.

While some sacraments, such as Baptism or the Eucharist, operate well enough independent of any jurisdiction of any kind, other sacraments, primarily Penance and Marriage, depend also on a quality known as jurisdiction, or "priestly faculties." Normally, jurisdiction comes from the Pope through the regular diocesan bishops in union with him. Unfortunately, much of the Church today finds Herself obliged to function either in a seeming defiance of a pope who obstinately refuses to grant them regular faculties, or even in the complete absence of any living pope at all. Even that portion of the Church which has purportedly been granted regular faculties from their pope share those rather questionable "regular faculties" with the non-Catholics of the Novus Ordo religion.

How can a Catholic priest, operating outside the confines of the Vatican organization, absolve from sin or solemnize a marriage? Fortunately, the Church has long ago provided a simple, and for many, and elegant, solution. There are several varieties of jurisdiction provided by the Church. The first and most important and common of these is ordinary jurisdiction which is granted by the Pope to the diocesan bishops, and by the diocesan bishops to the priests of their diocese. There are some things which can only take place within the realm of ordinary jurisdiction, such as a formal condemnation of some new heresy, an excommunication, certain reserved absolutions, or the granting of an Imprimatur for a religious book.

The authority the Pope has to grant such regular jurisdiction can be delegated to others, as for example the Patriarch of some particular Rite, and under exceptional circumstances, involving a prolonged status of having no access to a Pope, such ability can devolve to other legitimate bishops, as has occurred in rare cases where a group of Catholics were stranded or confined with no communication with the Pope, or also during a several year period in the 1200's when the Church had no Pope. But these exceptional cases had not been explored, and in any case were difficult to discern in today's confusing circumstance. Other than that, with the exception of some small scattered handful of elderly priests (and for a shorter time, some very few bishops) who were given permanent regular faculties and assignments by the Church before the changes, and who have faithfully served the true Catholics in their community in the name of their parish or diocesan assignment, ordinary or regular jurisdiction seemed to have vanished off the face of the earth. Many assumed it actually did.

At the opposite end of the scale is what is called supplied jurisdiction. This jurisdiction comes about as a result of a doubt regarding either a law or a fact pertinent to a cleric's jurisdiction or faculties to perform a given act requiring jurisdiction, and it often arises to serve a need on the part of the faithful for the Sacraments which are not otherwise available. In Canon Law, the Latin expression *Ecclesia supplet* which translates, the Church supplies, is used to refer to this type of jurisdiction. In the Code of Canon Law (both old and new), several types of this jurisdiction are explicitly recognized including "common error," where for example a Catholic approaches a priest for confession not realizing that the priest is only visiting in that diocese and has no faculties there, "doubt of law," where the benefit of the doubt is given to priests whose legal situation is too complex to know if he retains jurisdiction, and "doubt of fact," where some material fact(s) relevant to a priest's jurisdiction is not established. Different canons within the Law also grant such supplied jurisdiction for all cases of "immediate danger of death," where a person is dying and desperately needs the Last Sacraments, or for a couple to be free to marry should they lack access to any proper ecclesial minister for a given duration of some months.

Another expression often used in reference to supplied jurisdiction is *epikeia*, Latin for equity, otherwise known as a sense of proportion, or common sense. If one's child is dreadfully ill and the only physician readily available is not licensed to practice medicine in that state, would a responsible parent refuse to allow that doctor to see the child merely because of his lack of an applicable medical license? Jesus said the same thing about healing on the Sabbath. Such laws were intended only to help us and where they fail to do so they no longer apply. Certainly, at the very least, any traditional priests as do not possess some further and more direct source of their jurisdiction nevertheless do indeed have access to this form of jurisdiction as a fallback lawful basis to perform their services.

Between those two types of jurisdiction, Canon Law also speaks of another type of jurisdiction known as delegated jurisdiction. This is provided in the case where an individual with jurisdiction delegates to another individual the authority, the jurisdiction as it were, for some specific task, but this is of little relevance to the present situation.

Jurisdiction is normally granted personally by a Pope but in various circumstances the ability to grant jurisdiction can be delegated or in extreme necessity and lack of access to a Pope or any such designated authorities, can also devolve to any minister whose jurisdiction stems ultimately from the Pope (through a "legal" or "tacit" will of the Pope). And for those without such canonical jurisdiction, such as others stemming from historically schismatic lines but genuinely willing to "pinch-hit" for the Church, or even those with a legitimate mission but ministering to someone not of their own flock, there exists supplied jurisdiction to fill in that gap.

This canonical jurisdictional authority applies to most priests and bishops from all portions of the Church today ranging from some few of those permitted to function as such by the Modernist Vatican organization as well as those functioning outside that organization, the SSPX, the SSPV, the CMRI, Trento Priests, Istituto Mater Boni Consilii (founded 1985 by several SSPX priests who accepted the Sede Vacante finding), and in fact all priests and bishops all over the world who hold and teach the Catholic Faith. All of this is despite a variety of different theories regarding the nature of the current problems in "the Church" ranging from the abuse of authority position (whether they resist it or obtain an excuse from its non-Catholic demands) to the various sedevacantist positions, and even to some few who have gone to some other less known theories as long as they have avoided non-Catholic associations or beliefs. These groups taken together comprise the canonical hierarchy of the Church today.

As indicated earlier, sedevacantism comes in two basic flavors. There are the absolute sedevacantists such as the Mexican Thục-line bishops or Dr. Coomaraswamy who believe that through heresy the Vatican organization has lost all special claims whatsoever that it had once enjoyed as the Roman Catholic

Church before Vatican II. To them, the loss of authority on the part of the Vatican leadership is complete and total. The absolute form is much simpler to understand but appears to have been developed by those who have not considered issues of Church visibility etc. The other kind of sedevacantists are those of the materialiter/formaliter (or "sedeprivationist" or "Cassiciacum") variety such as Bps. des Lauriers, McKenna, Sanborn, and Fr. Noel Barbara who believe that the leader of the Vatican organization is a material pope (he was validly elected), but not a formal pope (he never validly accepted or understood the office), thus granting the Vatican organization some special privileges particularly with regard to its potential to become once again the visible hierarchy of the Church, if only its leadership repents of its false new religion, retracts its false liturgical rites, abrogates Vatican II, and condemns the heresies of modernism, ecumenism, and religious liberty. To them, the loss of authority on the part of the Vatican leadership is partial, not total, since it leaves it up to them to fix everything, should they ever choose to do so.

Just as the SSPX refuses to entertain or harbor any theories as to what happened to the Church, the SSPV correspondingly does not entertain or harbor any theories as to whether the loss of authority on the part of the Vatican leader means that he is a material pope or what. The CMRI, having been under the tutelage of Bp. Musey who was an absolute sedevacantist and then Bp. McKenna who was a materialiter/formaliter, has had to learn about these two different kinds of sedevacantism. Though the clergy of the CMRI are now all absolute sedevacantists, both kinds of sedevacantists can be found among the ranks of their attached lay faithful.

If the SSPV held a low opinion of the CMRI because of their origins with Schuckardt and also because of their continued association with Thục-line bishops, the CMRI also had a low opinion of the SSPV on account of their Pharisaical rejection of many fellow Catholics, although they are far too polite to air that opinion publicly. In all the years that they had endured Schuckardt's troubling ways, and all seemingly based on Holy Orders received from an Old Catholic schismatic, there was never so much as a kind word or encouragement from the SSPV. Indeed, the SSPV easily and even rashly passed along details out of context, all to smear the reputation of these fine and reputable and utterly Catholic clerics. It is amazing just how much a careful and selective concealment of particular details can make the most innocent series of events sound so sordid or even sinister. The only reprieve as can be given them is that they are not the authors of it but merely passed it along unexamined and unchallenged.

Ugly rumors circulated about Bp. Schuckardt and Abp. Thục. It was claimed that Schuckardt eagerly sought the episcopacy and had no compunctions about obtaining it from an Old Catholic, that he acted alone when in fact he consulted with several older, wiser, and experienced regular diocesan or religious order priests. And again, similar rumors circulated about Abp. Thục, claiming the Palmar de Troya group was already totally wacko when he arrived at them, and that he consecrated bishops for them knowing how wacko they were, and then claiming that he consecrated bishops for the various Old Catholic sects some men had turned to for Orders, much as Schuckardt had turned to such for his, that he had lost his mind, that the absurd ideas ventured at Vatican II were seriously intended by him rather than merely thrown out there in response to a request for suchlike, that there were no witnesses to any of his consecrations and no documents. Really, such accusations belong to the same category as the accusation made of the early Church that they practice cannibalism (all because they sacramentally eat the Body and Blood of Christ in the Eucharist). There are those who dig and do careful research and find all the things I have found and present here, and then there are those who simply accept the gossip at face value, and seem to have no compunction about passing it along and committing the grave sin of calumny against sacred persons. Thus exists one of the more painful divisions in the Church.

Despite their overall disapproval with the Thục-line bishops, the priests of the SSPV otherwise managed to avoid anything as could even look strange or invite malicious gossip, making them seem comparatively clean. Certainly, there is merit in avoiding even the appearance of evil, and that they did quite well. They also understood the need for bishops to continue the Church and had truly wished that Lefebvre would consecrate a bishop in order to continue the apostolic succession. Little did they realize how soon their wish would come true.

10

THE BISHOPS OF MARCEL LEFEBVRE AND ANTONIO DE CASTRO MAYER

On June 29, 1987, while Abp. Lefebvre was ordaining some priests he had trained in his seminaries, he gave a sermon which got Catholics up in arms all around the world. In it he said:

> What will become of souls if no one any more proclaims the divinity of Our Lord Jesus Christ? What will come of them if we do not give them the real grace which they need for their salvation?
>
> It is a question of obvious necessity. We must be convinced of this. This is why it is likely that I will give myself successors in order to be able to continue the work, because Rome is in darkness. Rome can no longer now listen to the voice of Truth.
>
> What echo have our appeals received?
>
> There you have twenty years that I have been going to Rome, writing, speaking, sending documents to say: "Follow Tradition. Come back to Tradition, or else the Church is going to her ruin. You who have been placed into the succession of those who have built up the Church, you must continue to build her up, and not demolish her." They are deaf to our appeals!
>
> The last document that we have received proves this fully; they are closing themselves up in their errors. They are locking themselves into darkness. And they are going to lead souls into apostasy, very simply, to the ruin of the divinity of Our Lord Jesus Christ, to the destruction of the Catholic and Christian Faith.
>
> This is why, if God asks it of us, we will not hesitate to give ourselves auxiliaries in order to continue this work; for we cannot think that God wants it to be destroyed, that He wills that it not continue, that souls be abandoned, and that by this fact itself the

Church have no more pastors. We are living in an age completely exceptional. We must realize this. The situation is no longer normal, quite particularly in Rome.

Read the newspaper, *Si, Si, No, No*, put out by the dear sisters who have come to see Ecône and to find here an encouragement to pursue the work that they are accomplishing. This newspaper gives some very precise indications about the Roman situation. A situation that is hard to believe, such that history has never known one like it. Never has history seen the Pope turning himself into some kind of guardian of the pantheon of all religions, as I have brought it to mind, making himself the pontiff of liberalism.

Let anyone tell me whether such a situation has ever existed in the Church. What should we do in the face of such a reality? Weep, without a doubt. Oh, we mourn and our heart is broken and sorrowful. We would give our life, our blood, for the situation to change. But the situation is such, the work which the Good Lord has put into our hands is such, that in the face of this darkness of Rome, this stubbornness of the Roman authorities in their error, this refusal to return to the Truth and to Tradition, it seems to me that the Good Lord is asking that the Church continue. This is why it is likely that I should, before rendering an account of my life to the Good Lord, perform some episcopal consecrations.

By saying that, he was promising to provide the SSPX with at least one bishop to succeed him, regardless of whether the Vatican establishment wishes to accept it or not, and that this would be how the Church would continue Her very existence!

In July, Abp. Lefebvre wrote a letter to Cardinal Ratzinger explaining what he meant and what he intended to do. The SSPX, which had long been ignored or just sent an occasional letter or two along the way, suddenly received top priority in Vatican circles. Many thought this would almost certainly be a schismatic act, unaware that the "schismatic act" happened back at Vatican II when practically the entire hierarchy of the Church got snookered into mandating that She would henceforth be divided between those who are inside the Vatican organization and those who are outside, and that the Vatican organization would no longer be the Church.

The Vatican establishment, once aware of this promise (being so completely out of touch with reality, they actually regarded it as a "threat") to do his part to continue the Church through making bishops, they at once began trying to placate him. Cardinal Ratzinger responded by proposing that an Apostolic visitor would be sent to Ecône in order to inspect the SSPX and see what is required to establish normal relations with the Vatican. Perhaps this might include allowing the priests of the SSPX the use of the 1962 liturgical books and the training and formation of priests using the traditional methods. On the other hand, Cardinal Ratzinger pointed out, if relations between Rome and Ecône were to break down after these negotiations, he would do whatever he could to make it look as if the break was all Lefebvre's fault. After quite some deliberation Lefebvre wrote back requesting that the Apostolic visitor be Cardinal Edouard Gagnon. That request was granted.

From November 11 through December 9, 1987, Cardinal Gagnon did visit the seminaries and various parishes of the SSPX in Europe and on December 8 even concelebrated Mass with Abp. Lefebvre. Perhaps without realizing it, this concelebration of the Mass constituted an acknowledgment on the part of Rome that Lefebvre had never been suspended nor had any lawful penalties applied against him. Since that time the law has been changed to read that such a concelebration **lifts** all previous penalties, but alas for the Vatican organization, the law on the books at that time read that such a concelebration could only take place on the condition that there are no existing penalties. On the basis of what he saw, Cardinal Gagnon wrote a report (which again almost no one ever saw just as no one ever saw the report of the 1974 visitors) which appears to have been quite favorable towards the SSPX, and which landed on the Pope's desk on January 5, 1988.

In the months of February through April of 1988, negotiations took place between Lefebvre and Ratzinger on the basis of the favorable report Cardinal Gagnon had written. It was in the course of the correspondence of these negotiations that Cd. Benelli first used the phrase "Conciliar Church." Though intended merely as a private correspondence between himself and Lefebvre, this one phrase took on a life of its own, owing to its unintended exactitude. The "Conciliar Church" is properly defined to be that brand new "church" defined into existence during the Vatican II Council, at *Lumen Gentium*, to be precise, and consists of that significant majority of the Vatican organization (everything of it other than that tiny traditionalist fringe) which that new religion enshrined in the Novus Ordo service (it ought NOT be called a Mass), and in all other official organs of spreading their new alien doctrines, morals, and liturgies. Cardinal Gagnon himself however, was permitted no further role in the negotiations. Lefebvre would ask for at least one bishop, or better yet several, to continue his work and Ratzinger would respond by saying that once the SSPX was regularized, their priests could be ordained by any of the thousands of bishops all around the world, so there is no need for any bishop.

Ratzinger seemed to have been unaware that many of these "bishops" he was suggesting could be used for ordinations of the SSPX priests once the SSPX is regularized were not really validly consecrated bishops at all, and furthermore most would undoubtedly use the new ordination rite which was also of doubtful validity even in the case where the bishop doing it was validly consecrated. This put Lefebvre in a tight spot. He did not dare voice his suspicions regarding the validity of the new rites since that would have brought the whole set of negotiations to a grinding halt. It was a very delicate situation for Lefebvre to find a way to request a bishop to succeed him without admitting to his belief regarding the doubtful validity of the new rites.

Another thing which would have taken place had the SSPX been regularized would have been the setting up of a secretariat or commission to oversee the relationship between the SSPX and the Vatican. This secretariat or commission would exercise tremendous authority over the SSPX but would also have some pull with John Paul II and his successors. Lefebvre was greatly concerned that a clear majority of its voting members be taken from Tradition, whether as members of his society or as sympathetic persons (such as Cardinal Gagnon) within the Vatican. Ratzinger was just as determined to see to it that the secretariat or commission (towards the end of the negotiations they settled on having it be a commission) would have at most only a minority of voting members take from, or sympathetic to Tradition. This is the one area they never agreed upon.

The problem was this: Lefebvre wanted to ensure that Tradition could continue indefinitely, which having a majority of voting members of this commission be traditional would greatly increase the odds of. Ratzinger (representing modernist Rome here more than his own private opinion) was trying to ensure that once Lefebvre was gone it would be a simple matter to shut down Tradition and be done with it once and for all. It is this one point which made it impossible from the very start that Rome and the SSPX would have ever made their peace so long as Rome adheres to Vatican II. How does one compromise between life and death?

Finally, there came the infamous "May 5th Protocol." This one document, which in the opinion of this writer has been greatly overrated in its importance, was really just one more minor step in the doomed negotiations between Lefebvre and Ratzinger, between the Catholic Church and the Vatican organization. For the first time, Rome formally acknowledged Lefebvre's need for a bishop to succeed him, even if only "for practical and psychological reasons." In other words, "if we allow Lefebvre his bishop which he seems to want so much, although he doesn't explain why, then he may be easier to control." Feeling that getting one out of the two things he needed (a bishop but no majority on the commission) might be the best he could get, he signed the Protocol on May 5.

Immediately after that, Fr. Klemens, Cardinal Ratzinger's secretary, gave a draft of a letter to Lefebvre which was supposed to be the sort of letter Lefebvre was being expected to write to John Paul II. This letter to John Paul II would have had Lefebvre apologizing for his "mistake" in defending tradition. This gave

Lefebvre cause to question what was going on. He thought about the Protocol he had just signed. Not only did it not place a majority of Traditional Catholic voting members on the commission, it was uncommonly vague about the promise of a bishop to succeed him. No mention or commitment had been made regarding who this bishop would be, what authority he would have, or even when, where, how, and by whom, he would be consecrated.

Only then did it sink in: He'd been had! The Vatican leadership was not turning from their errors by taking a small step back towards Catholic Tradition; they were just treating him with kid gloves. They never intended to give him that bishop, not while he was alive anyway, and certainly not consecrated by a reliable sacramental form, or by a reliably consecrated bishop! From the very beginning of the negotiations the sole interest of the Vatican in all of this had been to prevent Lefebvre from consecrating any bishops at all. They were hoping that he would just die off soon, thereby abandoning the flock of God that had been so long entrusted to his care. For that reason, he wrote back informing of his intention to consecrate bishops on June 30 no matter what. If the Vatican wanted to grant those bishops official recognition, great. If they wanted to grant that recognition on the condition that a candidate more to their liking was chosen, Lefebvre was still quite willing to work with them on that, but if not, he already had his candidates picked out.

Cardinal Ratzinger's response unmasked the Vatican's true aim to these negotiations when he wrote, "I have attentively read the letter which you just addressed to me, in which you tell me your intentions concerning the episcopal consecration of a member of the Society on June 30th next. Since these intentions are in sharp contrast with what has been accepted during our dialogue on May 4th, and which have been signed in the Protocol yesterday, I wish to inform you that the release of the press communiqué has to be deferred." The press communiqué spoken of was yet another carrot the Vatican was dangling in front of Lefebvre, which was basically a public statement to the effect that an agreement has been reached, and that Lefebvre was going to be recognized by "the Church." If the Protocol contains an agreement to permit Lefebvre a bishop, how can Lefebvre's stated intention to consecrate one be "in sharp contrast" to that same Protocol? Do you smell a rat?

While one might reasonably concede that technically this last break was Lefebvre's doing, those who would fault him for that lack a true understanding of what was really going on then. For a clearer picture, imagine Lefebvre standing firmly on the rock of St. Peter, utterly solid and immovable because he is precisely where God wanted him (and all bishops including John Paul II). The remainder of the Vatican leaders, including John Paul II and represented here by him, is sinking in a nearby pool of quicksand. John Paul II reaches out a hand, apparently asking for help.

As Lefebvre reaches out to take John Paul II by the hand, the sedevacantists who are also standing on the rock are saying, "don't do it, Your Excellency, it's a trick!" But Lefebvre, being more charitable in his appraisal of John Paul II than they, ignores them and takes his hand and starts trying to pull him up out of the quicksand. Suddenly, John Paul II seems a great deal heavier than he should be, so Lefebvre starts pulling harder. Just as it starts looking as if Lefebvre might succeed in pulling John Paul II out of the quicksand, John Paul II's other hand is seen reaching for and grabbing something under the quicksand so as to keep Lefebvre from pulling him out. In a horrifying moment of truth, Lefebvre realizes that John Paul II wasn't trying to get rescued from the quicksand but trying to pull Lefebvre off the rock and into the quicksand. In what was almost as much a reflex as a carefully considered act, Lefebvre let him go.

Cardinal Ratzinger then offered the date of August 15 as the date to consecrate the bishop, but soon it came out that none of the candidates were considered acceptable by John Paul II, and so more dossiers on other candidates were requested of Lefebvre. It was quite obvious that that was only a dodge since that date could come and they would say, "we were all set to have an episcopal consecration today, but regrettably none of your candidates meet with our approval, so I guess we will just have to reschedule and try again some other time." They were just stringing Lefebvre along and would have continued doing so until he died, which everyone knew wasn't very far in the future. Meanwhile, the SSPX continued its expansion by moving its American seminary from Ridgefield to Winona, Minnesota, and by opening a new seminary in Australia.

Allow me to dispel a certain myth which began to be circulated at about this time. Many say that John Paul II actually forbade Lefebvre to consecrate those bishops. Having inspected in detail all correspondence which took place between Rome and Ecône during this crucial period June, 1987, to June, 1988, this writer has not found anything from either the mouth or the pen of John Paul II actually forbidding Lefebvre to consecrate bishops. He begs, he pleads, he suggests alternatives, and others, presuming to speak on his behalf, come quite close to forbidding it, but nowhere is the man himself ever on record saying to Lefebvre, "By virtue of my authority as a successor of St. Peter, I hereby forbid you to consecrate these bishops!" or words to that effect.

Is that merely an accident, or could it be that he as yet had at least some vestigial or partial claim to the papacy? Any legitimate claim to the papacy on the part of John Paul II would have required the Holy Spirit to prevent him from saying the exact opposite of what the Holy Spirit wanted the Pope to say, namely "I hereby grant you the mandate to consecrate those bishops." Being unwilling to say what the Holy Spirit wanted him to say and being unable to say what the Holy

Spirit prevented him from saying, he said nothing, apart from a lot of vacuous begging, pleading, and cajolery. In his only known demonstration of malice, John Paul II refused to grant the desired papal mandate. Lefebvre, knowing that John Paul II would later regret not having granted that mandate (if not in this life then certainly in the next), proceeded precisely as if John Paul II had said what the Holy Spirit wanted him to say.

On June 29, 1988, Lefebvre ordained a large group of seminarians to the holy priesthood. He was joined by Bp. Antonio de Castro Mayer who had come from Brazil in support of his friend and of the true Faith and the Church. Then on June 30, 1988, Abp. Marcel Lefebvre, with Antonio de Castro Mayer as co-consecrator, consecrated four men to the episcopacy. At the time he said:

> …this ceremony, which is apparently done against the will of Rome, is in no way a schism. We are not schismatics! If an excommunication was pronounced against the bishops of China, who separated themselves from Rome and put themselves under the Chinese government, one very easily understands why Pope Pius XII excommunicated them. There is no question of us separating ourselves from Rome, nor of putting ourselves under a foreign government, nor of establishing a sort of parallel church as the Bishops of Palmar de Troya have done in Spain. They have even elected a pope, formed a college of cardinals… It is out of the question for us to do such things. Far from us be this miserable thought to separate ourselves from Rome!
>
> On the contrary, it is in order to manifest our attachment to Rome that we are performing this ceremony. It is in order to manifest our attachment to Eternal Rome, to the Pope, and to all those who have preceded these last Popes who, unfortunately since the Second Vatican Council, have thought it their duty to adhere to grievous errors which are demolishing the Church and the Catholic Priesthood…
>
> You need this Life of Our Lord Jesus Christ to go to Heaven. This Life of Our Lord Jesus Christ is disappearing everywhere in the Conciliar Church. They are following roads which are not Catholic roads: they simply lead to apostasy.
>
> This is why we do this ceremony. Far be it from me to set myself up as pope! I am simply a bishop of the Catholic Church who is continuing to transmit Catholic doctrine. I think, and this will certainly not be too far off, that you will be able to engrave on my tombstone these words of St. Paul: "*Tradidi quod et accepi*—I have transmitted to you what I have received," nothing else…
>
> It seems to me, my dear brethren, that I am hearing the voices of all these Popes—since Gregory XVI, Pius IX, Leo XIII, St. Pius X, Benedict XV, Pius XI, Pius XII—telling us: "Please, we beseech you, what are you

going to do with our teachings, with our predications, with the Catholic Faith? Are you going to abandon it? Are you going to let it disappear from the earth? Please, please, continue to keep this treasure which we have given you. Do not abandon the faithful, do not abandon the Church! Indeed, since the Council, what we condemned in the past the present Roman authorities have embraced and are professing...We have condemned them: Liberalism, Communism, Socialism, Modernism, Sillonism. All the errors which we have condemned are now professed, adopted and supported by the authorities of the Church...Unless you do something to continue this Tradition of the Church which we have given to you, all of it shall disappear. Souls shall be lost."...

It is not for me to know when Tradition will regain its rights at Rome, but I think it is my duty to provide the means of doing that which I shall call "Operation Survival," operation survival for Tradition. Today, this day, is Operation Survival. If I had made this deal with Rome, by continuing with the agreements we had signed, and by putting them into practice, I would have performed "Operation Suicide." There was no choice, we must live! That is why today, by consecrating these bishops, I am convinced that I am continuing to keep Tradition alive, that is to say, the Catholic Church...

Unfortunately the media will not assist us in the good sense. The headlines will, of course, be "Schism," "Excommunication!" as much as they want to—and, yet, we are convinced that all these accusations of which we are the object, are null, **absolutely null and void**, and of which we will take no account. Just as I took no account of the suspension, and ended up being congratulated by the Church and by progressive churchmen, so likewise in several years,... we will be embraced by the Roman authorities, who will thank us for having maintained the Faith in our seminaries, in our families, in civil societies, in our countries, and in our monasteries and our religious houses, for the greater glory of God and the salvation of souls.

His Excellency Bishop Antonio de Castro Mayer, who had kept his diocese of Campos, Brazil totally Catholic until his forced retirement in 1981, and who continued providing inspiration and leadership to the more than a couple dozen priests of that diocese in the time of their exile during the time of bishop Navarro, also had something to say:

> My presence here at this ceremony is a matter of conscience: It is the duty of a profession of the Catholic Faith before the entire Church and, more particularly before His Excellency Archbishop Lefebvre, before all the priests, religious, seminarians and faithful here present.

> St. Thomas Aquinas teaches that there is no obligation to make a public Profession of Faith in every circumstance, but when the Faith is in danger it is urgent to profess it, even at the risk of one's life…
>
> Because of this, since the conservation of the priesthood and of the Holy Mass is at stake, and in spite of the requests and the pressure brought to bear by many, I am here to accomplish my duty: to make a public Profession of Faith…
>
> I wish to manifest here my sincere and profound adherence to the position of His Excellency Archbishop Marcel Lefebvre, which is dictated by his fidelity to the Church of all centuries. The two of us have drunk at the same source, which is that of the Holy Catholic Apostolic and Roman Church.

The four new bishops consecrated by Abp. Lefebvre and Bp. de Castro Mayer were: Bernard Fellay, a Swiss born in 1958, ordained by Abp. Lefebvre in 1982, and at the time of his consecration the Bursar General of the Society of St. Pius X; Bernard Tissier de Mallerais, a Frenchman born in 1945, ordained by Abp. Lefebvre in 1975, and at the time of his consecration the Secretary General of the Society; Richard Williamson, an Englishman born in 1939 to Anglican parents, converted to Catholicism in 1970, ordained by Abp. Lefebvre in 1976, and serving as the Rector of St. Thomas Aquinas Seminary in the United States; and Alfonso de Galarreta, an Argentine born in 1957, ordained by Abp. Lefebvre in 1980, and at the time of his consecration the Superior of the South American District.

Fr. Franz Schmidberger, the Superior General of the Society, was not consecrated a bishop, in order to emphasize that the bishops being consecrated were not to be considered as having regular (territorial) jurisdiction which Fr. Schmidberger and the other District Superiors possessed over the Society. As one Society priest once crudely put it, the new bishops were merely "sacrament-machines." More properly and exactly, they were consecrated as auxiliaries to the SSPX. Their primary job, apart from the offices they already held, was to ordain priests, perform other functions normally reserved to bishops, and preserve an unquestionably valid apostolic succession. It was felt at the time that having a Superior General who was not sacramentally a bishop (though possessing a bishop's jurisdiction over the SSPX, with Abp. Lefebvre as its "Bishop Emeritus") might help with any further negotiations with Rome. As it turned out, it never made any difference, so they later dropped that policy.

There was a rich variety of reactions to these consecrations on the part of many different people. The secular press loved it. "A new Church was born at Ecône today," NBC said. Just as Lefebvre had warned that they would in his consecration homily, they all said "Schism," "Excommunication," and all the rest of that nonsense.

The Conciliar Church (note Lefebvre's making that phrase public in his homily) wasted no time in exhuming such long-forgotten concepts as "schism," and "excommunication," which apart from attacks on Traditional Catholics, are no longer in use. John Paul II, obviously writing in very hot blood, composed his Motu Proprio which is ironically named *Ecclesia Dei*. In this very strange and choppy document he attempts to condemn that portion of the Catholic Church which subsists outside the Vatican organization, most especially Abp. Lefebvre by name and the SSPX. However, that portion of the Church is not under his jurisdiction, so his condemnation of it makes no sense at all. He even managed to display an astonishing ignorance of his own Code of Canon law which he himself promulgated a scant five years previous, and in so doing ended up using this "official document" of his church to calumniate a private individual, Marcel Lefebvre, by name, which marks the first time such a document has ever been used for such a purpose. Yet also within this same document is a great broadening and strengthening of a permission given to those Catholics subsisting within the Vatican organization to obtain Catholic sacraments and teaching. Truly, this is a case of foul water and fresh emerging from the same spring!

Since this is the only known document of the post-Vatican II Vatican organization which deals with the entire Catholic Church both inside and outside the Vatican organization (albeit with serious flaws regarding the portions of the Church subsisting outside the Vatican organization), and with nothing else, there is an almost providential appropriateness in the fact that it was named *Ecclesia Dei*, the Church of God. The Traditional Catholics who celebrate the Tridentine Mass, both those who do so within the Vatican organization with their blessing and those who do so outside its confines with their cursing, together constitute the entirety of the Roman Catholic Church, the "Ecclesia Dei." And that includes the sedevacantists as well, though not explicitly named, for they too are part of the "Ecclesia Dei."

Another point of interest is the fact that Abp. Lefebvre is never actually excommunicated within this document or any other, but merely spoken of as having incurred an automatic excommunication according to various Canons. When the Canonists later examined the Lefebvre case closely, what they found is that Abp. Lefebvre had **NOT** incurred an automatic excommunication by the standards of Canon Law. If certain Canons are quoted out of context they can be made to sound as if the Archbishop managed to excommunicate himself. However, other Canons in that same Code of Canon Law make it quite clear that an emergency, **even if imagined and not real**, can make it licit to perform certain actions (including such actions as episcopal consecrations without a Papal Mandate) which would otherwise not be licit.

In this case, the emergency is very real, and the old Code of Canon Law (which sedevacantists abide by) differs from the new Code of Canon Law on this point by requiring the emergency to be proven real. Abp. Lefebvre would have had no difficulty doing that. The Canonical issues are explored in detail in the volume, *Is Tradition Excommunicated?* (See Bibliography). The key point of all of that canonical analysis is that Their Excellencies Marcel Lefebvre and Antonio de Castro Mayer are not schismatic and have never been excommunicated, even if one would have to grant that relations between the SSPX (and Bp. de Castro Mayer's order of priests, the St. John Vianney Society as well), and the Vatican establishment became quite strained and somewhat distant. Any claim to the contrary can be, and ought to be, properly dismissed as nothing more than idle, vicious gossip, of no substance whatsoever. Pay it no mind.

The response of a clear majority of those adhering to the SSPX was one of great rejoicing, whether as the formal members, the priests, or affiliated, as in the case of many religious houses of men and women around the world, or as those laypeople who regularly attended the totally Roman Catholic Masses of the SSPX and often sent their children to any one of their many schools. Since they had been following these events closely, they knew the issues and knew that Abp. Lefebvre spoke for all true Catholics and all reliable popes of the Church when he argued for a bishop to succeed him and for a clear majority on the commission. Many, not only in the SSPX but even in the various sedevacantist groups, had been afraid that the membership of the SSPX and their Faithful would get swallowed up by Modernist Rome.

All said and told, about 15 percent of their priests, religious orders and houses, and lay parishioners left off following Abp. Lefebvre and the SSPX, some directly transferring to the Novus Ordo religion, but most transferring to the Vatican organization on the condition of some special permission (then commonly called an "Indult") being granted to them to keep the Catholic Faith and worship. The first of such transfers to the Vatican organization was Dom Gérard Calvet, the Superior of a Benedictine monastery in Le Barroux, France, known as the community of St. Madeleine. Where the previous shakeup from 1983 through 1985 had gotten rid of the hardliners within their ranks, the episcopal consecrations of 1988 got rid of the softliners in one fell swoop. More will be said of those transferring to the Vatican organization in the next chapter.

As for Marcel Lefebvre himself, having the consecrations behind him was a monumental relief. Where before there had always been the possibility of a reconciliation with the Vatican which would have had to have been dealt with, perhaps by consecrating someone who would be bad for his order, or at least not being able to consecrate four new bishops, now there was the *fait accompli* of the four bishops of his choosing already consecrated. The stress of trying to make those negotiations weighed on him so heavily that anyone looking at him wouldn't have given him six months to live. Afterwards, he was so serene and lighthearted over the sure knowledge that he had done the right thing that his health improved, and he went on to live for almost another three years. During that time, he wrote and published a couple more books, *They Have Uncrowned Him*, and *Spiritual Journey* (See Bibliography). When interviewed one year later, he admitted that his name no longer appeared in the *Annuario Pontifico*, a directory of all "approved" bishops, but being so close to the grave he was quite serene in the certain knowledge that his name was still on "the Annuario of the good Lord,…and that is what matters."

The fears of those who thought he would compromise with the Vatican and hand all of his priests, religious, seminarians, and lay supporters over to the Modernists, had proven to be unfounded. Indeed, Abp. Lefebvre's own policy on this was really quite simple and straightforward. The policy he set for the SSPX to consider any future negotiations with the Vatican organization was, "Do you agree with the great encyclicals of all the popes who preceded you? Do you agree with *Quanta cura* of Pius IX, *Immortale Dei* and *Libertas* of Leo XIII, *Pascendi* of Pius X, *Quas Primas* of Pius XI, *Humani Generis* of Pius XII? Are you in full communion with these popes and their teachings? Do you still accept the entire Anti-Modernist Oath? Are you in favor of the social reign of Our Lord Jesus Christ? If you do not accept the doctrine of your predecessors, it is useless to talk! As long as you do not accept the correction of the Council, in consideration of the doctrine of these popes, your predecessors, no dialogue is possible. It is useless." So, there it is. Until such time (if any) as the Vatican leadership should decide to embrace the teaching of the above listed popes and papal encyclicals (which would constitute a total rejection of Vatican II), the policy Lefebvre has set is that there can be no talk of the SSPX returning to the Vatican organization.

This resolve on the part of Abp. Lefebvre surprised many within the sedevacantist camp since they were sure (some had joined the sedevacantists on this belief) that Lefebvre would return the SSPX to the Vatican organization, lock, stock, and barrel. They might indeed do that someday, but not without either making the Vatican organization totally Catholic and Anti-Modernistic first, or by forsaking the wise policy of the Archbishop. The Lefebvre consecrations substantially altered the playing field. When he went ahead and consecrated those bishops, the SSPV had very high praise for that action, if not for the reasons given by Lefebvre to justify it. The fear that Abp. Lefebvre might yet cut a deal with the Modernists was again voiced. The CMRI took a somewhat dimmer view of Abp. Lefebvre's action stating that he should either obey his pope or become a sedevacantist. Ironically, it was the SSPV which was much more adversely affected by the Lefebvre consecrations than the CMRI.

By this point, most sedevacantist priests had some bishop to whom they could turn to for Confirmations of their children, and seminaries and Holy Orders to train and ordain their young men interested in the priesthood. For example, the CMRI had the services of Bp. McKenna, and only a couple months later, of Bp. Oravec (consecrated by Bp. McKenna) as well. The SSPX in going from one bishop to five gained very little against the CMRI. The CMRI on the other hand continued its own rehabilitation, now knowing that they had a rival in one of the largest SSPX priories on the West coast, Immaculate Conception Church in Post Falls, Idaho, only thirty miles away from Spokane. Further events, most notably the consecration of Fr. Mark Tarcisius Pivarunas to the episcopacy by Bp. Carmona of Mexico (one of the three original Catholic Thục bishops) in September of 1991, were already helping to bring the CMRI through the last lap of its rehabilitation.

The SSPV, on the other hand, was at last having its share of hard times. Unlike the SSPX which had just gained four new bishops, and the other sedevacantists who had the services of the Catholic Thục bishops, the SSPV had no bishops and this began to weigh heavily on their lay faithful who desired the sacrament of Confirmation for their children. Some of these families began turning to Bp. Williamson of the SSPX who, in response to that, soon began requiring those about to be confirmed to sign a document, written mostly in Latin and well beyond the reach of the typical ten-year-old being asked to sign it, which stated that they were not sedevacantists. This cost the SSPV many of their lay faithful families. Others were getting their children confirmed by Bps. McKenna and Oravec and other Thục bishops.

Fr. Kelly was adamant in his refusal to have anything to do with the Thục bishops. He felt that the scandals which so many of them had been involved in (by which he meant the Palmar de Troya and other non-Catholic Thục bishops, apparently unaware of the sequence of events that had led to these) made their reputation hopelessly irretrievable in his eyes. To many, it seemed as if he were willing to let the SSPV pass into oblivion with the death of its last priest sometime in the future rather than have anything to do with any known traditional bishops. In reality he was merely holding out for a Thục-like consecration without any scandal. While that was the main source of tension within the SSPV, and which was greatly exacerbated by the Lefebvre consecrations, there were a couple others brewing. Not all of the SSPV priests felt that he was running the SSPV in the best manner possible, and in particular, some of them were coming to feel that the SSPV should be less critical of the CMRI in view of the great strides they had made towards becoming a respectable Catholic organization.

It started in April of 1985 when Fr. Sanborn of the SSPV paid a visit to Bp. de Castro Mayer in Brazil to see about the ordination of priests, or at least some helpful advice. Bp. de Castro Mayer advised Fr. Sanborn to "go to Guérard (des Lauriers)!" Fr. Sanborn objected, stating his (and the entire SSPV's) doubts concerning the validity of the Thục consecrations. It was at this point that Bp. de Castro Mayer stated that "if it's valid for Guérard, it's valid for me." This put Fr. Sanborn, assisted by Fr. Cekada, on a search to discover the truth of the Thục consecrations.

As they studied, they began to realize that a) the ceremonies took place, as proven by not only the two lay witnesses present (Drs. Kurt Hiller and Eberhard Heller), but also Thục himself who never wavered from his claim of having validly consecrated them, and those consecrated themselves who were all entirely satisfied as to the validity of the sacrament they received, and that b) validity of a sacrament performed in accordance to the traditional rites must be presumed until proven otherwise. As early as 1987, Fr. Sanborn felt willing to recommend the recently made Thục bishop, Franco Munari (who was one of the four Italian priests who left the SSPX at the end of 1985), to a prospective seminarian.

In February of 1988, Frs. Kelly, Sanborn, and Jenkins went to Germany to talk to Drs. Hiller and Heller so as to gain first hand testimony (under oath) as to exactly what happened at the Thục consecrations. They don't appear to have paid a visit to Bp. des Lauriers himself who was in the last few days of his earthly life. Frs. Kelly and Jenkins had gone along rather grudgingly, hoping that the testimony would discourage any further involvement with the Thục bishops, but Fr. Sanborn was secretly hoping to affirm their validity. The Doctors were adamant as to the actual performance of the ceremonies in accordance with the book, but their having misspoken as to whether that book had been a Missal (which contains the Mass, but no episcopal consecration ceremony) or the Roman Pontifical (which contains the episcopal consecration ceremony) alarmed Fr. Sanborn who came away from that interview visibly shaken, much to the reassurance of Frs. Kelly and Jenkins.

Over the next few months it dawned on Fr. Sanborn that even if Drs. Hiller and Heller could not tell a Roman Missal from the Roman Pontifical (1908 edition, Rome), des Lauriers himself would have noticed immediately, as would Carmona, Zamora, and Thục. In September of 1988, the results of the study, now completed, were presented to the fathers of the SSPV. Fr. Kelly still wanted nothing to do with them on account of their bad reputation, and he was the Superior of the SSPV. Therefore, nothing was said to the religious and lay supporters of the SSPV at that time, but further troubles were on their way for the SSPV.

In July, 1989, shortly after writing the article "Feed Him a Guilt Cookie" in defense of Fr. Kelly and the SSPV after a recent lambasting by Michael Jones of

Fidelity magazine, Fr. Cekada left the SSPV along with Fr. Dolan, one of the others of the original nine priests. What followed over the next four years was not so much a schism as a brain drain as various fathers, more or less one-by-one, departed from the SSPV, starting with Frs. Cekada and Dolan in 1989 and ending with Fr. Zapp in 1993. Dr. Coomaraswamy also left, somewhere in the summer of 1990.

In late 1988, the SSPV started a television show called "What Catholics Believe" which featured the various priests of the SSPV discussing aspects of the Catholic Faith and current events and issues. The show, despite its limited airings, garnered very positive reviews and soon gained such famous guest speakers as Presidential candidate Pat Buchanan. It was Fr. Sanborn's heavy involvement with this television show along with his close friendship with Fr. Kelly which motivated him to stay in the SSPV and sit tight on what he knew about the Thục consecrations clear until sometime late in 1991.

Meanwhile, Fr. Cekada began taking a good hard look at the CMRI and MSM since some of their parishioners were attending his parish in Ohio, and he hadn't felt right about carrying out Fr. Kelly's policy of refusing them the sacraments. On getting to know these parishioners personally, he found them to be devout, sincere, zealous, even quite doctrinaire Catholics, not at all what he had been led to believe about them. Finally, in 1991 he wrote a ground breaking article entitled "The First Stone," in which he defended the CMRI and MSM. While passing along unexamined the supposedly sordid past they had had under Schuckardt (and things certainly were less than ideal, as explained), he had to concede (and showed that others are morally obliged to concede) that whatever problems there might have been had been cleaned out and the CMRI was fully worthy to take its place among respectable Traditional Catholic organizations.

There were some who claimed that Fr. Cekada had merely defended MSM with the hopes of obtaining for himself or his friend Fr. Dolan a chance at being consecrated a bishop. That claim is nonsense for the simple reason that if any of Frs. Cekada, Dolan, or Sanborn had wanted to avail themselves of the then current MSM bishop, Robert McKenna, they could have approached Bp. McKenna directly. There was no need to involve MSM in any way. Or else they could have approached Bp. Carmona in Mexico. Fr. Cekada defended MSM simply on account of its own merits.

Most of the priests who left the SSPV over the 1989 to 1993 period, along with Dr. Coomaraswamy, then coalesced into an unnamed loose knit association. In late 1991, Fr. Sanborn began publishing his own magazine, *Catholic Restoration*, and soon after, its companion for priests, *Sacerdotium*. Many of the priests of this association have worked with the priests of MSM on certain projects of mutual interest. Two of Fr. Cekada's articles were published in *The Reign of*

Mary, but the most important project was the consecration of Fr. Dolan to the episcopacy on November 30, 1993. Bp. Dolan served by ordaining the seminary graduates from Fr. Sanborn's seminary, Most Holy Trinity, in Warren, Michigan Dr. Coomaraswamy continued to be frequently seen at various MSM functions including their Fatima conferences which they hold every October.

In 1994, Fr. Ahern (one of the three priests who joined the SSPV in 1984) published in both *Catholic Restoration* and *Sacerdotium* a very long article in which he examined the canonical status of the CMRI and MSM in great detail. His conclusions were:

1. MSM is not and never has been a sect through heresy,

2. MSM may possibly have been once a sect through schism, though probably not, and it is highly unlikely that it could be properly considered such today,

3. MSM almost certainly cannot be regarded as an Old Catholic sect,

4. While MSM had once been a psychologically destructive cult, it is not so now, and

5. that MSM was very scandalous in its past, and continues to be at least somewhat scandalous since he felt that it needs a stronger repudiation of its past (such as by changing its name or the color of the nuns' habits) than has as yet taken place.

So well researched was his article that it is generally accepted by most traditional priests and bishops of the sedevacantist position. The one exception this writer takes to Fr. Ahern's conclusions is the issue of scandal. There really is no scandal which can be rightly derived from the CMRI having retained its name and the blue habits for its nuns, even if some of those details might have been Schuckardt's invention. That is because those details stem from an early period in his career as leader of the CMRI which came well in advance of any later problems.

Meanwhile, the SSPV, having passed through its great brain drain and the hard times it had faced being the only sedevacantist group without a bishop, finally got its own problems sorted out. This was made possible by a bishop by the name of Alfred Francis Mendez y Gonzalez. Bp. Mendez was born on June 3, 1907, ordained on June 24, 1935, and served as a priest in Austin, Texas where he got to know many of the Mexican Catholics who were escaping to the United States as refugees from the persecutions they were having in Mexico at that time. Fr. Mendez worked side-by-side with the Mexican immigrants building churches and ever since harbored a great love for the Mexican people. He helped form the diocese of Austin, Texas and in 1948 was promoted to an administrative post at Notre Dame and in 1956 named Director of Province Development for the Congregation of the Holy Cross.

On October 28, 1960, he was consecrated bishop and set over the newly created diocese of Arecibo in Puerto Rico. In February of 1974 he found himself obliged to retire, due to ill health. In his retirement he began to study the issues which were tearing up the Church. In time he came to repudiate the liberal ways he had learned from the ultra-liberal Holy Cross Fathers and began to do what little he could to foster tradition. He wrote letters to the Vatican and to Lefebvre recommending some sort of Tridentine Ordinariate, but lacking Lefebvre's clout and visibility the Vatican merely wrote back saying in effect, "You better stay out of this!" Lefebvre at least wrote him a kindly letter thanking him for his support. Shortly after that, in 1988, two seminarians approached him in order to see about being ordained.

After two years of attempts to negotiate a regular ordination for these seminarians with some as yet active bishop, Bp. Mendez came to realize that the only way these two men could be ordained was for him to do it himself. Being retired, this meant doing it without any clear incardination (though in fact they were ordained in order to serve the SSPV, in effect at least, their incardination. On September 3, 1990, Bp. Mendez ordained Joseph Greenwell and Paul Baumberger to the priesthood for the SSPV. For the first time in over six years, the SSPV finally had newly ordained priests of its own. Thereafter, Bp. Mendez took a more active role in the support of the SSPV. He supported them not only with his prayers, but also provided quite generous financial support which covered many of the costs of producing the television show, "What Catholics Believe," and which would also assist in the purchase of seminary grounds.

Finally, in late 1993, he voluntarily made the offer to Fr. Kelly to consecrate him to be a bishop. He was at that time still concerned about being excommunicated on account of the rule in the Code of Canon Law about that. To put his mind at ease, he was shown an article in *The Latin Mass* magazine, in which Canon lawyer Count Neri Capponi was interviewed and expressed his view that Lefebvre was not really excommunicated after all. After another month of thinking it over and also enduring a bout of chronic illness, he recovered and made specific plans for the consecration. On October 19, 1993, Bp. Mendez consecrated Fr. Clarence Kelly to the episcopacy in the presence of the five other remaining priests of the SSPV.

At his request, this consecration was kept secret from the world until his death. All five priests and Bp. Kelly kept quiet about it, saying only that "God will Provide," to any who expressed concern over the future of the SSPV. Being a private ceremony, this very much resembled the Thục consecrations, but with one important difference: where Abp. Thục had been the one blazing all the trails and making all the mistakes, Bp. Mendez had been laying in the sick bed doing a whole lot of nothing. This at least had kept him out of trouble. Such a life of having done nothing was sufficiently free of scandal so as even to satisfy Fr. Kelly.

While there were some questions raised as to his sanity and clarity of mind, these questions have been raised only by those who stood to gain from casting doubt on Bp. Mendez' last wishes and episcopal consecration. There was the Novus Ordo group who wanted to win his corpse in the courts by deceiving the judge into thinking that his last request had not been made with a sound mind, and there were those at the CMRI and others who had had enough of **Fr.** Kelly's criticism against them and were not about to put up with **Bp.** Kelly's criticism. However, under closer investigation, the claims against the sanity and soundness of mind on the part of Bp. Mendez fall flat on their face. Those who knew the bishop in his last days knew him to be alert and quite sound in his mind.

On July 9, 1994, Bishop Mendez wrote this message of encouragement to Traditional Catholics:

> Vatican Council II and the changes that followed from it have proven a disaster for our dear Roman Catholic Church. I fell ill during the first session of Vatican II and did not return until the closing days of the council. I was surprised and saddened by what I saw when I returned.
>
> Since the conclusion of Vatican II, there has been an epidemic of marriage annulments. The liturgy has become a kind of show rather than the true Holy Sacrifice of the Mass. Once-Catholic institutions, such as colleges and universities, have lost their Catholic character altogether. The number of religious sisters fell away sharply, and many of those remaining no longer lived like religious. Many priests left the priesthood. The number of seminarians fell drastically and many of those who remained hardly lived like seminarians.
>
> As first bishop of Arecibo, Puerto Rico, I had to send my seminarians away to study at various diocesan seminaries on the continental U. S. After visiting them later, I pulled them all out of those seminaries because their training was so liberal and so contrary to what a Catholic seminarian and Catholic priest should be.
>
> The constant change has confused the Catholic faithful. Some have held firmly to the traditional Roman Catholic faith in these troubled times. I encourage them and I long for the restoration of the traditional Catholic faith, the traditional Latin Mass and the sacraments.

On January 28, 1995, Bishop Mendez died, leaving behind his legacy, Immaculate Heart Seminary, for the priests of the Society of Saint Pius V which was made possible by his generous financial gifts towards the purchase of the seminary grounds at Round Top and by the episcopal consecration he performed in 1993. Even having done this, an ugly court battle ensued over the question of who should bury his body and what sort of funeral there should be. The judge, in rank defiance of the bishop's own formal and written request, although perhaps also in deference to members of the bishop's immediate family who were all in the Novus Ordo, gave the body over to the Novus Ordo clergy.

On March 25, 1991, His Excellency Archbishop Marcel Lefebvre died somewhat unexpectedly in a hospital. Exactly one month later, on the 25th of April, His Excellency Antonio de Castro Mayer also died in Brazil. The Lefebvre bishops, in carrying out a promise made by Abp. Lefebvre to de Castro Mayer and the priests of the Brazilian Society of Saint John Vianney, consecrated a bishop for those priests. On July 28, 1991, Bp. Bernard Tissier de Mallerais, with Bps. Richard Williamson and Alfonso de Galarreta as co-consecrators, consecrated Licínio Rangel, one of the priests of the Saint John Vianney Society and Rector of the Seminary in Campos, to the episcopacy.

So, despite some unpleasant events along the way, everything turned out quite well. In the period from 1987 to 1993, Lefebvre got his bishops without actually consummating any "schism" and yet also without compromising with the Modernists, the CMRI up at MSM truly came to have earned their respectability and even got their own bishop, the priests who stayed with the SSPV got their bishop, and that without turning to the Thục line (it is good to have multiple lines of the episcopacy), and the priests who left the SSPV also got their own bishop. In time, even Fr. Sanborn would also come to be raised to the episcopacy, by Bp. McKenna on June 19, 2002. Finally, as will be discussed in more detail in the next chapter, the Indult also got a tremendous shot in the arm with the promulgation of *Ecclesia Dei* and would even come to have a traditional bishop of their own. None of the fears each group had regarding the future of the others proved to have any substance to them. Since all of these priests and bishops who hold and teach the Catholic Faith have been given jurisdiction by the Church, all have been protected from error and from oblivion. It is the indefectibility of the Church which has kept all of these traditional priests and bishops on the straight path as they labor under the One Lord Jesus Christ, in union with all the reliable popes, and ultimately in union with each other and the next reliable pope.

11

THE ADVANCE OF THE INDULT PRIESTS AND THE FSSP

There once was a man who had long been an upstanding citizen, well known and widely respected, who suddenly decided to become a very mean and nasty individual. Everywhere he went he began hurting everyone he met. If he saw a person carrying a load down the street, he would knock it out of his hands making the poor fellow have to pick it all up again. If he passed an older lady in the street, he would greet her by saying, "Hello, Mrs. Woodruff. My, you're looking old today! Is that a new wrinkle?" If he passed a child on the way to school, he would say, "Hey pip-squeak! I'm bigger than you! Know what that means?" and then take his lunch money (or his lunch). In general, he spent all of his days going around making enemies of everyone he had known and everyone he met. It didn't take long before he had no friends, no respect, and no admiration. People never smiled in his presence except to laugh at him. But mostly they stayed away from him as much as they could.

Sometimes, he got lonely as he wished that someone would be his friend, but he enjoyed his own meanness so much he knew he had no intention of parting with his mean and nasty ways. Then one day he found a mask which he bought so that he could put it over his face, and no one would know who he was. It only helped for a little while since his mean and nasty ways would still shine through and people quickly came to hate the face on his mask as much as they hated the sight of his own face.

Then one day he encountered a magician who promised to make it so that he could make friends. The magician put a spell on the mask and made it come alive. It worked! People began treating him with respect and counting him as a friend. The problem he had was the things that this mask did to make it so. Where he wanted to say mean and spiteful things to people the mask would open its mouth and say something nice, cheery, and pleasant. When he saw someone drop a load they were carrying, where he wanted to say, "serves you right for carrying around such stupid things," the mask would say, "may I help you with those?" and next he knew he would actually be helping that person pick the things up.

Sometimes he so hated the way this enchanted mask was forcing him to be kind where he wanted to be mean that he was tempted to rip off the mask so that he could go back to making everyone around him feel miserable and unhappy. But then someone would come along to whom his mask had been particularly kind and would be very glad to see him and he would be glad he had the mask. And so things went, until one of two endings of this story occurred.

The first ending is thus: Despite his frequent temptation to tear off his mask he kept it on and little by little the mask became his entire self and the part of him that wanted to be mean and nasty faded away as he found great fulfillment in his new friends, the respect and admiration of the community, and finally his place as a great community leader, and so he lived happily ever after, the end.

The second ending is thus: One day it was too much for him. All of that niceness that mask had encouraged in everyone he had met and worst of all the nice person he saw himself becoming made him nauseous as he thought, "how sickening, all of this pathetic niceness!" And so, in one angry moment he ripped the mask off his face which was suddenly peeled raw and bleeding and hideous. Once again everyone he met was horrified and disgusted at meeting him, and his rude and cruel treatment of everyone he met rapidly cost him every friend his mask had gained him. He ended up a lonely, horrible monster, ugly inside and out, and so he died miserably ever after, the end.

Vatican II did state that part of the Roman Catholic Church would "subsist in" the Vatican organization. The creation of the Fraternal Society of Saint Peter almost immediately after Abp. Lefebvre consecrated those four bishops is so very like having the mask the man in that story wore come to life. What remains to be seen is whether the Vatican will keep this mask known as the FSSP and the other Indult priests and priestly associations until it is converted to the Faith, or whether it will one day rip off the mask and lose even what vestigial appearance it retains of having anything to do with Roman Catholicism.

As mentioned several chapters ago, there were three basic ways to continue Catholic worship within the Vatican organization after the changes were being imposed on all of the unfortunate members of what was rapidly becoming a brand new religion, the Novus Ordo, and a brand new church, the Conciliar Church. One had been for faithful priests and bishops to "drag their feet" about making any changes to their parish or diocese, and perhaps also invoking the "Perpetual Indult" granted by Pope Pius V for the use of his Missal. The last such bishop in the Latin Rite had been Bp. de Castro Mayer until his forced retirement in 1981, and the last known such priest was Fr. Schoonbroodt who had been forced out in 1988. Word was put out that there was to be a gradual transition to the new service, but that the old Mass would only be permitted until Advent November 28, 1971. The "Congregation for the Doctrine of the Faith" stated on June 14, 1971, "Starting from the day that the translations must be adopted for services in the vernacular, those who continue to use Latin must only use the renovated texts of the Mass and the Divine Office." On October 28, 1974, Paul VI published this truly reprehensible "note" saying:

> This sacred congregation, in a Note published on 14 June, 1971, and approved by the Supreme Pontiff, defined the role of episcopal conferences in the preparation of vernacular versions of liturgical books and set out the regulations for obtaining their confirmation by the Holy See. Gradually, the employment of the vernacular versions spread everywhere to such an extent that, enough time having elapsed, it is clear that the work is almost complete.
>
> With regard to the Roman Missal: when an episcopal conference has determined that a vernacular version of the Roman Missal—or a part of it, such as the Order of the Mass—must be used in its territory, from then on Mass may not be celebrated, whether in Latin or in the vernacular, save according to the rite of the Roman Missal promulgated by the authority of Paul VI on 7 April, 1969.
>
> With regard to the regulations issued by this sacred congregation in favor of priests who, on account of advanced years or infirm health, find it difficult to use the new Order of the Roman Missal or the Mass Lectionary: it is clear that an ordinary may grant permission to use, in whole or in part, the 1962 edition of the Roman Missal, with the changes introduced by the Decrees of 1965 and 1967. But this permission can only be granted for Masses celebrated without a congregation. Ordinaries may not grant it for Masses celebrated with a congregation. Ordinaries, both religious and local, should rather endeavor to secure the acceptance of the Order of the Mass of the new Roman Missal by priests and laity. They should see to it that priests and laity, by dint of greater effort and with greater reverence comprehend the treasures of divine wisdom and of liturgical and pastoral teaching it contains. What has been said does not apply to officially recognized non-Roman rites, but it does hold against any pretext of even an immemorial custom.

One can see in this horrific "note" part of one and the other of the remaining two of the three ways to sustain Roman Catholic worship within the Vatican organization even under the dark days of Paul VI. One of them was by a special permission of the local functionary, or "bishop" (in this chapter I will just say bishop everywhere to avoid numerous and cumbersome circumlocutions for them, and in any case those as allow the Catholic Mass could be seen, at least in that moment, as functioning enough like bishops to be properly so called), or "Indult," at this point officially allowable only to aged and infirm priests who are saying Mass "without a congregation," which, contrary to what the language seems to imply, does not exclude the presence of others (obviously an Altar server at least, and by extension others as well as might be visiting with the priest at his own private home or other such nonpublic venue where he is saying this unofficial Tridentine Mass), but forbids the use of it officially or publicly, in his parish church for example. Although not mentioned within this document, another Indult had been issued to Cardinal Heenan (of parts of England and Wales) in 1971 to allow the occasional use of the 1967 Missal for public worship, the last to retain something of the Tridentine basis. This was sometimes called the "Agatha Christie" Indult, because the famous author was among those who had requested it. This indult read:

> His Holiness Pope Paul VI, by letter of 30 October 1971, has given special faculties to the undersigned Secretary of this Sacred Congregation to convey to Your Eminence, as Chairman of the Episcopal Conference of England and Wales, the following points regarding the Order of the Mass:
>
> 1. Considering the pastoral needs referred to by Your Eminence, it is permitted to the local Ordinaries of England and Wales to grant that certain groups of the faithful may on special occasions be allowed to participate in the Mass celebrated according to the Rites and texts of the former Roman Missal. The edition of the Missal to be used on these occasions should be that published again by the Decree of the Sacred Congregation of Rites (27 January 1965), and with the modifications indicated in the Instructio altera (4 May 1967).
>
> This faculty may be granted provided that groups make the request for reasons of genuine devotion, and provided that the permission does not disturb or damage the general communion of the faithful. For this reason the permission is limited to certain groups on special occasions; at all regular parish and other community Masses, the Order of the Mass given in the new Roman Missal should be used. Since the Eucharist is the sacrament of unity, it is necessary that the use of the Order of Mass given in the former Missal should not become a sign or cause of disunity in the Catholic community. For this reason agreement among the Bishops of the Episcopal Conference as to how

this faculty is to be exercised will be a further guarantee of unity of praxis in this area.

2. Priests who on occasion wish to celebrate Mass according to the above-mentioned edition of the Roman Missal may do so by consent of their Ordinary and in accordance with the norms given by the same. When these priests celebrate Mass with the people and wish to use the rites and texts of the former Missal, the conditions and limits mentioned above for celebration by certain groups on special occasions are to be applied.

The remaining way to sustain Catholic worship within the Vatican organization, as mentioned at the end of that note, was to be a member of the Non-Latin (mostly Eastern) Catholic rites, whose worship was not in any way legally affected by the promulgation of the 1969 Missal and other new sacraments. Over the years from 1974 to 1984, all but the barest handful of foot-draggers had been brought in line with the new religion or else forced to continue their Roman Catholic worship outside the Vatican organization, either as "independents," or aligned with any of the new Catholic associations as were beginning to arise, such as the SSPX, and these severely limited Indults and the Non-Latin rites were the only remaining bastions of Catholic worship within the Vatican organization.

The negotiations between Abp. Lefebvre and John Paul II did finally bring the latter to rescind that horrible note (which since seems to have been deleted from every place they have control over, to the point that many now deny it ever existed) and introduce the following bit of legislation on October 3, 1984:

> Four years ago, at the direction of the Holy Father, John Paul II, the bishops of the entire Church were invited to submit a report on the following topics:
> - the manner in which the priests and the people of their diocese, in observance of the decrees of Vatican Council II, have received the Roman Missal promulgated by authority of Pope Paul VI;
> - difficulties arising in connection with the implementation of the liturgical reform;
> - opposition to the reform which may need to be overcome.
>
> The results of this survey were reported to all bishops (see *Notitiae*, No. 185, December 1981). Based on the responses received from the bishops of the world, the problem of those priests and faithful who had remained attached to the so-called Tridentine Rite seemed to have been almost completely resolved.
>
> But the problem continues and the Holy Father wishes to be responsive to such groups of priests and faithful. Accordingly, he grants

to diocesan bishops the faculty of using an Indult on behalf of such priests and faithful. The diocesan bishop may allow those who are explicitly named in a petition submitted to him to celebrate Mass by use of the 1962 *Editio Typica* of the Roman Missal. The following norms must be observed:

- A. There must be unequivocal, even public evidence, that the priests and faithful petitioning have no ties with those who impugn the lawfulness and doctrinal soundness of the Roman Missal promulgated in 1970 by Pope Paul VI.

- B. The celebration of Mass in question must take place exclusively for the benefit of those who petition for it; the celebration must be in a church or oratory designated by the diocesan bishop (but not in parish churches, unless in extraordinary instances, the bishop allows this); the celebration may take place only on those days and in those circumstances approved by the bishop, whether for an individual instance or as a regular occurrence.

- C. The celebration is to follow the Roman Missal of 1962, and must be in Latin.

- D. In the celebration there is to be no intermingling of the rites or texts of the two Missals.

- E. Each bishop is to inform this Congregation of the concessions he grants and, one year from the date of the date of the present Indult, of the outcome of its use.

The Pope, who is the Father of the entire Church, grants this Indult as a sign of his concern for all his children. The Indult is to be used without prejudice to the liturgical reform that is to be observed in the life of each ecclesial community.

I take this opportunity of extending my cordial good wishes in the Lord to your excellency.

As one can see, while the Catholic Mass was at last being permitted on a grudging basis, at least it was being permitted. The conditions imposed were quite unreasonable, from a truly Catholic standpoint, but typical of the way most dioceses were run, such unreasonable rules were seldom followed or taken seriously. In a few early attempts to conduct a Mass on this basis, only those who personally wrote a satisfactory letter to their bishop would be sent an acknowledgment letter admitting them to the diocesan Indult Mass. At the Mass a guard would be posted at the door so as to allow through only those who brought their "Boarding Pass to Tradition."

The attempts to follow such ridiculous rules were so farcical that by the end of the first year they were no longer followed or taken seriously. Some bishops who had allowed this Indult Mass simply stopped allowing them anymore; other bishops continued the Indult Mass but stopped bothering about who may or may not attend. Abp. Lefebvre, though expressing some appreciation for the step, refused to commit the SSPX to working under its terms. In 1986, a committee of nine cardinals was called together to review the results of the application of the Indult to what few places it was granted. Cardinal Stickler, who was one of the nine, confirmed that all unanimously agreed that, "the Mass of Saint Pius V (Tridentine Mass) has never been suppressed." Even so, on the eve of the Lefebvre consecrations in 1988 only a dozen dioceses in America had Indult Masses on a weekly or better basis. Throughout the rest of the world there were fewer than that regular Indult Masses. The problem here was that it was up to the diocesan bishops, nearly all of whom had fully converted to the new religion, to allow the Catholic Mass if they so choose. The vast majority did not so choose. Certainly, such permission was a great source of rejoicing among those conservatives who were still Catholic at heart. They saw in this new Indult a beginning of a return to common sense. Those who were in a position to attend these Masses were even happier. For many, it had been a long time since they had felt truly at home in their worship of God. Still others merely took advantage of these Masses to show their children the Church they grew up with, the Roman Catholic Church. At a January 19, 1985 Mass in Louisville, a mother was seen, "huddled with her two daughters in the back corner of the Church, following along the Latin in her missal with her finger, guiding her young daughters."

There certainly were problems, mostly resulting from the many years that so very many of those attending had been kept away from Catholic worship and had grown unfamiliar with the Tridentine rite. Women came in dressed much as they had come to dress in the Novus Ordo, with their heads uncovered; many others repeated the altar boys' responses aloud, and the priests showed signs of being quite rusty at doing the old rite. On the greater front, many liturgists (whose job it was to invent new material for Fr. Bozo's sing-a-long stand-up comic act) reacted almost violently against the new Indult. To them it was as if the Pope was letting them down, and the fear the old rite inspired (despite the extreme restrictions placed upon it) was quite well founded. In the Indult Mass (the Vatican approved Catholic Mass), they saw what could be the end of their little mad kingdom. Instinctively, these liturgists all know that they and their works of darkness are the kind that will get swept up in the morning with the trash, and the 1984 Indult was like the first glints of the morning sun.

The Traditional Catholics outside the Vatican organization also rejoiced at the new Indult, even though few of them were in any position to take advantage of it. At long last, the Pope was publicly stating that the Tridentine Mass was not evil or forbidden. This even brought a small number of people to the traditional chapels operating outside the Vatican organization. Probably about the same number of people transferred from long standing Traditional Catholic chapels to their local Indult Mass. Many other persons who had not gone to Church at all for years and years returned by attending the Indult Mass. But, in many respects, it was business as usual for the traditional priests operating outside the Vatican organization.

Finally, while the Indult Mass was a cause of rejoicing among the Catholics-at-heart who were trapped in the new religion, many of them deeply sensed that more was needed. In particular no priests were being trained to say the Tridentine Mass (at least within the Vatican organization) and those capable of saying the Mass in accordance with the Indult were all old-timers whose days were numbered. For nearly four years, this was all they had to make do with, and then came the Lefebvre consecrations.

Meanwhile, other events were happening in the arena of the Lay Faithful who wanted the true Catholic Mass but were for the most part unable to get it from their local Vatican representatives, and unwilling to get it from their local Catholic parish outside the Vatican organization. When (in 1967) Walter Matt departed from the editorial staff of *The Wanderer* and founded *The Remnant*, he soon found himself speaking for a truly vast number of Faithful who were quite determined to bring the Vatican organization back to its senses and back to the Catholic Faith and Worship. In time, these Faithful began to organize into groups, such as *Una Voce*.

In 1982, a group of Traditional Catholics made the pilgrimage on foot, a distance of about 70 miles, from Paris to the Cathedral in Chartres. The doors of the Cathedral were barred shut and the pilgrims had to celebrate their Mass (with the help of an SSPX priest) somewhere outside nearby. This pilgrimage came to be repeated each year, with more and more participants each time. Even more impressive, a significantly high percentage of the participants are young people, teenagers, and young adults both with and without families and small children.

On account of the new Indult, starting in 1992, the Bishop began opening the doors and letting the Traditional Catholics come in and have the Tridentine Mass. The pilgrimage continues to this day, starting on the Vigil of Pentecost at Notre Dame Cathedral in Paris and going all 70 miles on foot. Many other pilgrimages have taken place, some such as the pilgrimage to Rome to ask Paul VI to bring back the Catholic Mass were unique occurrences, but others taking place at regular (usually yearly) intervals, such as the Corpus Christi processions started by Abbe Louis Coache (the suspended village curate), and the annual Pilgrimage for Catholic Restoration which is made to the Shrine of the North American Martyrs in Auriesville, New York, each autumn. Yet another such pilgrimage was the "Spirit of Chartres" Phoenix Pilgrimage from the State Capitol (in Arizona) to St. Mary's Basilica, in which all sorts of Traditional Catholics participated from 1998 to 2002.

Dietrich von Hildebrand was one of the earliest of the Laity to get involved with writing about "the crisis." In his book, *The Devastated Vineyard*, he wrote, "Truly, if one of the devils in C. S. Lewis's *The Screwtape Letters* had been entrusted with the ruin of the liturgy, he could not have done it better."

A number of great laymen, such as Hamish Fraser, Dietrich von Hildebrand, Evelyn Waugh, and Michael Davies also rose to the occasion. Hamish Fraser, having converted from Communism to Catholicism in 1945, became one of Communism's most articulate opponents, and in that, an opponent of the Spirit of Vatican II. Dietrich von Hildebrand, called a "Twentieth Century Doctor of the Church" by no less than Pope Pius XII, and going on to enjoy the respect of Benedict XVI (posthumously), had written a book in 1965 titled *The Trojan Horse in the City of God* which predicted in detail the logical consequences of Vatican II. When, by 1973, the nightmare scenarios he had described had all taken place, he then wrote a follow up volume titled *The Devastated Vineyard* in which he analyzed the philosophical and other weaknesses of the new Church.

Evelyn Waugh was yet another literary figure who used his writing to warn the people against the liturgical changes and all that went with them, including the Welfare State, and so forth, the book, *A Bitter Trial: Evelyn Waugh and John Carmel Cardinal Heenan on Liturgical Changes*, is a testament of his love for the traditional Mass and teachings, and of the shocking indifference of a typical Vatican II prelate, none other than Cardinal Heenan himself, who nevertheless went on to request and obtain the first of these Indults. Heartbroken over what he was seeing, and the direction it was taking, Evelyn Waugh passed away on April 10, 1966, at least being spared having to see the full manifestation of the new religion.

Michael Davies was also a convert, in his case from the Church of England. Educated at London University and St. Mary's Catholic College, Twickenham, and a very intelligent and capable speaker, he entered the fray in 1976 with his book, *Cranmer's Godly Order*, the first of his Liturgical Revolution series (See Bibliography). Being very familiar with both the Catholic and Anglican Churches, he was in a position to document in detail the striking parallels between the liturgical changes Thomas Cranmer made to those of Annibale Bugnini. After that followed numerous books, booklets, articles, and public addresses devoted to the Traditional Catholic cause. Through these earlier works he acquired a reputation for reliable scholarship and almost relentless gentlemanly politeness.

His Liturgical Revolution series continued with *Pope John's Council* and *Pope Paul's New Mass*, and meanwhile he began his next series, his *Apologia Pro Marcel Lefebvre* books in which he explains to the world what Abp. Lefebvre was doing and why. Yet throughout all of this, despite his obviously traditionalist sympathy, the basis of his theological outlook remained firmly rooted in the Novus Ordo. It was Novus Ordo theologians he would turn to, virtually never any of the great pre-Vatican II theologians, for the backing of his "points," for example that the Novus Ordo, though undeniably inferior, would nevertheless be valid and lawful for a Catholic to attend in performance of his "Sunday duty."

In the beginning, Michael Davies pretty much seemed to speak for all Traditional Catholics, but in 1981, sedevacantism began to express itself in the public forum in the form of Dr. Coomaraswamy's book, *The Destruction of the Christian Tradition*. In a review of that book published shortly thereafter in *The Angelus*, Mr. Davies slammed that book, having discerned its sedevacantist leanings, by comparing the sedevacantists' attempt to explain the current Church crisis to an attempt on the part of a High School Science teacher in Arkansas to build a moon rocket. Regrettably, *The Destruction of the Christian Tradition* in its original 1981 edition actually did possess a few minor scholastic weaknesses which made it an easy target for Michael Davies. Coomaraswamy's book had been based on his correspondence with Mother Teresa and notes taken and used for that correspondence. At times, his notes had failed to delineate where some pope's words stopped and some commentator's (such as Patrick Omlor's) words began, and similar mistakes. The good doctor was to do much better next time with *The Problems With the New Mass*, and in time his first book would be revised and expanded, correcting these minor mistakes.

As if to balance that off, Michael Davies also similarly slammed a conservative Novus Ordo publication, *The Pope, the Council, and the Mass*, by James Likoudis and Kenneth D. Whitehead, at about the same time in a review published in *The Remnant*. That book, a compendium of almost every known complaint against the Traditional Catholic Faith, is totally deficient, both scholastically and theologically. In it, every unverified rumor against tradition (such as Cardinal Ottaviani's supposed "letter to Dom Gérard Lafond") gets free rein. (Against the "Lafond" letter in particular, Michael Davies wrote an excellent response which appears in his booklet, *The New Mass* (See Bibliography).) Michael Davies shows that the Likoudis & Whitehead book is little more than a point-by-point attempt at refutation of some of his own earlier works, even though his name never occurs in it even once, as if Likoudis and Whitehead were afraid their readers might learn who Michael Davies is, read his books, and see the strength of his arguments and the weakness of theirs.

But meanwhile, what of Michael Davies' own scholastic and even doctrinal errors? At the root of it all (besides basing his works on Novus Ordo "theologians"), he had a flawed understanding of the nature of the Church's magisterium, the teaching authority and that of the body of bishops united to him. Having come from one of the many non-Catholic churches that have no concept of a fixed, infallible, and irrevocable magisterium, let alone express papal teachings, he had seen, for example in the change made to the Novus Ordo's General instruction, or even the correction of the Arian prayer associated with "Eucharistic Prayer #4," that the Novus Ordo demonstrably had no compunction about withdrawing or changing a magisterial teaching, even a papal teaching, as having been wrong, to say nothing of the withdraw or change made to numerous pre-Vatican II teachings, even papal ones, by outright replacing such teachings with new teachings. From this he concluded that the "Ordinary Magisterium" is fallible and revocable, something that, while observably true with the Novus Ordo Church of the People of God, runs directly against Catholic dogma. Where he came from (and again as observed in the Novus Ordo and its "theologians"), it is perfectly fine to publish something as true, and then when found false, to go "oops" and issue a new version that corrects it. No FIAT of irreformable doctrinal definitions, only the "FIAT" of "Fix It Again, Tony."

Having then expressed for the first time, and repeatedly sustained ever since, his disagreement with the sedevacantist position (he had long since disowned the conservative Novus Ordo position), Michael Davies became at that point a speaker only for the SSPX and the (as yet to appear) Indult crowd. A long stream of articles on various liturgical and historical topics written by him appeared on the pages of *The Angelus* and *The Remnant* in roughly equal numbers over the next eight or nine years, and some in other periodicals as well. When Abp. Lefebvre consecrated the four new bishops in 1988, Michael Davies began to distance himself from the SSPX, although he continued to hold Abp. Lefebvre himself in high esteem and to defend his 1988 consecrations, albeit primarily from the standpoint of Lefebvre's legitimate desire to protect the traditions and priesthood of the Church, and how much that encouraged the Vatican leadership to grant somewhat wider permissions for Catholic worship.

In his later years, his defenses of the SSPX grew rather sparse as he then focused his attention almost exclusively on the concerns of the Indult Catholics. In 1994, Dr. Saventhem stepped down from his long-held post as President of Una Voce, and Michael Davies succeeded him and continued as President of Una Voce for many years, and then passed away September 25, 2004. This role of his as President of Una Voce placed him in a diplomatic position in which he frequently met with various high officials of the Vatican organization, and that is another reason he further distanced himself from the SSPX.

Indeed, as it turned out, there were a number of serious weaknesses in his scholarship, as uncovered by John Daly in his book, *Michael Davies—an Evaluation*. Though fairly detailed and painstaking in his historical details, his understanding of theology suffered significantly from his reliance upon sources that cover things with extreme brevity, sources tainted with Modernism and at times even outright post-Vatican II, sources not meant as academic theological works but meant for general reading, dictionaries, and of course, his small team of nameless "experts" who were in any case aligned with the Modernists. But at least many received a first light introduction to the traditional Faith from a reading of some of Michael Davies' works, and much was documented by him for which the original sources are now quite difficult to trace.

Another prominent Catholic who was all too ready to jump into the arms of the Vatican organization once they would be allowed to keep the traditional Rites was Dom Gérard Calvet, Superior of the Benedictine Monastery in Le Barroux, France. Even before the Lefebvre consecrations, he had approached Abp. Lefebvre and told him that he and his monks were not overly concerned about the Vatican II documents, ecumenism, and religious liberty, but were primarily concerned with being able to continue their Benedictine traditions and liturgy. He discussed the possibility of accepting the hand of the Vatican, should one be offered, with the Archbishop who had formed and ordained his priests and performed many other episcopal functions for the brothers. Lefebvre advised against it, however conceding that the danger was probably a great deal less for a religious order because their subjects were grouped together. Dom Gérard wanted that admission published, but Lefebvre never saw any need to do so. It remained a private understanding between the two men.

There were certainly reasons to be concerned, even for the future of a religious order. In 1985, Dom Pére Augustin had made a deal with the Vatican to be recognized, after years of running his monastery outside the Vatican organization. He gained it, only to be told that the Indult (of 1984) would not be applied to him. Prior to his agreement with the Vatican, they had promised that he would be granted it. Again, on October 15, 1986, the Vatican organization had opened a seminary in Rome called Mater Ecclesiae for those seminarians who dropped out of Ecône and others who thought like them. Although promised that they would be allowed the Latin Mass and a sound priestly formation, the Modernist staff and faculty there treated the Catholic seminarians with such contempt that the situation had quickly deteriorated, and the seminary had to be quietly abandoned and closed down less than two years later. Such precedents did not bode well for those Catholics who threw themselves into the arms of the Vatican organization.

Despite those events, Dom Gérard felt that the risk was justified and began negotiations with the Vatican leaders shortly after Lefebvre changed his mind about having signed the May 5th Protocol. In June, several secret discussions took place between the Vatican (in the person of Cardinal Augustin Mayer and Monsignor Camille Perl) and his monastery. Dom Gérard's presence at the 1988 consecrations was his last public show of unity with Abp. Lefebvre.

On July 7, the Commission spoken of in the Protocol was officially set up, with Cardinal Mayer (not to be confused with the heroic Bishop de Castro Mayer of Brazil) as its President, Camille Perl as its secretary, and seven other faceless bureaucrats none of whom were known for having any particular sympathy for Catholic tradition, nor for any other reason to have any business being on such a Commission. On July 8, Dom Gérard wrote to Cardinal Ratzinger formally requesting to be granted the new Indult and placing his order at the feet of the Vatican leadership. Two conditions he imposed, however, were that the Vatican leadership is not to consider this reconciliation as a discredit to Abp. Lefebvre, and that they would not be required to make any doctrinal nor liturgical concessions nor be prevented from preaching against Modernism.

On July 23, the decision was made by Cardinals Ratzinger and Mayer to accept Dom Gérard's offer and grant full "regularity" to his monastery and yet also permit them the use of the traditional liturgical books (of 1962) for both private and public worship. When asked why he was willing to do this, Dom Gérard gave as his main reason that he felt that fearful Catholics might prefer to attend the Masses of his monastery if they were "approved," and thereby more Catholics could be returned to their roots by being exposed to it again. This is what he felt his own contribution to the restoration of Tradition could be. Thereafter, the monks of Le Barroux no longer had the services of Abp. Lefebvre.

The new Indult given in *Ecclesia Dei* was actually quite impressive in its generosity. The key parts are the portions which read, "To all those Catholic faithful who feel attached to some previous liturgical and disciplinary forms of the Latin tradition I wish to manifest my will to facilitate their ecclesial communion by means of the necessary measures to guarantee respect for their rightful aspirations. In this matter I ask for the support of the bishops and of all those engaged in the pastoral ministry in the Church," and "respect must everywhere be shown for the feelings of all those who are attached to the Latin liturgical tradition, by a wide and generous application of the directives already issued some time ago by the Apostolic See, for the use of the Roman Missal according to the typical edition of 1962."

On the 5th and 6th of July, eight priests from the SSPX paid a visit to the Vatican. They were quite keen on founding a society modeled on the SSPX, but in a modified form to represent the vision offered to them in the May 5th Protocol, and answerable to the Commission so being set up. The society they had in mind was officially set up and founded as a Society of Pontifical Right on July 18 with 12 priests and 20 seminarians at the Cistercian Abbey of Hauterive in the Canton of Fribourg. It was called the Fraternal Society of Saint Peter, or FSSP for short.

Fr. Joseph Bisig, a Swiss who had been the Rector of the SSPX Seminary in Zaitzkofen and First Assistant to the Superior General of SSPX, was elected to be the Superior General of the FSSP. His assistants were Frs. Gabriel Baumann, a Swiss who had been the Vice-Rector of Zaitzkofen, and Denis Coiffet, a French priest who had also been a member of the SSPX. Before long, the FSSP opened its first seminary in Wigratzbad, Germany. All in all, about 25 priests ordained by Abp. Lefebvre had joined the FSSP by the end of 1988. Certainly, the FSSP faced a number of risks. Would the diocesan bishops allow them in? One of the first invitations came from Cardinal Decourtray of Lyons, France, one of the many French modernist bishops who had long been hostile to Lefebvre, and who believed in teaching out of the heretical French Catechism, "*Pierres Vivantes*," (which means "Living Stones"). They would be allowed to say Mass in a few locations, and to hear confessions of anyone who might collar them in the street, nothing more.

But they had their seminary in Wigratzbad, and many other young men eager to be trained in the traditional manner. Given time, they did grow, and would even come to have opportunities to administer the other sacraments at the various (Novus Ordo) parishes at which they were permitted Masses. In 1991, the FSSP first entered the United States in Dallas, Texas. Before long, with Fr. Arnaud Devillers as the District Superior for the United States, they would come to be accepted in over 15 American Dioceses, be able to open and run three Traditional Catholic schools, and even have an order of nuns, the Oblates of Mary Queen of the Apostles.

In 1994, Bishop James Timlin of Scranton, Pennsylvania, allowed the FSSP to open a seminary in Elmhurst, called Our Lady of Guadalupe, for English speaking candidates. Even Tridentine ordinations of their priests happen on occasion. For example, on June 15, 1996, Charles Van Vliet was ordained by Archbishop Marcel Gervais (Ottawa Archdiocese) using the traditional Ordination Rite. Fr. Vliet presently ministers in a "quasi-parish" called St. Clement's in Ottawa, Canada. In this true parish (by Catholic standards), all sacraments are routinely given in their traditional forms!

The Fraternal Society of Saint Peter is the first, but by no means the only traditional order of priests allied with the Vatican organization. As the new Indult became common knowledge, many approached Fr. Gilles Wach, a member of the Opus Sacertodale, an association of diocesan priests approved many years previous by Cardinal Siri, to draft a constitution for what would become the Institutio Christi Regis (Institute of Christ the King, "ICR"), a Society of Apostolic life when given final approval and recognition on September 1, 1990. Starting in the diocese of Mouila in Gabon, Africa, and extending soon to Florence, Italy, the ICR is a missionary order devoted to spreading the Catholic Faith through the traditional sacraments and other missionary activities. Sixteen American dioceses and several in Canada have welcomed priests of the ICR. Other more recent orders include Opus Mariae founded by Fr. William Ashley, Servi Jesu et Mariae which is also a Society of Pontifical Right just like the FSSP, and the Society of St. John the Evangelist, based in the diocese of Scranton. All but the last of these continue on to this day, albeit watered down with full-blown bi-ritualism (meaning that they do both Catholic Tridentine Masses and Conciliar Novus Ordo services). The last however entered scandal and before long was shut down. Carlos Urrutigoity had dreams of being a priest, but his puerile interest in Altar boys exceeded his interest in the Catholic Faith. In the 1980's he had applied for the SSPX seminary in La Reja, Argentina, and though a priest named Andres Morello had known him well and of his proclivities, Bp. de Galarreta ignored his warning owing to his prejudice against Fr. Morello owing to his being a sedevacantist. In time, the SSPX also had to eject him, after which he approached Bp. Timlin. The SSPX warned Bp. Timlin of by then Fr. Urrutigoity's proclivities, but again the warning was ignored, and after some years the Society of Saint John also had to be closed down.

On May 12, 1991, a Mass was said in the Diocese of Richmond, Virginia which officially inaugurated the first Traditional Catholic parish in union with the Vatican organization. Saint Joseph's Villa, technically an "affiliate" or "ministry" of the Novus Ordo parish of St. Benedict, was primarily brought into existence through the efforts of two laymen, Charles Furlough and Michael Reardon. They had organized an "independent" Catholic parish, utilizing the priests of the Orthodox Roman Catholic Movement, until that support was withdrawn in the Summer of 1990 due to Fr. Francis Fenton's declining health (he passed away on August 3, 1995). After a failed attempt to bring in a priest of the SSPX, they took the radical step of approaching the liberal (to the extreme) diocesan bishop, Walter Sullivan.

To their immense surprise, the arrangement worked, and this bears some explanation since Bp. Sullivan was probably one of the most radically liberal Pro-Abortion, Pro-Homosexual bishops there are. He also tolerated all of the most unbelievable liturgical abuses, including a "Wizard of Oz Mass." Why should such a raving liberal be suddenly willing to go to such lengths on behalf of true

Catholics? What it comes down to is that there are two basic kinds of liberal. One kind is the "agenda liberal" whose agenda is to eliminate the truth and spread any heresy he can. Such liberals are only generous in any effort which is based on, or which supports, errors and confusion and disarray. These liberals often pretend (but the disguise is necessarily a very thin one) to be the other type of liberal, the "consistent liberal," or the "sincere liberal." This second kind of liberal, a rare breed indeed, is everything all liberals pretend to be.

Bp. Walter Sullivan was truly this second kind of liberal. To him, all persons have rights, whether they want truth, falsehood, or anything else. If Fr. Holy at St. Joseph's Villa wants to teach that 2 + 2 = 4 in his Catholic parish while Fr. Bozo teaches that 2 + 2 = PURPLE in his Wizard of Oz "Mass," that's all perfectly fine with him. In his diocese, Anything Goes, even the Truth! The agenda liberals are not like that. For them, 2 + 2 can equal anything (*sotto voce*: except 4!) that anyone wants.

This was the very first of what has expanded to nearly 120 Catholic parishes (Western Rite) worldwide (as of 2019) inside the Vatican organization. At this parish and those like it to follow, all teaching and sacraments are performed according to the Catholic (Tridentine) forms. This same bishop would soon bring about the second Indult parish. In 1992, St. Benedict's Chapel in Chesapeake was acquired by Bp. Sullivan after existing for years as a parish ministered to by priests of each of the SSPX and SSPV in turn. For his own curious reasons, Bp. Sullivan preferred not to have his traditional parishes serviced by the FSSP, but merely by older Benedictine priests to whom he has granted the Indult.

The Archdiocese of Atlanta introduced its Traditional Catholic parish in 1995. Archbishop John Donoghue granted full permission to Traditional Catholics to have their very own parish, not merely an "affiliate," or "ministry" of some nearby Novus Ordo parish, but free-standing. It is serviced by priests of the FSSP and called St. Francis de Sales. The next traditional parish which was to appear is the one previously mentioned in Ottawa, Canada. Next, Bishop Thomas Tobin of the Diocese of Youngstown, Ohio took over a parish which had functioned outside the Vatican organization since its founding in 1979 and brought in a priest of the FSSP to provide for the needs of that flock. Then, Bishop Thomas G. Doran of the Diocese of Rockford, Illinois allowed his growing indult community to take over a dying parish, St. Mary's Catholic Church, and put Fr. Brian Bovee of the Institute of Christ the King in charge as parish priest. And so on things just grew from there.

Why all this emphasis on these Indult parishes? Because outside of these Indult parishes, only the Mass had been made readily available using the Indult, or by what would come along later on, a Motu Proprio, to be discussed below. Out of over 1,400 locations worldwide that the Catholic Mass is said with permission from the Modernists (as of 2019), barely one in twelve comes with the full complement of all sacraments and a traditional parish-like setting. While there had been some weddings and funerals according to the Tridentine Rite in other places within the Vatican organization, such instances are still somewhat rare. What the devout Catholic truly needs is not merely the Mass, although that is what he needs most frequently, but an entirely Catholic community in which all teaching and sacraments follow the Catholic norms. While such numbers can look impressive, the fact remains that they represent but a 0.003% fringe on a society which is otherwise non-Catholic. Apart from these relatively few parishes and maybe what little (if any) has not been yet corrupted within the Eastern Rites, that is simply not available anywhere else within the Vatican organization.

Further difficulties of the FSSP are that while the Order had staunchly maintained a firm commitment to using the traditional liturgy exclusively (despite rumors to the contrary), pressure continued to be put on them to go to Bi-ritualism, namely that they would be open to both the Catholic Mass and the Novus Ordo "Mass." In theory they are Bi-ritualists since their membership in the FSSP does not place them in a position to criticize the Novus Ordo. They are forced to pretend that the Novus Ordo is all right, and that it is only the rules of their own Order which forbid them to do it personally. Although the freedom granted to perform the Catholic Mass had gained Cardinal Decourtray's apparent sympathy, he was still circling around out there like a vulture saying, "I would like all the priests of the Fraternity of St. Peter to have the authorization to celebrate both Rites, that of Paul VI as well as that of Pius V, and that the seminarians of the Fraternity might be equally prepared to celebrate in the Rite of Paul VI." With enough time and patience, he would get his wish.

Notice that this statement carries with it an admission that the priests of the Fraternity had no authorization to "celebrate in the Rite of Paul VI." That should have put to rest the rumor spread by some their rivals that priests of the FSSP would sometimes say the new "Mass." However, a number of German diocesan "bishops" have given priests of the FSSP an ultimatum: Either say the new "Mass" at least once in a while or else I don't let you in my diocese at all. As a result, while the FSSP itself did not operate in those dioceses, a tiny handful of its priests have left that order so as to meet that unreasonable demand and enter those dioceses. Since priests of the FSSP only enter into dioceses to which they have been invited, it will be interesting to see what will become of their excess priests when they run out of friendly dioceses.

An important point must be made here: There are many of those among the Modernist-approved groups who make the mistake of supposing that by coming under the auspices of the Vatican organization they are returning to the Church. If it is a question of having not bothered to attend Church at all since "the changes" until hearing that the Mass of their youth is being offered once again, or alternatively settling for going to Novus Ordo services for all that duration, that could be a fair enough assessment, albeit somewhat oversimplified. But if the individual is already attending a Traditional Catholic parish operated by a priest of the SSPX, the CMRI, or any other equivalent society in other parts of the world, or even what few remaining "independent" Traditional Catholic priests as exist, such a transfer to the Vatican organization at best gains that individual precisely nothing. Indeed, he actually distances himself from the Church somewhat if he transfers from a full-fledged Catholic parish to a mere traditional Mass sponsored by some Novus Ordo parish. He also runs a significant and increasing risk that the Masses, though often performed quite beautifully, may actually fail to be sacramentally valid Masses at all.

A layman contemplating such a move ought to consider: Is the priest of the Traditional Catholic parish he presently attends validly ordained? Does he perform all the sacraments in accordance with their Tridentine norms? Does he teach out of all the standard catechisms of the Church and adhere to the entirety of the Magisterium? If so, then his priest is as regular as any priest could ever be. A person who transfers from a parish, say of the CMRI, to (let us suppose the best possible scenario) one of the few Modernist-approved Tridentine parishes, is making a mistake if he thinks that such a move is a return to the Church. In truth, he was already fully in the Church and fully in union with the Pope before his move, and at best he would simply be continuing to be so after his move, assuming its priest is also valid. There might also be however the Novus Ordo "programs" going on or being pushed, which would have to be systematically ignored or resisted, a continual distraction. Worst of all, though the Indult would continue (albeit in a new form) for the immediately foreseeable future, it is my observation that particular Indult situations often prove to be rather fleeting, and at times, even "fly-by-night," or else start out idealistically enough, but gradually get worn down into Novus Ordo forms and activities, mostly by a kind of erosion. The Indult parishes, with their buildings usually more suited architecturally for Catholic worship, may be more permanent.

So what about this issue of a valid priest? Someone not actually ordained will, despite using the ceremonial of the Catholic Mass with all proper form, matter, and intent, will end up merely simulating the sacrament – the bread and wine merely remain bread and wine in his case, and the grace of the Sacrament is absent, leaving its recipients to rely on their own mere human goodness and not on God's grace. The invalidity itself may well be invisible, but the absence of grace within the congregation might be much more evident. In the sedevacantist orders such as CMRI or SSPV, the risk of that is exactly zero. The risk of invalidity might well be about 5% to 20% for those clerics serving as Catholics but tracing their Orders merely to any of the various schismatic lines. In the SSPX, the risk is about 1% since most of their priests were ordained by Abp. Lefebvre himself or by one of his traditional succession, some of the oldest were ordained before the ordination ceremony was rendered questionable (though these are dying off and already rather few), a very few may have transferred over from other Rites, but there are also some who transferred over from the Novus Ordo. Of these, most were conditionally ordained by the SSPX, or otherwise able to verify the validity of their ordination in some other substantive way. But there remains a small few, less than half a dozen, such as Philip Stark, who transferred from the Novus Ordo to the SSPX, were not ordained or at best only most doubtfully ordained, yet directly pressed into service by the SSPX without the conditional ordination they offer to such transfers.

The risk is far greater with the Indult. Many of the new priests training for the Indult Mass are getting ordained with the doubtfully valid Novus Ordo ceremony, or else ordained by bishops whose episcopal consecration is even more unreliable. One might allow that a priest being ordained for the purpose of saying the Indult (Catholic Tridentine) Mass instead of the Novus Ordo may have better chances of obtaining a valid ordination because the intent meant in the traditional form in that case could still be to confer priestly powers where in the other it is not. Even so, a number of these priests (as led by the Holy Ghost) have secretly approached traditional bishops outside the Vatican organization for a conditional ordination, in case the Novus Ordo "ordination" they received already was not valid. In many cases those bishops are very accommodating in this, despite their leadership of rival groups.

In all cases, it remains key to realize that Vatican II moved the visible boundaries of the Church so that they no longer coincide with the boundaries of the Vatican organization. Being inside the visible boundaries of the Church is essential for all Catholics; being inside or outside the Vatican organization is entirely immaterial, from the standpoint of a soul's individual salvation. As Abp. Lefebvre put it, "But the 'Church' against her past and her Tradition is not the Catholic Church; this is why being excommunicated by [such] a liberal, ecumenical, and revolutionary 'Church' is a matter of indifference to us." It is ironic to see the way God works here. Many in the Vatican organization fondly imagine in their ignorance that the traditionalist permission is bringing Traditional Catholics into "the Church." But actually, God is using that permission to bring many who are already inside the Vatican organization into Tradition, and thereby into that which is **really** the Church.

That being the case, is there any reason that an offer made to a Catholic parish to enter into an Indultarian basis with the Vatican organization should ever be accepted? There could be one. While a parish stands to gain absolutely nothing from a doctrinal, spiritual, or canonical standpoint by accepting such an offer on the part of a diocesan bishop to be "regularized," the bishop himself stands to gain a great deal. Take Bishop Sullivan for example. His open toleration and even advocacy of Abortion, Homosexuality, and liturgical abuses of every kind means that he would have to have had a great deal to answer for when he went to meet his Maker. His creation and generous support of the very first two Indult parishes in the whole world could very likely have been the beginning of the salvation of his soul. Might it not be that his good deed has moved his soul from hell to purgatory?

What price is a human soul? Pope Pius XII surrendered many treasures of the Church to help rescue as many Jews from Nazi Germany as he could. As a result, Eugenio Zolli, the Chief Rabbi of Rome (and many other Jews) converted to the true Faith. Who could deny that his soul (and theirs) was worth more than all those treasures? If a bishop has the generosity to extend the Indult within his diocese, then that could be Grace at work in the heart of that bishop, which may lead him back into the Church, or at least be a real grace in his life.

Regrettably, that is not always the case. Some Indult Masses are simply opened up so as to try to put some local Catholic priest out of business. If enough of his parishioners stop attending his Mass and go to the Indult Mass instead under the mistaken impression that they are "returning to the Church," the priest may indeed be forced to close his doors, fold up shop, and go away. But after that, the Indult Mass may then be stopped since it has done its job and is "no longer needed." In many cases, it may be hard to tell what the diocesan bishop's intentions are. The best response would be that enough parishioners remain with the "independent" priest to keep him going, and yet some <u>few</u> others should go to the Indult so as to make it worth the bishop's while. As long as there is a balance, both those who stay with the "independent" priest and those who transfer over to the Indult are doing good. Such a reason would not, however, counterbalance the evil such a congregation being subjected to an invalid "priest," or worse, a propagandist for the Novus Ordo religion from the pulpit.

Sometimes it's quite obvious that the bishop has no good intention. For example, diocesan bishops have been known to offer "regularization" at an enormous cost, as follows:

1. Property and assets are to be in the bishop's name,
2. Acceptance of the leadership and pastoral care of the (Novus Ordo) diocesan bishop,

3. Acceptance of the services of the (Novus Ordo) tribunal for annulments,

4. Acceptance of the services of the department of religious education and pastoral services (Novus Ordo schools, religious education, youth ministry, marriage and family life),

5. "Faithful celebration of the sacraments according to the rites of the Catholic Church" (without any clear provision as to whether by "rites of the Catholic Church" they mean the Catholic traditional rites or the Novus Ordo rites and providing no guarantee of the right to have all Sacraments administered in the traditional Rites),

6. "The pastor and parish are to provide faithful teaching of the Catholic Faith", but with the acceptance of Vatican II understood, since the right to refuse its errors is not admitted,

7. Financial support of the Novus Ordo diocese: 9% cathedral tax, 7% high school tax, support of Bishops appeal and all special collections approved by the Bishop's Conferences,

8. Follow diocesan policies on religious education (such as by using the new catechism) and preparation for the sacraments (e.g. delaying baptisms of infants, elimination of First Confession before First Communion and delaying the age of Confirmation),

9. Encourage subscription of the diocesan Catholic newspaper, and

10. Participate in the "pastoral" life of the diocesan church, e.g. by assistance at such Novus Ordo functions as Millennium celebrations, the Chrism Mass, and so forth.

Obviously, such a list of rules is intolerable, and a Catholic parish can by no means be obligated to accept such an offer, even in the off chance of the hope of saving the bishop's soul. However, it so happens that the fifth of those Indult parishes, the one in Ohio, did just that, and at least for some several years or more, to what may have been to good effect.

While the motives for providing an Indult Mass can be varied and often evil, there are good motives for accepting any such offer which must be respected, chief of which is the desire to establish a beachhead of real Catholicism on the shores of what is, in all other respects, a patently non-Catholic society. While the likelihood of resuscitating the fallen Vatican organization is almost certainly quite futile, the intent and goal is nevertheless well-intended, and I don't see where those who attempt such a futile goal can be in any way blamed for pursuing it. Other problems with the Indult Mass are that some of them are not full Tridentine Masses, but a mixture of Tridentine and Novus Ordo, such as by using the Novus Ordo readings of the day, or even being a Latin Novus Ordo done with Tridentine rubrics. Such admixtures of the Tridentine and Novus Ordo Rites violate even the rules published in the 1984 Indult which state that "In the celebration there is to be no intermingling of the rites or texts of the two Missals." Sometimes the Indult Mass gets used as a place to preach "What a wonderful thing Vatican II is!" There may even be a mingling of hosts consecrated at the Tridentine Mass with those hosts doubtfully consecrated in the Novus Ordo "Mass."

Despite all of these problems, a vast majority of Indult Masses are performed in a very reverent manner by priests who obviously care about what they're doing. Such Masses frequently have the added blessing of having very high-quality Gregorian chant used for the music, as sung by professional choirs, and they might even take place in beautiful and large Churches and Cathedrals. Oftentimes, the presence of a Novus Ordo supper table poses some minor inconvenience for the priest as it may keep him from using the high altar, and that he may have to walk some distance around it to get to the tabernacle. A lay association known as the Coalition in support of Ecclesia Dei does much to promote the spread and use of the Indult Mass by publishing informational materials on the Tridentine Mass and by listing all Mass locations approved by the Vatican organization. Not to be confused with this Coalition (which still functions to this day), the Ecclesia Dei Commission was a Vatican-run Commission first run by Cardinal Mayer as its first President.

Cardinal Mayer did not remain President of the Commission founded in 1988 for very long. In 1991 he stepped down and was replaced by Cardinal Antonio Innocenti, a man who obviously didn't want the job. Cardinal Mayer had been a man who went to bat for all of those groups and individuals who sought Tridentine worship, regardless of whether they had been affiliated with any traditional order, or even whether their diocesan bishop approved. He would grant "celebrets" to priests wanting to say the Indult Mass, even over the bishop's head. He saw the 1988 Indult as being a right granted to the faithful who desire it, where the 1984 Indult had been merely a right extended to what few diocesan bishops desired it. Unfortunately, Cardinal Innocenti had no such zeal. For him it is all he would do to rubber-stamp an approval for an Indult Mass where everything else had already been worked out. Nonetheless, the Indult continued to be extended during his term. Some attribute this to some amount of behind-the-scenes activity on the part of Cardinal Innocenti, though his main behind-the-scenes activity appears to have been an attempt to sow the divisiveness of bi-ritualism within the FSSP, culminating in Protocols 1411 and 512. In 1995, Cardinal Innocenti was replaced by Cardinal Angelo Felici, who was later succeeded by Cardinal Castrillon Hoyos, who has been the first openly active President of Ecclesia Dei since Cardinal Mayer. Cd. Hoyos next served as the Commission's President. Monsignor Perl remained as its Secretary throughout those transitions. But in January of 2019 this Commission, then headed by Abp. Guido Pozzo, was shut down.

In 1994, Fr. Schmidberger's twelve-year term as Superior General of the SSPX came to a close, and Bp. Bernard Felay, the only one of the four original Lefebvre-de Castro Mayer bishops who was neither a consecrator nor a coconsecrator of Bp. Rangel of Campos, Brazil, was elected to succeed him as Superior General. With the SSPX maintaining its delicate balance between the Indult crowd and the sedevacantists, they placed the Vatican organization into a rather strange position. On the one hand the Vatican leaders want to write the SSPX right out of the Church, but on the other they wish to be able to treat the SSPX problem as an internal matter. They can't have it both ways.

In 1991, Bishop Joseph Ferrario of Hawaii attempted to excommunicate six prominent members of a small SSPX chapel in Hawaii. The six parishioners (sometimes called the "Hawaii 6") appealed to Rome, and "Rome," in the person of Cardinal Ratzinger, responded in 1993 that the Hawaii 6 were in no way schismatic, and that there were to be no penalties applied. Bp. Ferrario backed down, retreated in disgrace, and retired soon after.

One would think that should be enough. But in 1996, Bishop Fabian Bruskewitz of Lincoln, Nebraska attempted the very same thing with the parishioners of St. Michael Archangel Chapel, along with a host of other groups which really are properly condemned by the Church. Once again, there was an appeal to the Vatican and once again the erring bishop was told (albeit this time privately) to back off. In reality there is nothing he can do. Deprive the SSPX parishioners of the (Novus Ordo) "sacraments?" Even if any real Catholics who attend the SSPX parishes should desire to receive such Novus Ordo "sacraments," the rules for "Eucharistic Ministers" voted on at bishops' conferences render withholding them totally impossible by requiring that such "Ministers" give them to all who come forward, even notorious sinners. Realistically, there is no way for such lay "Eucharistic Ministers" to be "in the know" as to who might be under some disciplinary sentence among those approaching them. In reality, all Bp. Bruskewitz did was lose a great deal of face in public.

In 1998, with the quarters for the FSSP in Scranton getting more and more cramped for the increasing number of bright young seminarians, and also their bishop Timlin being hounded over the scandal of the Society of Saint John which he had also authorized, Bp. Bruskewitz, perhaps in a gesture that goes a long way towards making up for his former public mistake, invited the FSSP to begin construction of a new seminary in his diocese, so that their seminary is now based in Denton, Nebraska rather than Pennsylvania as before. Certainly, this is a commendable action on his part. Even more to his credit, when Fr. Devillers of the FSSP actually went so far as to deny that the administration of tonsure, the minor orders, and the Subdiaconate carried any validity (since the Novus Ordo had swept them all away), Bp. Bruskewitz continued to administer all degrees of Holy Orders on his FSSP seminarians as a point of honor, making it abundantly clear at the ordination Mass that tonsure, the minor orders, and the Subdiaconate most certainly do carry force and validity.

Yet another flap occurred when Fr. Gerald E. Murray, a priest from the Archdiocese of New York who was working on his doctorate in canon law at Rome's Gregorian University, prepared a licentiate thesis titled *The Canonical Status of the Lay Faithful Associated with the Late Archbishop Marcel Lefebvre and the Society of St. Pius X: Are They Excommunicated as Schismatics?* In it he concluded that the 1983 Code of Canon Law did not actually cause them to be excommunicated, according to a strict and literal application of that Code. All of those Vatican liars who wanted to deceive many into thinking that the SSPX is schismatic, or that anyone involved becomes excommunicated were in a total uproar.

He was interviewed by *The Latin Mass* and quoted in the literature published by the SSPX. His case had been absolutely devastating, an open and shut proof fit to lay to rest once and for all the claims made by those vocal liars at the Vatican (and other places) who say that those of the SSPX are schismatic or excommunicated. Unfortunately, the man himself since then refused to stand by his own work. In a letter to *The Latin Mass* (and published therein) he stated that he had changed his mind. The reasons he gave were stupid ones which any first-year student of canon law would at once see as fallacious. He got away from the literal meaning of the text of the law and went into the dark and murky territory of speculating on the intentions of the lawgiver and attempted to reinterpret the 1983 Code of Canon Law in terms of these mysterious and undocumented "intentions."

Perhaps they threatened him. Perhaps he was bought off. Perhaps he is simply very ambitious, his thesis being written so as to show that he knows precisely where a major Achilles' heel of Vatican organization is, and his subsequent backing down done so as to show that he is willing to play political ball with them. Whatever the scenario, in any case he is just like a surgeon who successfully performs open-heart surgery (and the patient lives), but who subsequently expresses remorse over having used a knife during the surgery since "knives are dangerous things which can hurt people."

As the SSPX continued to expand into many other countries, another event in 1996 was the conversion of Bishop Salvador Lazo of the Philippines. Kept incessantly busy by his diocesan duties clear up until his retirement shortly before that year, he had never before questioned the guidelines by which he had run his diocese and brought it in line with the new religion. Once he retired however, he finally had time to do a lot of reading and research on the issues. Once he had read a number of the books listed in the Bibliography of this book, he at last came to understand the need to take a stand for the Tridentine Mass.

Bp. Lazo soon exclusively said the traditional Mass in the SSPX parishes, primarily the Chapel (and priory), Our Lady of Victory, located in the metropolis of Manila. With his help, the SSPX has opened a house of Formation for priests, a place where seminarians receive the first part of their training for the priesthood. On April 11, 2000, Bp. Lazo passed away and was given a Traditional Catholic funeral by the priests of the SSPX in the Philippines. It was only after his death that his bishop friend Bp. John Bosco มนัส จวบสมัย (Manat Chuabsamai) of Thailand finally found the time to read the same books Bp. Lazo had read and come to the same conclusions. Bp. มนัส (Manat) regularly brought in SSPX priests for retreats and similar functions and encouraged a return to tradition within his diocese. Sadly, the man soon reached the age of 75 and had to retire, and in that, also resigned.

The Indult increasingly came to be used for various special event public Masses which have drawn yet more attention to the Traditional Catholic community. In 1995, Cardinal Ratzinger visited the Monastery at Le Barroux and celebrated a Tridentine Mass there, and yet that was nothing compared to the Mass at St. Patrick's Cathedral in New York celebrated in 1996. Cardinal Stickler had come to perform this Mass himself, and the Cathedral, despite its enormous size, was absolutely packed, with hundreds more standing outside. Many of those present, and the Cathedral itself, had not seen a Catholic Mass in over thirty years. Abp. O'Conner of New York declared that he would "never again" permit such a Mass in his Cathedral, and he never did. He soon died. Sadly, the piety and enthusiasm he saw not only failed to impress him, but even offended him.

In the first half of the 1990's a number of interesting things came from the pen of John Paul II. In 1992, there was the new catechism, which despite its theological weaknesses in the areas of Ecumenism and Religious liberty, its extreme verbosity, and underdeveloped Sacramental Theology, was nevertheless maintaining a firm and conservative stand in many controversial areas of public morality, which the Novus Ordo religion had eroded to a considerable extent. In 1993 came the encyclical, *Veritatis Splendor*, in which the one small and particular heresy of moral relativism was condemned.

Veritatis Splendor proved to be a source of seeming trouble for the SSPX. Some anti-traditional publications were claiming that this encyclical had split the SSPX, with one bishop agreeing with it and another rejecting it. The truth was far more innocuous than that. Bp. Fellay read *Veritatis Splendor* and commended John Paul II for at last taking a stand against one particular heresy which was tearing the Vatican organization apart (although also failing to address many others which were also tearing it apart). His praise for it was based strictly on its contents, so far as they went, which provided some limited but real basis to applaud John Paul II. Bp. Williamson on the other hand commented rather cynically that this encyclical was merely a bone being thrown to the traditionalists in order to pacify them, and that its fine counsel to forbid moral relativists to teach in the seminaries would never be enforced. As it turned out, Bp. Williamson's assessment proved correct. Its provisions were never carried out.

As can also be seen, there was no real disagreement between the two men. One commented strictly on its contents and praised it on that basis; the other commented strictly on the way it was being used merely to pacify Traditional Catholics but not for any real reform and spoke against it on that basis. The next major document from the pen of John Paul II would prove to be much more significant. In 1994, as if to balance off the new permission he gave to have Altar girls, *Ordinatio Sacerdotalis* was published on May 22. Unlike almost everything else the man has written, this document is extremely short, succinct, and bases its evidences directly on Scripture and Church Tradition.

The orthodoxy of this document was so complete and stunning that even the sedevacantists stood up and took notice of it. For five minutes, it was as if the whole Church had a Pope again. Hear what a leading sedevacantist publication, *The Reign of Mary*, has to say about it: In an article exposing the heresies of the Ste. Jovite Church (the followers of the "Pope" of Canada) it says, "As we set about the task of rebutting the claims of Ste. Jovite to a female priesthood, it is interesting to note that the very arguments of the modernist, John Paul II, could be used here to accomplish the task. Before turning to more traditional sources, let us—for curiosity's sake anyway—sum up his arguments (*Ordinatio Sacerdotalis*, May 22, 1994) regarding the impossibility of a female priesthood. Among his arguments are references to Paul VI's own opposition to it.

"a. It is the *constant* tradition *and* teaching of the Catholic Church, observed also in the Eastern Churches.

"b. Christ chose His Apostles only from among men and this choice 'did not proceed from sociological or cultural motives peculiar to the time.' The Church must conform to the Lord's way of acting.

"c. Not even the Blessed Virgin Mary herself was admitted to the mission of the ministry of the Apostles. This in no way lessened her inherent dignity, nor does the non-participation of women in the priesthood detract from their inherent dignity or role in the Church.

"John Paul II concludes: 'Wherefore, in order that all doubt may be removed regarding a matter of great importance, a matter which pertains to the Church's divine constitution itself, in virtue of my ministry of confirming the brethren, I declare that the Church has no authority whatsoever to confer priestly ordination on women and that this judgment is to be definitively held by all the Church's faithful.' Well said—this time anyway—John Paul! With this proclamation we wholeheartedly agree."—*Reign of Mary*, No. 81, Summer 1995, page 9.

If John Paul II were truly a pope, and if as such he ever resorted to the Extraordinary and Infallible Magisterium of the Church, it would be in that document. The very next year, he used the same strong language in the long and verbose encyclical *Evangelium Vitae: The Gospel of Life*, in his formal condemnation of abortion and "mercy-killing," though it also condemned the State's right to execute a criminal convicted of a serious crime, an evident heresy, but since they had relinquished any authority to guide any secular State in anything, that really didn't mean much of anything. The new catechism was finally translated into Latin, and a slightly revised edition (the biggest change seems to be the addition of a very large and detailed index) based on that Latin edition was finally published in 1999. While one would have hoped that it might fix some few small details, as could have been readily predicted, the fundamental theological weaknesses of the original edition were largely unaffected. Still, it was another instance of "Oops, we didn't get that one quite right, so let's Fix It Again, Tony."

In the 1980's and 1990's, another important trend which has appeared is the publication of directories of traditional Latin Masses. For quite some time, both the SSPX and the SSPV had been publishing listings of Mass locations in their magazines, *The Angelus* and *The Roman Catholic*, but of course only the Masses which each respective society approved. The Coalition in support of Ecclesia Dei also began publishing their own listing of "approved" Indult Masses. Mr. Radko Jansky first pioneered a directory for all Mass locations, regardless of affiliation. After 1989, Fr. Francis LeBlanc, of Our Lady of the Sun International Shrine, in El Mirage, Arizona, took over the preparing and publishing of this directory. Unfortunately, in Fr. LeBlanc's version, the SSPX and the SSPV listings were omitted since they both had declined to allow their Mass locations to be published in the directory. In order to fill up space and provide more Mass locations, Fr. LeBlanc took the odd step of including the locations of Old Catholic parishes, though this seems to have been priests functioning as Catholics but merely tracing the source of their orders to the Old Catholics. That oddness is significantly mitigated by his having included information in each listing as to whether the Mass is Indult, sedevacantist, some other Mass somehow in union with "Rome," or one of these (few) "Old Catholic" Mass locations.

In 1994, Fr. M. E. Morrison of the Fisherman's and Seaman's Memorial Chapel in San Francisco, California, picked up the ball by publishing a directory of his own. In it, he excluded the Old Catholic ordained clergy, but included the SSPX and SSPV Mass locations despite their opposition to being listed. On September 29 of that year he inaugurated the first Traditional Catholic Website. Like Fr. LeBlanc's directory, Fr. Morrison's directory, the *Official Directory of Traditional Latin Masses* (See Bibliography or On-Line) also indicates the affiliation (if any) of all the Mass locations listed. This now remains thus far the best and most thorough directory of the Catholic Church today.

At last, a single volume managed to lay out with an explicit list all Catholic parishes truly in union with Eternal Rome. This lends a significant amount of visibility to the real Roman Catholic Church as She exists today, and as a result of that great service, it was only a matter of time before Fr. Morrison would come under the most vicious attack. He had always had the policy of never discussing personalities (other than public figures in the news) by name, neither members of his e-mail list nor anyone else. As a result, he never mentioned the name of the bishop who had ordained him, and with time it emerged that this detail was not for public consumption. Eager to attack him, some clever rumor monger invented the fanciful claim that Fr. Morrison had been ordained by some shady Old Catholic bishop. Conveniently, some such Old Catholic bishop in the area had just passed away and so was therefore in no position to deny being the bishop who ordained him (as he doubtless would have, had he still been alive). Despite such unjust treatment, the always noble and gentlemanly Fr. Morrison continues to publish and update his directory to this day. Basically, the directory has only gotten bigger and bigger as new traditional priests are ordained, others join in, or are found, and as other information about religious orders, publications, suppliers, and even foreign countries are increasingly being included.

In 1999, the "In the Spirit of Chartres" Committee finally produced and began distributing its *tour de force* film, "What We Have Lost," which both exposes the Novus Ordo religion for the non-Catholic religion that it is, but also shows visually and persuasively where the true remnant of the Church is to be found today. "What We Have Lost" is quite literally the movie to which this is the book. It remains a classic despite some of the same limitations as beset the previous editions of this book. Another event of interest in this year (at least to this writer) is that "the-pope.com" finally went on-line, including the entire text of this book and many other articles dedicated to helping people understand what is going on and to take the right action.

The FSSP came under a terrible attack and compromise to its integrity. Sixteen of its priests had somehow converted to the new religion and desired "permission" to say the Novus Ordo "Mass" on at least some few rare occasions. Rather than simply leave the order and found some new bi-ritual order (of which at least one already existed, anyway), they wrote some letter exaggerating the FSSP's hostility for the modernism at the Vatican, and that resulted in the infamous "1411 Protocol" document which effectively mandated that the FSSP give tolerance to all its members to do the Novus Ordo, at least for the yearly chrism "Mass" performed by the local bishop. Cardinal Hoyos had arrived in his new office at Ecclesia Dei with a spirit of great friendliness towards Traditional Catholics, and yet one of his first official actions was this shameful protocol (and another less well-known one, numbered 512, as well) which did much to make the SSPX fear to have any dealings with the present Vatican.

As part of this change, Fr. Joseph Bisig, who had faithfully served as Superior General of the FSSP for almost twelve years, was ousted from his post a mere six months prior to its natural termination, and an American, Fr. Arnaud Devillers, one of the sixteen priests who had pushed for having a "Protocol 1411" (and the only American among them) was installed in his place. Fr. Bisig had always maintained that the FSSP was a mono-ritual order, that is, ordained only to do the Traditional Mass and Sacraments. Fr. Devillers on the other hand had pushed the idea of saying a Novus Ordo at least once a year in one's local bishop's Chrism "Mass." Obviously this is the camel's nose in the tent, and if I had been an FSSP priest at the time, my response to any local bishop requesting that I join him in such a sin would be "Tell you what, do the Chrism Mass using the 1962 Missal and I will be most honored to participate in it with you." This is again to emphasize that we are to be traditionalists, **NOT** separatists. Invite us to help you feed the hungry and we accept. Invite us to help you rob a bank and you do that alone without us. We avoid sinful actions (such as the Novus Ordo), not sinful people. Alas, no FSSP priest seems to have thought of that, and many Traditional Catholics saw this as signaling the possible end or termination, and certainly a watering-down, of the Indult. As described in the Vatican II document *Lumen Gentium*, a part of the Church thus far seems to "subsist in" the Vatican organization, and at least for a while, would soon gain a considerable boost (and a significant modification) under Benedict XVI.

However, come the Jubilee year of 2000, the SSPX made a great pilgrimage to the Vatican to pay homage. So impressed by their devotion and seriousness and piety, Cardinal Hoyos quickly invited Bp. Fellay to come and begin negotiations about trying to end the separation between the SSPX and the Vatican organization and so some major negotiations began. Bp. Fellay decided that all he would ask for would be a formal lifting of all seeming "penalties" supposedly imposed on the SSPX, permission to operate in any diocese regardless of whether permission was granted or not, and for all priests to be formally permitted to say the Mass. It was this last point which became the occasion for the termination of the negotiations as the SSPX could never compromise on this point, and the Vatican organization's recognition of the fundamental right for all priests to say the Mass (as promulgated as eternal law by Pope Pius V) would (correctly) be seen as the beginning of the end, or even removal of, Vatican II. Unfortunately, Vatican II is such a sacred cow of the Novus Ordo Conciliar Church that they utterly refused to permit anything that even looked like its inevitable revocation.

In keeping with that attitude of treating Vatican II like some sacred cow, they went through the motions of pretending to release the famous "Third Secret of Fátima" which unfortunately is not the Third Secret which John XXIII read in 1960, at which point the seer had claimed it would be clearer. That latter document consists of a single page (unlike the new one released which is four pages long), and begins with the phrase "In Portugal, the dogma of the Faith will always be preserved…" Because it went on to warn the Pope against holding a Council, one can see why that document shall never be released so long as Vatican II continues to be their sacred cow. Indeed, it was the children of Fátima who were the main "prophets of doom and gloom" that John XXIII was talking about in his opening remarks at Vatican II.

When Bp. Rangel's health began to give out, the Saint John Vianney Society of priests in Campos elected to continue the negotiations with the Vatican which the SSPX was unable to consummate, and the Vatican offered them quite a deal, which they finally accepted. In January of 2002, Bp. Rangel became (officially in the Vatican organization's eyes), the auxiliary bishop of Campos. Apparently, several of the Saint John Vianney priests were able to strike up quite a friendship with the recently appointed new bishop of Campos, Roberto Gomes Guimarães, who labored to secure hopefully as good a deal as they were willing to extend to the much larger SSPX.

One welcome start for Campos came on August 18, 2002 when Cd. Hoyos consecrated Fr. Rifan to the episcopacy. In case there are any questions of its sacramental validity (the traditional rite was used, but the validity of Cd. Hoyos' consecration could be challenged), one of the co-consecrators was the ailing Bp. Rangel (who then passed away on December 17, 2002). The other co-consecrator was Bp. Alano Maria Pena. It is not clear why Guimarães was not party to that ceremony. At last, there is that "traditional bishop" promised in the original 1988 protocol that Abp. Lefebvre had decided against. Now, if only Bp. Rifan could be elected to be the next Vatican leader! He is only one truly Catholic bishop in their organization, against thousands worldwide who are not. Can he win? Only

THE ADVANCE OF THE INDULT PRIESTS AND THE FSSP 241

if he has the strength to stand his ground and not be led into the sin of the Novus Ordo by the new "bad companions" he had surrounded himself with.

His first couple years or so as bishop for the Catholics in Campos were quite scandal-ridden however. More than once he was roped into actions which compromised him with Novus Ordo aberrations, things that as a priest of the St. John Vianney Society (the faithful priests of Campos under Bp. de Castro-Meyer) he had sworn to himself that he would never do. After that, it is not clear whether, suitably warned, he conducted himself much more circumspectly ever since then, or else there is simply a news blackout as the goings on in Campos simply seem to have fallen off the radar.

As John Paul II's health continued to decline, and he grappled with increasing age and advancing Parkinson's Disease, several modernist cardinals, acting like vultures in my opinion, and led by Cd. Lehmann (whom he had just made a cardinal), began pressing for his resignation, but he refused to entertain that request. I think it is interesting and worthwhile to point out that not one traditionalist, even any sedevacantist, has ever applauded this request. This is because we do know that the problems are bigger than merely the man himself, and furthermore we know that Cd. Lehmann and his cronies want someone quite a bit more to their liking, something which did not bode well for the Vatican organization's future or rehabilitation.

Hope was raised for many however with the election of Joseph Ratzinger who took the name of Benedict XVI. Though known at Vatican II as a "necktie priest," he had since then gradually acquired something of a conservative leaning. Many hoped that he would bring some real order and restore at least something of the old ways and values. By and large however, he mostly left things as they had been when he acquired the leadership office at Vatican City. But he did take one truly remarkable step, however.

On July 7, 2007, he promulgated his famous Motu Proprio, *Summorum Pontificum*. This document replaced all previous Indults and set up a brand-new arrangement by which the Catholic Mass would be tolerated within the Vatican organization. For at least about a decade or so, some sort of further "releasing" of the authentic Catholic Mass for use within the modernist Vatican apparatus had been bandied back and forth, variously hinted at, promised, delayed, shelved, reopened for reconsideration, shelved again, promised again and again, and only in 2007 had finally emerged out the end of the chute. All in all, it was a rather generous expansion of the rights of Catholics to be recognized as such and permitted their worship within the Vatican organization. This was not another Indult, but an outright ruling, or law, governed by the Motu Proprio. With that, even the word "Indult" practically fell into disuse, and with remarkable speed. So what were the provisions of this new official policy? The key parts read as follows:

> Art 1. The Roman Missal promulgated by Paul VI is the ordinary expression of the 'Lex orandi' (Law of prayer) of the Catholic Church of the Latin rite. Nonetheless, the Roman Missal promulgated by St. Pius V and reissued by Bl. John XXIII is to be considered as an extraordinary expression of that same 'Lex orandi,' and must be given due honour for its venerable and ancient usage. These two expressions of the Church's Lex orandi will in no way lead to a division in the Church's 'Lex credendi' (Law of belief). They are, in fact two usages of the one Roman rite.
>
> It is, therefore, permissible to celebrate the Sacrifice of the Mass following the typical edition of the Roman Missal promulgated by Bl. John XXIII in 1962 and never abrogated, as an extraordinary form of the Liturgy of the Church. The conditions for the use of this Missal as laid down by earlier documents 'Quattuor abhinc annis' and 'Ecclesia Dei,' are substituted as follows:
>
> Art. 2. In Masses celebrated without the people, each Catholic priest of the Latin rite, whether secular or regular, may use the Roman Missal published by Bl. Pope John XXIII in 1962, or the Roman Missal promulgated by Pope Paul VI in 1970, and may do so on any day with the exception of the Easter Triduum. For such celebrations, with either one Missal or the other, the priest has no need for permission from the Apostolic See or from his Ordinary.
>
> Art. 3. Communities of Institutes of consecrated life and of Societies of apostolic life, of either pontifical or diocesan right, wishing to celebrate Mass in accordance with the edition of the Roman Missal promulgated in 1962, for conventual or "community" celebration in their oratories, may do so. If an individual community or an entire Institute or Society wishes to undertake such celebrations often, habitually or permanently, the decision must be taken by the Superiors Major, in accordance with the law and following their own specific decrees and statues.
>
> Art. 4. Celebrations of Mass as mentioned above in art. 2 may - observing all the norms of law - also be attended by faithful who, of their own free will, ask to be admitted.
>
> Art. 5. § 1 In parishes, where there is a stable group of faithful who adhere to the earlier liturgical tradition, the pastor should willingly accept their requests to celebrate the Mass according to the rite of the Roman Missal published in 1962, and ensure that the welfare of these faithful harmonizes with the ordinary pastoral care of the

parish, under the guidance of the bishop in accordance with canon 392, avoiding discord and favouring the unity of the whole Church.

§ 2 Celebration in accordance with the Missal of Bl. John XXIII may take place on working days; while on Sundays and feast days one such celebration may also be held.

§ 3 For faithful and priests who request it, the pastor should also allow celebrations in this extraordinary form for special circumstances such as marriages, funerals or occasional celebrations, e.g. pilgrimages.

§ 4 Priests who use the Missal of Bl. John XXIII must be qualified to do so and not juridically impeded.

§ 5 In churches that are not parish or conventual churches, it is the duty of the Rector of the church to grant the above permission.

Art. 6. In Masses celebrated in the presence of the people in accordance with the Missal of Bl. John XXIII, the readings may be given in the vernacular, using editions recognized by the Apostolic See.

Art. 7. If a group of lay faithful, as mentioned in art. 5 § 1, has not obtained satisfaction to their requests from the pastor, they should inform the diocesan bishop. The bishop is strongly requested to satisfy their wishes. If he does not want to arrange for such celebration to take place, the matter should be referred to the Pontifical Commission "Ecclesia Dei".

Art. 8. A bishop who, desirous of satisfying such requests, but who for various reasons is unable to do so, may refer the problem to the Commission "Ecclesia Dei" to obtain counsel and assistance.

Art. 9. § 1 The pastor, having attentively examined all aspects, may also grant permission to use the earlier ritual for the administration of the Sacraments of Baptism, Marriage, Penance, and the Anointing of the Sick, if the good of souls would seem to require it.

§ 2 Ordinaries are given the right to celebrate the Sacrament of Confirmation using the earlier Roman Pontifical, if the good of souls would seem to require it.

§ 3 Clerics ordained "in sacris constitutis" may use the Roman Breviary promulgated by Bl. John XXIII in 1962.

Art. 10. The ordinary of a particular place, if he feels it appropriate, may erect a personal parish in accordance with can. 518 for celebrations following the ancient form of the Roman rite, or appoint a chaplain, while observing all the norms of law.

Art. 11. The Pontifical Commission "Ecclesia Dei", erected by John Paul II in 1988, continues to exercise its function. Said Commission will have the form, duties and norms that the Roman Pontiff wishes to assign it.

> Art. 12. This Commission, apart from the powers it enjoys, will exercise the authority of the Holy See, supervising the observance and application of these dispositions.
>
> We order that everything We have established with these Apostolic Letters issued as Motu Proprio be considered as "established and decreed", and to be observed from 14 September of this year, Feast of the Exaltation of the Cross, whatever there may be to the contrary.

The many and detailed provisions of this new policy create the novel idea of a Rite with two different "uses." If it had been a matter of Latin versus vernacular Novus Ordo then perhaps such a word might well have been appropriate. But the Catholic Rite (Tridentine) and Novus Ordo are not only two different rites, but even rites belonging to two different religions. Where the one is ancient and venerable, dating from the time of the Apostles, with only such minor tweaks as the addition of many particular newer saints and feasts to the calendar, rubrical and musical notations, and the like, but recognizably the same Mass, the other is a mere modernist synthesis of unbelief and irreverence, or as Ratzinger himself once quite properly put it, "a banal, on the spot product."

Now suddenly the two are put on equal footing, or at least very nearly so, with one being the hum drum "normal" or "ordinary" form (Novus Ordo), and the other being the "extraordinary" form (Tridentine). The only way that putting truth and error on (pretty much) the same parity could be considered a "good" thing would be in comparison to the previous state of affairs in which the truth has no right to exist, or at least no right to be transmitted at all (especially to new generations), or else was to be only most grudgingly tolerated on some few rare occasions as some sort of indulgent favor, while error enjoys all full status as being for everyone pretty much all the time. Viewed in that light, this Motu Proprio is undeniably real progress, a significant step unequalled by any of the somewhat similar steps documented throughout this chapter. Not only is the Catholic Mass given all clear freedom, but also all the other Catholic sacraments.

Unfortunately, by this time, we have reached the state predicted by the SSPX priest Fr. Carl Pulvermacher when he stated that "Just as soon as they have no valid priests left, you'll have all the Latin Masses you want." But so far, a few old timers do remain, and some have crossed over from traditional societies (mostly the SSPX) or else from some Eastern Rite, or quietly obtained a conditional ordination from a traditional bishop.

In this form of course, the FSSP and other such Vatican-tolerated orders, as well as any other deemed capable or fit to say the true Mass (whether validly or not) has seen some considerable growth, to the figures given above (nearly 120 such parishes and 1,400 other Mass locations, serviced by over 300 FSSP priests and at least a quarter as many more from the other traditional Vatican-tolerated orders, plus any number of others otherwise considered by them to be qualified, whether rightly or not. Such growth seems impressive, but it seems to have hit a peak. The problem is not interest on the part of the people who desire to assist or even insist upon it for themselves and their families, but at the top.

Pope Saint Pius X wrote of in Pascendi (Paragraph 38) that the Modernists "Regarding worship, they say, the number of external devotions is to be reduced, and steps must be taken to prevent their further increase, though, indeed, some of the admirers of symbolism are disposed to be more **indulgent** on this head." Joseph Ratzinger, as Benedict XVI, was undeniably quite generous about the Catholic Mass, one of the very few prelates within that organization who, despite being nevertheless a Modernist, is among what few "admirers of symbolism" as were inclined to be "more **indulgent** on this head," actually rather generous, all said and told. What could be more **indulgent** than an Indult?

But alas for the Vatican-tolerated Catholic Masses and those who say them and those who assist at them, a chill wind now blows in the opposite direction. Rather than waiting to die, Benedict XVI stepped down, making room for another to be elected, one Jorge Bergoglio ("Francis I"), who has made it quite clear that he has no affection for traditionalists. He has referred to real Catholics as being "Pelagians" (what a stunning ignorance of history that choice of epithet displays on his part!), or "rigid" and the like, and that refers to those attending Masses approved by his society under the terms of the "extraordinary form." One can only guess what choice and unprintable words he would have to describe those Catholics who openly resist his paganism and atheism (SSPX and like), to say nothing of sedevacantist Catholics who have nothing whatsoever to do with him. He first reduced the Commission Ecclesia Dei staffing in 2017, and then eliminated the Commission altogether in 2019, handing those "troublesome traditionalists" to some other, lesser, outfit. And now, having stocked the Vatican organization's "cardinalate" with a majority of like-minded individuals, he has already done everything in his power to ensure that his pathetic mad kingdom will never find its way back to God. Why is the 2007 Motu Proprio even tolerated by him? It is merely something he has inherited, and for which he has absolutely no sympathy, interest, or support. Perhaps it is human respect; Joseph Ratzinger is still alive as I write this. What will happen when he is not?

12

THE CATHOLIC LAY APOSTLATES AND FURTHER EVENTS

A married Novus Ordo "permanent deacon" once told me that "the Church will be saved by the "little people," by which he meant those who have no hierarchical place in the Church, namely the laity. The Novus Ordo Church of the People of God places great stock in their "laity," to whom they entrust quite a number of significant liturgical and parish council roles. Whole schools and retreat centers have been prepared in order to brainwash unsuspecting persons into thinking that they are being taken behind the scenes so as to see how things really work, and what sorts of great and wonderful things might one day be implemented if only more persons were willing to get involved and go along with it, and if only "stick in the mud" priests and conservative laity would just permit it. These ideas are of course merely the germ of yet further liturgical abuses, the Novus Ordo service itself not being sufficiently an abuse in at of itself to those of running the schools and retreats. Even the most archconservative priest or bishop would find himself overwhelmed with absurd requests (and even directions or commands!) from their "laity" so brainwashed. Though of course, it has to be obvious that the laity had nothing to do with this in the first place, and by and large simply wanted things to stay the same and not to become laboratory rats for the latest experimentation.

Interestingly, within Tradition, the Laity have also come to have something of a role not expected nor existing much previous to Vatican II. Confused, dismayed, disillusioned, disappointed, angered, appalled, and seriously upset by the continuous upset and instability that had come to their parish churches, they suddenly had to become interested in the Faith, in preserving it in their own life, and the lives of their family members. I have already mentioned the labors of Thomas A. Nelson who formed a publishing company dedicated to keeping in print (and returning to print) numerous Catholic books, the writings of saints, mystics, popes, theologians, ecumenical councils, historic catechisms and apologetic works, and so forth. Suddenly it was no longer appropriate to take the "Father knows best" approach of simply allowing "Father" to answer all their questions and tell them what to do, what is right and what is wrong. For the first time, "Father" really did not know best, as he struggled to balance his priestly formation and training with the alien new direction being imposed from what he sincerely mistook for authority.

Some of the earlier laity thus showing initiative have already been described in the preceding chapters, but there were many more to come, and their role in reaching out to fellow Catholic laity with explanations understandable to each other has also come to take on a legitimate role. Some have even read many vast and heavy theological tomes, perhaps even in Latin which they had to learn, so as to prepare works of theology themselves. Ordinarily such a step would be unprecedented, even often seen as forbidden to involve themselves with.

Yet Dr. Don Felix Sarda Y Salvany, with the approval and support of no less than the Sacred Congregation of the Index, wrote that "Yes the faithful are permitted and even commanded to give a reason for their faith, to draw out its consequences, to make applications of it, to deduce parallels and analogies from it. It is thus by use of their reason that the faithful are enabled to suspect and measure the orthodoxy of any new doctrine, presented to them, by comparing it with a doctrine already defined. If it be not in accord, they can combat it as bad and justly stigmatize as bad the book or journal which sustains it. They cannot of course define it ex cathedra, but they can lawfully hold it as perverse and declare it such, warn others against it, raise the cry of alarm and strike the first blow against it. The faithful layman can do all this, and has done it at all times with the applause of the Church. Nor in so doing does he make himself the pastor of the flock, nor even its humblest attendant; he simply serves it as a watchdog who gives the alarm. Oportet allatrare canes. 'It behooves watchdogs to bark' very opportunely said a great Spanish Bishop in reference to such occasions" in his book, *Liberalism is a Sin.*

In a similar vein, one such Traditional Catholic layman, John S. Daly, justifies the theological writings of educated laity in our unusual times, stating that "I am not, of course, denying that in some circumstances it may be permissible to publish without ecclesiastical approval. For instance, when it is likely to be of considerable benefit to souls that a particular truth be publicized and recourse to ecclesiastical authority is impossible, a writer who had taken sufficient steps to ensure the orthodoxy of his writings might prudently publish by virtue of epikeia. And indeed if it were to become wholly impossible to obtain an imprimatur by legitimate means it is possible that cessatio legis – the automatic cessation of the law – would indeed take place." That of course assumes that the impossibility comes from there being no Catholic bishops willing to review written works for imprimatur purposes, not from Catholic bishops reviewing a work and refusing it an imprimatur due to errors or other deficiencies.

The Traditional Catholic community has produced a number of significant books which have further defined the traditionalist cause. The first of these, by Romano Amerio, is *Iota Unum* which is a most detailed study of the changes made as a result of Vatican II itself. Several authors, Atila S. Guimarães, Michael J. Matt (of *Remnant* fame), John Vennari (publisher of *Catholic Family News*), and Marian T. Horvat, got together and assembled a work entitled *We Resist You to the Face*, which does what only the Abbe de Nantes has ever done before, namely confront the present Vatican leadership with their errors and demand an answer. So often we talk about them behind their back so to speak, but how seldom we think of what we would want to say **to** them. Atila S. Guimarães also began a series of eleven books, the set entitled *Eli, Eli, lamma sabacthani?*, and the first volume of which, published in 1997, is entitled *In the Murky Waters of Vatican II*. He completed this series in 2017.

As the old guard began to die off, new writers have emerged to take their place. One of the better-known writers is Christopher Ferrara, who seems to fill something of the gap left by Michael Davies. Both take (took) a mostly Motu Proprio (formerly Indult) position, but also hold Abp. Lefebvre in high regard, though Mr. Ferrara seems to be a little more sympathetic to the SSPX, much as Mr. Davies had been prior to the 1988 consecrations. When criticizing the Novus Ordo his books and other writings and talks are quite incisive, often truly brilliant commentary. But, like Michael Davies, he reacts against sedevacantists with extreme and clumsy prejudice, demonstrating for all to see the differences between a secular lawyer (his secular occupation in which he defends the rights of the unborn and other worthy causes) and a canon lawyer, to say nothing of the difference between either and anything at all of a theologian. His most important book relevant to the issues discussed herein is *The Great Façade*, co-authored with Thomas E. Woods, Jr. which many have found to be a splendid introduction to how the authentic Catholic traditions have been removed and destroyed by the Novus Ordo religion and Vatican II.

One somewhat surprising turn of events has been the creation of new paper newspapers at a time that newsprint has gone through a tremendous decline, owing to so many small town and other smaller newspapers going out of business, and even larger city newspapers shrinking. While other Catholic papers such as *The Remnant* and *Catholic Family News* have continued their press runs, new papers, dedicated to the Traditional Catholic Faith, have nevertheless also made their appearance. In November of 2005, Kathleen Plumb, with support from her husband Robert, began organizing a paper first published February 2006 titled *The Four Marks*. Despite Robert getting killed in a vehicular accident in June 2006, Kathleen has carried forth with publishing this excellent monthly paper from then until now. Robert's death had, however occasioned a tremendous outpouring of love and support and Masses offered for him. Yet another paper arising later on is *The Catholic Inquisitor* beginning with the December 2018/January 2019 issue.

It is a good thing that Catholic papers on physical newsprint, and books, have flourished since, ages hence, who knows what might remain on the Internet, and what might be deleted, even if only for space, to say nothing of things deleted in a malicious suppression of the truth. It is therefore appropriate that these old fashioned means of communication continue to thrive among Traditional Catholics while dying on the vine for all other categories, all of which will one day be as lost as all the numerous day-to-day events of the High Middle Ages as were only known verbally in their own time. Many sources, on the Internet today, or recently enough to have been captured, will certainly be totally lost, apart from what reference is made to them here and in other Traditional Catholic publications. Nevertheless, the Internet has served as a spiritual and intellectual lifeline to Catholics in this dire time. But now, with Catholics so few and so scattered in every part of the world, and in every country, some Catholics are many hundreds of miles from other Catholics and have only such contact as this to reassure them that the Church thrives on in other areas. We have reached a time when the Internet is easier to access than food in some areas.

One effect both good and bad however has been the serious discussion of serious theological questions. For practically everyone, dear old "Fr." no longer knows best, himself being altogether completely unrecognizable as a Catholic to any of the great Saints through the ages. The putatively "official" sources of information, as Catholics had long trusted with such complete serenity, have patently come to fail us. Here in America such publications as *Our Sunday Visitor* or even *The Wanderer* have failed us; even the catechisms provided to us give us no idea what the Faith is actually all about, and in other countries, the same has happened. Suddenly, if you want to find the actual Catholic Faith as known for over nineteen centuries, you have to dig, to find old (and often out of print) catechisms, theological works, saintly writings, and so forth, and though many of them are available in reprint, and sometimes also as old original volumes, you have to know what to look for by title, and even sometimes, by edition. How immensely helpful it is to have friends, even geographically remote, that you can contact through the Internet, to point you to the right materials, and to answer our questions that we would have put to our priests in better times, but dare not now for fear of being given a bum steer.

Our authentic Catholic clergy however seldom have much time to devote to the Internet, what with trinating the Mass every Sunday, often in different cities, bringing the Last Sacraments to individuals within a large distance, an area many times greater than your typical diocese, as there is no other priest to cover them (and perform other functions such as House blessings and the other Sacraments). They keep up with their Breviary, and many teach catechism, write books and articles, run the business end of having a parish church, visit the sick and in prison, go on priestly retreats and other meetings, and so forth. What this amount to is that the Traditional Catholic Internet is primarily dominated by the Laity (and perhaps by some few consecrated religious, if assigned such a task), as the Laity are more likely to have the time to pour into such an apostolate.

Ideally, such an apostolate would be supervised by at least some form of clerical spiritual advisor. But if any cleric had the resources of time etc. to supervise, then he would have the resources of time etc. to do it himself. The upshot of this is that laity, and self-appointed laity at that, are having to give out the sorts of advice that is the proper domain of priests. Such a thing becomes permissible precisely because there simply are not the sheer number of priests it would take. So we sort out our personal, or moral, or doctrinal problems as much as possible, and then, if any absolution or other priestly function is needed, guidance to the nearest real Catholic priest is given, and he doesn't have to explain the whole world to us.

As technology has changed, computers and the Internet have gone from "bulletin boards" to e-mail lists, and finally to blogs and forums, wherein Catholics from all around the world can discuss their theological questions, bounce ideas off each other, and (hopefully) the ignorant become informed. There are some on these places who are quite well-read, and sometimes seem to know everything, and undeniably they do know a lot. One cannot fault these self-appointed lay armchair theologians for attempting to make some sort of sense of the current crisis; ideas have to come from somewhere after all, and sometimes their knowledge may be more detailed than that of the clergy (being all read up on everything), but I have seen them wax dogmatic about some opinion or other which really does not hold up very well in the face of some doctrine or other.

Though I don't think very many such intend such an outcome of their efforts, one deleterious effect has been to usurp the clergy and their role, effectively becoming a kind of "mini-magisterium" of their own. Mere Canons of Canon Law may get cited as some sort of moral absolutes that cannot get interpreted, adapted, or set aside for particular cases, even where doctrine and practical circumstance demand it. This is the sort of thing that happens when one is well-read but not trained, and furthermore don't have the advantage of looking at these issues from the pastoral perspective. Without intending it, the Church is defined out of existence, clergy are re-interpreted as mere laymen with Holy Orders, subservient to their "mini-magisterium," and ultimately mere sacrament vending machines. Whatever may have been intended, the usurpation is real.

The worst effect is when the clergy, already subjectively uncertain as to their place in the Church, begin to doubt themselves or their ministries, pressing on in some deep sensing that what they are doing is important (and obviously so to those whom they minister to), yet rarely willing to exercise the pastoral authority which the Church has given to them. Objectively, their position is clear and integral to the Church, but subjectively, many have not had the time and resources to check the findings of these self-appointed lay armchair theologians and their mini-magisterium, and this is the greatest crisis facing the true Church today. In any case, it is not clear that many of these armchair lay theologians actually mean any harm. Among the best of these was John Lane, though he has largely fallen silent these past few years. But in his heyday he ran the Bellarmine Forum which at least kept the discussions at an incomparably high level. But perhaps his most prominent moment was his October 16, 2006 debate with Robert Sungenis over the Sede Vacante issue.

In 2003 and into 2004, a particular Traditional Catholic layman, Mel Gibson by name, after a richly successful career in the movie industry, came to be known for his authentic Traditional Catholic beliefs, and in union with these beliefs he produced and released a movie called "The Passion of the Christ" which for the first time truly shows the Divine cost incurred by our sins. Instead of doing the usual Hollywood "job" on it, he decided to follow the Gospel texts carefully and literally and take the account seriously and reverently. This was the first time since the 1988 consecrations by Abp. Lefebvre that Traditional Catholicism became an item for the daily news.

Suddenly, people wanted to know about him and the private chapel he had built on his property for the Mass, his own rejection of Vatican II, and his father's well-known sedevacantism. Almost immediately, certain self-appointed movie experts were falsely accusing him of being anti-Semitic. His movie was originally to include the line (taken from the Gospel of Matthew) "His blood be upon us and upon our children." Under popular pressure, he added subtitles (since it was filmed in two of the three original Biblical languages used at the time, Aramaic and Latin, Greek unaccountably left out), but left this particular line untranslated (it can still be heard in the original Aramaic).

So once again, Traditional Catholics find themselves the victim of calumny, but at the same time, a surprising number of others came to stand behind him and his film, including many Protestants. When John Paul II viewed an early cut of his film, his only response was "It is as it was." There was a bit of a flap when this was at first officially acknowledged, but then denied by the Modernist Vatican. The film itself however proved truly moving and profound, and with that Traditional Catholicism has also moved into the limelight. Most of all, with this movie, Mel Gibson truly stood up to be counted with the true Church of God.

The Home-aloner group has gone through an extraordinary (and very bad) transition from trying to create popes to giving up and defining the Church right out of existence. For a number of years, prominent Home-aloner and Conclavist Ken Mock had travelled the countryside seeking some sufficiently untainted clergyman to be made into a pope. After one failure with Michael Bawden ("Pope" Michael 1), he had moved on to making another with Fr. Lucien Pulvermacher (a brother of Fr. Carl Pulvermacher and other priests), who declared himself to be "Pope" Pius XIII. Since he was not a bishop, rather than await the arrival of some bishop (somehow), he invented a way (so he thought) to make himself a bishop, so that he could then be a "Bishop of Rome." He had found a rare instance (in fact discussed by Ludwig Ott and Msgr. Charles Journet) in which a mere priest, at the command of the Pope, had successfully ordained some several other men to the priesthood, and who were then simply pressed into service. Apparently, in such an extraordinary circumstance, a priest could ordain another man to the priesthood, as an "extraordinary minister." But for a priest to make a bishop, even at the command of a Pope, would mean that he is attempting to give to another that which he does not himself possess. One curious upshot of this is the possibility that even though "bishops" in the Novus Ordo today virtually never possess valid Orders as such (other than many in the Alternate Rites, some few very old timers (virtually all retired), and transfers from schismatic but valid successions), but may often be valid priests. As valid priests, directed to ordain seminarians by who they mistake for Pope, it seems quite possible that through supplied jurisdiction due to common error, such ordinations if performed by the authentic Catholic rite, might well be valid. So for example, if "bishop" (actually merely a priest, in terms of any valid orders) Bruskewitz were to go through the ceremonial motions of ordaining a seminarian of, say, the FSSP, given his creditable insistence on using the full traditional form, the result could be a valid priest. Despite the plausibility of this scenario to apply in some cases, the stern cautions regarding invalid "priests" loosely "ordained" by the Novus Ordo must remain.

After this failure, Ken Mock began scouring Europe and the rest of the world seeking his "untainted" bishop, and as it has become clear to the Home-aloner community that there remains none such to be found who would also be willing to reconstitute the Church, this is why Home-aloners now claim that the Priesthood and the Sacraments (other than Baptism and Marriage which can be done by laypeople) are simply scheduled to disappear. They have closed in on themselves and vanished into heresy, the heresy of really believing that the hierarchical Church has disappeared, and that no Catholic authority exists. As for the new "popes" themselves, their universal failure to provide credible alternatives to the Vatican claimant has quite well demonstrated that the Conclavist approach based on an attempt to create authority where none already exists will not work. But of course, that stands to reason. The electors, even were there no cardinals, must at least represent real jurisdiction over souls of the Church.

For clarification however, this is not to say that all, or even most, of those who might "stay at home" on a Sunday are necessarily entangled with either Ken Mock's conclave efforts, or else in total denial of the existence of all Catholic authorities whatsoever. There are many who "stay at home" on Sundays simply because they have no reasonable access to a Catholic Mass, being isolated by great distance, ill health, lack of transportation and so forth, making each Mass they do manage to attend something of a pilgrimage. And there are others who, while admitting hypothetically the existence of some real Catholic authority somewhere, nevertheless denying it in practice by claiming it cannot be found or might exist only in some forgotten part of the world. Where it most seriously goes astray is in cases where a Catholic Mass is avoided even where convenient on the specious claim that no Mass has any business existing at all, all (or at least all known or accessible Masses) being supposedly illegitimate.

Ironically, though the home-aloner position is most often associated with sedevacantism, one of the main, if unacknowledged, intellects behind the home-aloner group was neither a sedevacantist, nor (strictly speaking) a home-aloner himself. This would be the Abbé de Nantes. Despite the value of some of his works in demonstrating the heretical nature of the new religion and its main proponents, and in addressing directly the fallen Vatican leadership, the man himself had inexcusably become critical of Abp. Lefebvre and all other bishops and clerics who have heroically sustained the valid and lawful apostolic succession. Taking a grotesquely overscrupulous and pharisaical stance against performing any clerical function without Modernist Vatican approval, a number of young men in his group, though eminently qualified to serve as priests, go unordained. Thus, his followers have been effectively neutralized against taking any positive constructive action in this crisis. If it were up to him, the very future of the Church itself would be held hostage by the Modernist heretics.

Far better news in France is that on an unidentified Sunday somewhere in the year of 2002, the total number of Faithful in attendance at traditional Masses in that country finally surpassed the total number in attendance at Novus

Ordo ser- vices on that given Sunday, and pretty much every Sunday ever since, excluding only certain high holy days and weddings and funerals. The total number of Novus Ordo believers there still outnumbers the number of Catholics by quite a bit, but far, far fewer of them (percentage wise) care enough to bother to show up to church on a typical Sunday than do Traditional Catholics.

Events continue to unfold within the fold of Traditional Catholicism, many involving clerics more than laity, and so we begin focusing here on events to follow, and as known to be occurring in the many various parts of the world:

In the years since the previous edition (2004), there have been a couple transitions of power in the Modernist-usurped Vatican. John Paul II died only about a year after that edition, on April 2, 2005. I was not surprised that Joseph Ratzinger was the next person to take that role on April 19, but I was surprised that he took the name Benedict XVI instead of John Paul III. Slightly more conservative in personal moral matters, and generous towards the Catholic Mass and Sacraments, his high point was *Summorum Pontificum*, though his lifting of the fictitious "excommunications" of the SSPX in 2009 drew that Society a bit closer to his orbit (which at that time did not seem quite so bad as it does now), and was a mixed blessing, in that it sustains the fiction that the "excommunications" ever existed in the first place. On February 28, 2013, he stepped down (resigned) and Jorge Bergoglio, taking the name of Francis I, was elected on March 13. A stark raving communist, Francis I has gone out of his way to alienate Catholics from his now quite patently anti-Catholic society, and taken and continues to take measures to ensure that his society will never be bothered by real Catholics in any places of importance.

Meanwhile, the SSPX has gone through a few minor transitions. Bp. Fellay completed his first 12-year term as that society's Superior General in 2006 but was reelected to serve a second 12-year term which ended on July 11, 2018, and he was succeeded by Fr. Davide Pagliarani on that day. Though not really accepting the SSPX (and certainly not their religion), the Modernist Vatican has, to keep them pacified and undermine their actual canonical mission received through Abp. Lefebvre, by granting "recognition" of the lawful juridical acts of the SSPX, such as marriages, confessions, and even ordinations of priests, as if these actions did not already carry with them an obligation to be recognized by all Catholics from the get-go. Like the "lifting" of the fictitious "excommunications," these Modernist "recognitions" carry with them the subversive intent of getting the SSPX clergy to reject the true value of their juridical acts prior to these phony "recognitions." This threatens the validity of their real canonical mission from the real Church, by fooling them into thinking they have been given something when in fact it has been taken away. If taken in the best possible sense, "OK, they are finally recognizing the authority and jurisdiction we have always possessed as a lawful and canonical entity within the Church," well then no harm, no foul. But if taken as "Hey, now we finally have ordinary jurisdiction for the first time," then that is no better than if the Anglicans had granted canonical recognition to the authority and jurisdiction of the SSPX, and based on that the SSPX were to say the exact same thing. In that latter case, those taken in by that sophism really and truly have only supplied jurisdiction, where before they had regular and ordinary jurisdiction.

One other major event in the SSPX was the departure of Bp. Williamson on October 4, 2012. His peculiar and ahistorical claims, in particular about the scope and scale of the Nazi slaughter of Jews (he is generally considered a "Holocaust denier" by claiming their deaths number in the few hundred thousands, even though Yad Vashem has cataloged and documented about 3 million by name and family history, and it does appear that there remain quite a few more yet to be identified) created an unnecessary scandal for the SSPX back in 2009, resulting in his eventual expulsion in 2012. Since that time, he has consecrated Bp. Jean-Michel Faure on March 19, 2015, Bp. Tomás de Aquino Ferreira da Costa on March 19 2016 (with Bp. Jean-Michel Faure as co-consecrator), and Bp. Gerardo Zendejas on May 11, 2017 (with Bps. Faure and Aquino as co-consecrators).

Other societies started by SSPX priests include the Istituto Mater Boni Consilii (Institute of our Mother of Good Counsel, founded 1985), a sedevacantist society first headed by Bp. Munari, now currently led by Bp. Geert Jan Stuyver, and the Institute of the Good Shepherd (founded 2006) which joined up with the Vatican organization. On February 22, 2018, Bp. Sanborn consecrated Bp. Joseph S. Selway to assist in his ministry. On February 28, 2007, Bp. Kelly consecrated Bp. Joseph Santay, and on December 27, 2018 he consecrated Bp. James Carroll (with Bp. Santay as co-consecrator) to assist him in his ministry.

One sorrowful bit of news is the bankruptcy and ultimate takeover of TAN Books and Publishers, Thomas A. Nelson's brainchild, in 2008. He had been brave enough to publish key Traditional Catholic texts, that is to say, texts pertaining to the circumstances of Traditional Catholics today even by such authors as Dr. Rama Coomaraswamy, Fr. Cekada, Fr. Wathen, and the like, alongside the equally valuable works of the Saints, the Mystics, the Apologists, and the Catechists of the Church from the pre-Vatican II era. For some reason, coming into the new Millennium Thomas Nelson seemed to be having business difficulties. He had already found himself unable to pay royalties to Atila Sinke Guimarães for his edition of *In The Murky Waters of Vatican II*, and perhaps other works as well by current and living authors, and debts were accumulating until he was forced to file for bankruptcy in 2005. By 2008 the publishing concern had paid its debts by being bought out by a "St. Benedict's Press," a Novus Ordo publishing concern, which predictably ceased to print or distribute all traditionalist publications as had formerly been available. Some of the Saints and Mystics of the pre-Vatican II era are no longer available (though some have found other publishers), though a few of the most contemporarily popular remain in print, and new living authors, only some with a mild traditional leaning, and the rest in no way Traditional Catholic at all, have been added. But many of the pre-2005 books, though no longer printed by TAN (now rebranded to mean "Tuum Adoramus Nomen" instead of the initials of Thomas A. Nelson), have eventually found homes with other publishers.

On January 7, 2011, the Right Reverend Leonard Giardina, O.S.B. passed away. While there had been any number of Traditional Catholic religious communities set up over the course of the crisis, the Abbey of Christ the King in Alabama is probably a fairly typical case. Abbot Giardina was a Benedictine priest ordained May 26, 1949, and had departed from a formerly Benedictine religious house which, like virtually all others, was being forcibly converted to the Novus Ordo religion, thereby ceasing to be Catholic, back in the early 1980's. He took an ordinary job to support himself, and started saying Masses for a few friends, neighbors, and associates. With the generous help of supporters, he was able to establish a monastery in Cullman, Alabama in 1984, and this had attracted the attention and support of Traditional Catholics of all stripes. He had refused to comment publicly on any of the controversial questions, but merely got on with being (and leading his small community into being) a faithful follower of Saint Bernard. His actual leanings, which he kept to himself, were nevertheless evidenced by his turning to Bp. McKenna for ordaining anyone of the brothers as a priest, and who also canonically blessed and established him as a Benedictine Abbot on July 13, 1994. Up until that point he had been the only actual Benedictine, with his followers only living as Benedictines, but then he had the power to make them actual Benedictines.

Being quiet about his beliefs certainly had its perks, most notably the respect and support and approval of many Traditional Catholics, from sedevacantist through SSPX-Resistance-like clear to the Indult/Motu Proprio crowd. But this carried a price which exposed itself upon his death. As death grew near, one of the priest-brothers, untrained in Catholic theology (pointing up that even for a contemplative community a priest should always be properly and fully trained), managed to become in charge of the civil corporation owning the monastery property, and in the Abbot's final convalescence was able to initiation dealings with the Vatican Modernists, which meant that everything they had, the farms the brothers (and some sisters as well) had worked, the living quarters, and the church building where about 70 to 100 Catholics attended Mass regularly and others visited frequently also for Mass, all got handed over to the Novus Ordo

Conciliar Church, lock, stock, and barrel, by May 1, 2011. It had been easy for the softlining priest-brother to suppose that Abbot Giardina might well have intended such a action as he had never told them otherwise, or indeed, anything.

In May of 2011, the German government imprisoned Pater (Father) Rolf Hermann Lingen, a sedevacantist priest, as if they thought that it is up to the secular government (themselves) to decide religious questions rather than churchmen. He endured prison rather than renounce his Catholic Faith, which the German court insisted upon as a condition of his release. On August 2, 2012, he appealed to the International Criminal Court and to the Office of the United Nations and eventually his freedom was won. In the beginning of 2013, a petition was circulated for his freedom and recognition as the valid and lawful Catholic cleric that he in fact is. This petition drew widely diversified support from many countries, from Germany, many states of the United States, several provinces of Canada, also Greece, Hungary, Portugal, Argentina, Poland, Ecuador, India, New Zealand, Mexico, Romania, South Africa, Brazil, Russian Federation, United Kingdom, Philippines, Japan, Netherlands, and Sweden, demonstrating a Traditional Catholic (and even, specifically, sedevacantist, or at least sedevacantist-sympathetic) presence in countries of which I have no information as to what, in the way of Traditional Catholics, even exists there.

The Traditional Catholic community continues its exponential expansion in all its various groups and Orders. In the United States, the seminaries Immaculate Heart, Most Holy Trinity, and Mater Dei, operated by the SSPV, the priests associated with Bp. Sanborn, and the CMRI, respectively, are all training sedevacantist priests, to be ordained by Bishops Kelly, Sanborn, Dolan, and Pivarunas. Rather promptly, priests began to emerge from these new seminaries, beginning with Fr. Cyr ordained by Bp. Dolan and Fr. Santay ordained by Bp. Kelly (now bishop). These few priests were transfers from other seminaries who merely completed their training and formation with Bp. Dolan's group or the SSPV. There have since come to be many priests coming from these seminaries who have been there since the beginning of their priestly formation and at that point the pace of ordinations have greatly increased. Saint Thomas Aquinas Seminary continues to turn out priests for the SSPX, and Our Lady of Guadalupe Seminary (now relocated in Nebraska) turns out priests for the FSSP. Even "independent" and "Motu Proprio" priests are becoming more numerous. The Indult priests are easy enough to explain since more and more diocesan uses of the "extraordinary form" are learning that the best way to raise money is to grant the use of the Catholic sacramental forms to some old priest in their diocese and start having Latin Masses. Unfortunately, many young men are often doubtfully ordained for this purpose, as the old timers die off and validity becomes more and more of a concern, despite what could still be considered a profession of the Catholic Faith by virtue of their ceremonial forms.

The growth of the "independent" priests may seem a bit mysterious, but they are also becoming more numerous through several means. Many of the new "independent" priests were trained and ordained by the traditional Orders described in this book, but eventually left their Orders for various reasons. Priests often transfer in many different directions. For example, Fr. Gregory Foley who was trained and ordained by the SSPX left the SSPX in 1992, but continued to serve in his parish church in San Jose, California, until moving to Walnut Creek, also in California. Another priest trained and ordained by the SSPX, Fr. John Rizzo, transferred over to the FSSP in 1992. Another source of "independent" priests is regular diocesan and religious Order priests who are returning to the traditional Mass in their old age. A good example of this category was Fr. Frederick Schell who said Mass in quite a number of locations in the Southern part of California. Fr. Schell, originally a Jesuit, had to leave that order in 1967 when it became too liberal to qualify as Catholic anymore. After serving in a variety of diocesan positions for a number of years, he simply began saying the Tridentine Mass for those who need it, and continued until his declining health made it impossible, several months later dying on September 28, 2002. He was succeeded in his parish communities by Fr. Patrick Perez, who had been trained and ordained by and for the Institute of Christ the King, but somehow set aside and forgotten by them.

This wholesome trend has continued with more priests leaving the Novus Ordo, or at least refusing to do the Novus Ordo, and then letting what happens to them, happen. Fr. Zigrang, once the Saint Andrew's Church in Channelview, Texas (Novus Ordo), decided on the feast of Saints Peter and Paul (June 29, 2003) that he would cease offending those two great saints by saying a ruined "Mass" and instead began saying the Catholic Mass exclusively. Needless to say, he was soon put out and now serves with the SSPX. Not long after, another priest, Fr. Lawrence C. Smith made the same decision, and others are beginning to follow suit, all to God's glory. At last united to the Church, they are "sacerdos vagus" no longer! In most cases, when a doubtfully ordained Novus Ordo presider converts to the Catholic Faith, he receives a conditional ordination from a Traditional Catholic bishop, as in many cases of putative priestly ordination one cannot be certain as to the validity of the gravely doubtful act performed by them. But on June 29, 2011, Bp. Pivarunas ordained Father Michael Oswalt to the priesthood absolutely, that is NOT conditionally, despite his previous "ordination" by the Novus Ordo, which they had examined and ascertained to have been categorically invalid. This is the first time that a Novus Ordo ceremony has been categorically rejected as invalid in such a manner.

THE CATHOLIC LAY APOSTLATES AND FURTHER EVENTS   259

It appears that at least one, perhaps two additional lines of valid and canonical episcopal lineages may have joined us, or perhaps might eventually, both based on the official but barely-documented successions continued clandestinely in the Eastern European Communist nations. The one group most clearly with us begins with Bp. Pavol Hnilica, clandestinely ordained a priest on September 29, 1950 and consecrated on January 2, 1951 by Bishop Robert Pobožný for the Underground Church of Czechoslovakia. Seeing how the Communists operated, by setting up some few (very few) "clergy" approved by themselves (and therefore disapproved by Pope Pius XII) who agree to water down the religion and spread disbelief among the increasingly nominal "believers" who attend their services, and how the Novus Ordo, now effectively operating in exactly the same way, he came to see the Traditional Catholic community (and most especially the theologically astute sedevacantists) as being the Underground Church today. In that vein, he is believed to have given a conditional consecration and canonical recognition (mission) to one Emmanuel Korab in November 1999 (exact date uncertain), though Bp. Korab had already been validly consecrated (probably sans canonical mission, however) by Bp. Lopez-Gaston on June 26, 1994.

The other group of Uniate Eastern Rite Catholics, seeing the increasing modernism, finally reached "the last straw" as to what they could put up with and call "Papal" with the shenanigans of Francis I, whose heresies finally forced this group to accept the Sede Vacante finding and seek to make sense of the whole Church crisis morass. So far, I have focused almost exclusively on events in the Western Rite of the Church. The Eastern Rites have had a very different struggle throughout this same period. Unlike the Western Rites, extremely little Eastern Catholic worship takes place in the Church and yet outside the Vatican organization. Some few Uniate priests joined (and a few others even been ordained by) the Transalpine Redemptorists, a small group founded in 1988 with the blessings of Abp. Lefebvre and Cd. Gagnon. In 2008, the Transalpine Redemptorists threw in with the modernists and were hopelessly compromised, however a similar society, again aligned with the SSPX, called the Priestly Society of Saint Josaphat, also now exists which preserves the original and authentic Ruthenian Rite traditions including the use of the Slavic language for their liturgy.

But regarding this other group of Uniate Easter Rite Catholics, Bp. Mychajlo Osidach, ordination information unknown, but believed to have been consecrated by Archbishop Volodymyr Sterniuk (ordained by Bp. Basil Vladimir Ladyka on September 21, 1931, consecrated by Bp. Vasyl Vsevolod Velychkovsky (no co-consecrators) on July 19, 1964, and deceased September 29, 1997) on September 6, 1989, assisted by Bp. Philemon Kurchaba (ordained by Bp. Mykola (Nicola) Czarneckyj (Charnetsky) on July 25, 1937, consecrated by Bp. Volodymyr

Sterniuk on February 23, 1985, and deceased October 26, 1995), consecrated Anthony Elias Dohnal (ordination details unknown but believed to have been ordained for the diocese of Litoměřice in Czechoslovakia in 1971 or 1972) on March 3, 2008, and laid hands on him as elected Byzantine Catholic Patriarch on 5 April 2011 through the election and the imposition of hands by the Synod of Bishops headed by Archbishop Mychajlo Osidach. They declared their belief that the Church is in a Sede Vacante condition, but as it stands however, some of the historical background still needs to be established, and it is not clear whether or if this Byzantine Catholic Patriarchate will be willing and able to work with other Traditional Catholics, or normalize their procedures. They have unilaterally (that is, without the cooperation of other Traditional Catholic societies) elected one Archbishop Carlo Maria Viganò to be a Pope on October, 2019 (regnal name unknown), who had formerly served in the Vatican organization as Apostolic Nuncio to the United States from 2011 to 2016, but in a post-retirement assertion of integrity criticized the Novus Ordo leader (Francis I) for his open heresies. As I write this, it remains to be seen what will come of this.

With regards to the Eastern Rites, the overlap between that part of the Vatican organization and that part of the Church had continued to be much larger, but they too became different, only more slowly. New liturgies began to be prepared in the early-to-mid 1990's but by now have been imposed practically everywhere. It has not been as yet studied as to whether the changed rites being introduced to the Eastern Rite Churches have as yet imperiled the validity of their sacraments. But validity aside, one of the main currents of liturgical change has been to remove those details that show unity with the Catholic Church in differing from schismatic East Orthodox practice, details sometimes called "Latinizations" but in fact the preservation of such things as the filioque part of the Creed (by which the Holy Ghost is professed to proceed from the Son as well as the Father, an important component of the Creed rejected by most Eastern schismatics), and the sinless conception and assumption of our Lady. Presumably there may still remain some Eastern Rite worship that remains truly and specifically Catholic, but serious abuses and compromises to the integrity of the authentic Eastern traditions are becoming more frequent.

As a result of this, the Eastern Rites have largely ceased to be the safe haven they had previously been. And there are other pressures laid upon them, though of course always with the possibility of some faithful priest (or even bishop?) recalcitrantly opposing and even rejecting these pressures, not only to use the liturgies of the schismatic churches, but even to engage in communicatio in sacris with them, to form Kyivan (or "Kievan") churches, localized unions of putatively "Uniate" with schismatic "East Orthodox" congregations, more united to their particular place of origin, language, and customs than to either the Catholic or even Orthodox Churches for that matter.

Persecution continues to be a major problem. The liberation of Russia in 1991 brought about only a very temporary reprieve. Already, what was once the Soviet Union is once again deporting Eastern Rite Catholics to the Gulags of Siberia. Adding insult to injury, the Vatican organization uses its policies to drive Catholics directly into the schismatic East Orthodox churches, since, as they see it, the services of schismatic ministers is all that is needed in such regions as where the Eastern schismatics hold sway. In this, they display their formal renunciation of the Mark of Catholicity, such that now, the only Catholics operative as such in such regions would be those, such as the Society of Saint Josaphat, as are seen as "dissident" for not following the new fallen Vatican organization into the local Eastern schism.

As it now stands, Eastern Catholics have to disobey their Vatican leadership in order to remain Catholic. In the Declaration of Balamand, their leadership has formally relinquished their apostolic mission to the Schismatic East Orthodox, and one only demonstrates their Catholic Mark of Apostolicity by disobediently continuing to proselytize among the Schismatics. To some as yet limited extent, this necessary disobedience puts all true Eastern Rite Catholics outside the Vatican organization, though some few faithful clergy might still be tolerated for various historical and localized reasons. I would not be surprised to learn some day that the Schismatics should come to be counted as being in full union with the Vatican organization while those "rebellious" Uniates are formally kicked out of that organization on account of their desire to be loyal to Rome and the Pope.

Before long, Eastern Rite Uniate Catholics, who already suffered heroically under Communism, but who at least had been spared having to know about Modernism, in the 90's, finally encountered Modernism's left foot of fellowship. Father Stehlin (of the SSPX) reported 12 parish priests and seven Basilian sisters, who have been persecuted by their own particular leadership, on account of their refusal of any kind of ecumenism with the Orthodox, and of the "renewal" of the liturgy, with the use of the vernacular and shortened prayers, etc. In October of 1999, one of these priests was thrown out of his parish and suspended a divinis. The next Sunday, together with five other priests and 500 faithful, this priest entered the church from which he had been ejected, occupied it, celebrated the three hour long traditional liturgy, and left in a solemn procession of the Blessed Sacrament to a small chapel, where the banished priest now celebrates the traditional Byzantine liturgy. And so, with cooperation between these and other faithful Uniate clergy and the priests of the SSPX, the order known as the Transalpine Redemptorists, followed by the Society of Saint Josaphat were founded for the preservation of the authentic Eastern Rites and beliefs, and thus was the traditional community officially resumed in the East.

The stand is becoming clear, and even showing the true strength of the real Catholic Church to the Eastern schismatics. Several such clerics, seeking finally to re-unify themselves with the Church approached the Vatican asking to be received, and all were turned away, with a statement to the effect that "We prefer that you remain with the schismatic East Orthodox, since we need more of their kind to reach out to us and participate in our ecumenical agenda." Very disgusted with this, one such bishop, Yuri Yurchyk of Donetsk, Ukraine, came to realize that if he wanted to return to the Catholic Uniate Church, it is to the traditional Church he must go. But going to the SSPX would accomplish nothing since all they could do is refer him back to the Vatican. So, Bp. Yurchyk took the extraordinary step of approaching the sedevacantist Bishop Pivarunas of CMRI and MSM fame, and on October 24, 2002, he formally abjured the errors of East Orthodoxy and became reconciled to the Church. He was the first Uniate Eastern Rite sedevacantist cleric. When he returned to his own country, the persecution was turned on and he was not permitted to speak again since. No reliable information exists as to his present whereabouts and status.

In India, Fr. Blute reported of a certain Fr. Pancras Raja, of the city of Periasamipuram (meaning "Big Priest Town") who refused to perform the Novus Ordo. For this he had been forcibly retired and he lived a monastic existence, saying Mass by himself, for many years. When however, a new bishop of the (Novus Ordo) "diocese" of Tuticorin desired to put an end to a longstanding tribal feud in that town, Fr. Raja was the only person they could find who had gained the trust and admiration of all sides of the feud. However, they so hated the Catholic Mass that even though they finally agreed to make him the regular parish priest in Periasamipuram, it was with the unique and previously unheard-of proviso that he does not say Mass for anyone in his own parish. Instead, an ill-trained young pup Novus Ordo presider was sent in to keep the Novus Ordos going while Fr. Raja was supposed to deal with the individual citizenry and their quarrels and feuds. Before long, the young pup gave up and abandoned the town for "greener pastures," and after a brief period of borrowing a Novus Ordo presider from a neighboring town (which spread that man far too thin), finally consented to allow Fr. Raja to say Mass in his own parish Church for his congregation.

As the priest (presider?) from the neighboring area became less and less able to fill in, Fr. Raja's Catholic Masses became seen more and more frequently by the town's populace, and before long, attendance grew mightily. Whole youth gangs were converted, with a gang leader since becoming the lead altar boy. Attendance continued to grow, as did lines at the confessional, and the common people were even learning all their prayers in Latin! With consummate skill and diplomacy, the "Big Priest" (Fr. Raja was large in stature, beard, and legend) earned the esteem of all the police chiefs, and political leaders of the District. He put in motion the starting of a High School, so the children of the village would no longer have to go to Tuticorin where so many lost their faith. Programs are being started for a water supply, and for employment for the handicapped. Needless to say, the Novus Ordo "bishop" could tolerate none of that, but so severely outvoted by the village citizenry, there is nothing he can do but permit Fr. Raja to continue doing his job there. At one point, Fr. Raja explained to the bishop's face that, in his opinion, "the new sacraments are invalid, give no grace, and are incapable of sanctifying souls." The Bishop "was left blinking." The fruits which the traditional faith can bring forth in that place are evident: virtue and vocations abound. But in 2018, with his health fading, Fr. Raja left India and works with the SSPX in the United States. Other priests may attempt to continue in his role, but none seem strong enough to sustain a clear Catholic stance.

In China, the situation had grown truly bizarre. The Chinese Patriotic Church, a product of that nation's Communist government, and utterly beholden to it, had remained traditional in their liturgical manner despite their schismatic condition, while those Catholics of the underground had (for the most part) gone Novus Ordo. At first, they thought they were merely being obedient, and that such a different rite would also distance them from the Patriotic Church. For a season, saying a Novus Ordo was itself a kind of protest against the Chinese government, and almost enthusiastically engaged in on that basis by many who were nevertheless truly "Catholic-at-heart," while the schismatics had, ironically enough, the blessings of the authentic Catholic sacraments (but not teaching).

This was too much, even under the present confusing times, for Providence to permit. Over the course of the 1990's and into the early 2000's, the issues again became clear. First, in the early 1990's, the Chinese government finally "Novus Ordo-ised" their services. They did this, because they had come to realize what a disaster to the Catholic Faith of the people the Novus Ordo is, and so this was a technique for them to begin killing off Catholicism. Next, the new Vatican began to "recognize" the Chinese Patriotic Church bishops and clergy, even seeing to their "training" and "formation" and even, at times, their "ordination." Needless to say, the Chinese underground Catholics, finally seeing the truth of their situation, have been returning to the authentic traditional Mass of the Church, while the Novus Ordo Conciliar Church and the Patriotic Church waltz off into the sunset together, arm-in-arm. They now work together so closely that today's fallen Vatican sees no reason to continue having its own representatives in China, since those of the Chinese Patriotic Church are now considered perfectly capable of tending all Chinese souls, once again a renunciation of the Mark of Catholicity.

Finally, there is also some sad news from the years since my 2004 edition, namely the passing of so very many persons named in this account, heroes of the Faith, or at least key to the life of the Church through this difficult period, plus a few who had passed away without my knowledge prior to that edition, but this list omits those whose deaths have been given elsewhere in this volume:

Fr. Urban Snyder - 25 January, 1995
Fr. Denis Chicoine 10 August, 1995
Fr. Malachi Martin - 27 July, 1999
Fr. John Keane - 28 February, 2001
Fr. Paul Wickens - 08 July, 2004
Michael Davies - 25 September, 2004
Fr. Grommar DePauw - 06 May, 2005
Fr. Carl Pulvermacher - 29 May, 2006
Fr. Dr. Rama Coomaraswamy - 19 July, 2006
Bp. Pavol Hnilica - 08 October, 2006
Fr. James Francis Wathen - 07 November, 2006
Fr. Lawrence Brey - 26 November, 2006
Fr. Roy Randolph - ?? January, 2008
Bp. Roberto Martinez - ??, 2008
Bp. Lopez-Gaston - 05 May, 2009
Abbe de Nantes - 15 February, 2010
Bp. John Bosco มนัส จวบสมัย (Manat Chuabsamai) - 20 October, 2011
Fr. Paul Schoonbroodt - 26 May, 2012
Bp. Louis Vezelis - 01 January, 2013
Archbishop Mychajlo Osidac – 21 February, 2013
Jean Arfel (also known as Jean Madiran, and Jean-Louis Lagor) - 31 July, 2013
Bp. Oliver Oravec - 09 July, 2014
Fr. Nicholas Gruner - 29 April, 2015
Bp. Robert Fidelis McKenna - 16 December, 2015
Fr. Gabriele Amorth - 16 September, 2016
Bp. John Hesson - 26 August, 2017
Rev. Terence R. Fulham - 19 November, 2017
Bp. Frank Slupski - 14 May, 2018

If any of these heroic and faithful souls failed to enter Heaven directly so as to have already joined that great cloud of witnesses over our head (Hebrews 12:1) who intercede for the restoration of all things in Christ and the Triumph of the Immaculate Heart of Mary, it could only be for secret sins unevidenced by anything known to anyone in their lives. We appeal to those who have entered Heaven to pray for us, and for any as might be yet awaiting any purgation, we intercede with our prayers and sacrifices for their speedy release into Heaven. And we ask of all of these to intercede for the restoration of all things in Christ, and for a real, Catholic Pope and the Voice of Peter to be returned to the Church.

Now that the Traditional Catholic community has grown so strong and aware of itself, it should be ready to cope with the many changes that are bound to come in the years to come. In the original edition of this book (2002), I concluded the penultimate chapter with the words, "Already, speculation is rife as to who will succeed him [John Paul II] and what that person will do. Will he be a conservative who is sympathetic to the Traditional Catholic community, such as Cardinal Thiandoum or Cardinal Ratzinger, or will he be a raving liberal, and if that, a consistent liberal like Bp. Sullivan or an agenda liberal, or what?" Interesting to have been able to include within a list of only a mere two names the name of Karol Wojtyła's successor! But now Joseph Ratzinger, as once Benedict XVI, is no longer in control, having been succeeded (while still alive) by Jorge Bergoglio as Francis I, an agenda liberal if ever any such existed.

Indeed, who knows what will become of the Vatican organization in the years to come? How far will it degrade, and will things come to such a clear pass that large numbers will finally abandon it for the real and Traditional Catholic Church, or will it be a long slow battle of attrition as Traditional Catholicism grows slowly and the Conciliarists shrivel up over some several centuries or more? Perhaps there may come a time when the Vatican organization is unable to agree upon anyone to lead them, and they just decide to get along without any "pope" at all! If, by such a time, the Traditional Catholics have not elected a Pope, then at least that would be a good time for Traditional Catholics to select their own pope. Then, who would be most obviously in union with the Pope? The same people who were truly in union all along.

CONCLUSION

◆

WE BELONG TO ROME, AND ROME IS OURS

The supernatural Mystery of the Incarnation is sustained until our time, and until the end of time, in the Mystery of the Church. The Church of Christ was never merely a haphazard cluster of disciples (although under Vatican II, it has come as close to being like that as ever to be possible in all history), but an institution, a hierarchical establishment meant to have an earthly leader who possesses that leading Apostolic role that Peter himself occupied during his lifetime. Although composed of sinful, fallen mankind, the Church exists as a single entity, the Mystical Body of Christ, which is incapable of moral or doctrinal error, through which all graces flow from Christ Himself in Heaven to all here on earth who receive His Graces, and outside of which there is no salvation.

The Church, being His Mystical Body, lives and exists with His Divine Life, and not merely the limited life of any other institution or establishment. His Resurrection Life is so powerful that not even death can swallow it up. It is not only He Himself personally who possesses this life, but also His Church. Those of us who have lived through the last sixty years or so (or more) have seen the events described in this volume. We saw His Mystical Body, the Church, condemned to be scourged and then driven outside the walls, displaced by the changed religion imposed in the name of aggiornamento, and then abandoned and left to die in the hands of the lynch mob back at Vatican II. We saw Him scourged and beaten with the ever-increasing disrespect the Novus Ordo shows towards real Catholics and the reduction of their society to its barest threadbare outlines. The Novus Ordo religion, not suffering at all, says, "I sit a queen and am no widow: and sorrow I shall not see," while the traditionalists say, "Father, forgive them, for they know not what they do." We saw Him dead and buried with the vacuous "excommunications" of every faithful bishop, along with many priests and their attached lay Faithful as well, and the frequent inability of most Catholics to discern the authority and unity of the traditionalist Catholic clergy.

But those who remember the Gospel account will also remember that the story does not end here.

Since the Church must always exist, the characteristics and marks of the Church also must always exist. Furthermore, their existence can never be merely as philosophical ideals, but must always be living and material realities, attached to some corporate entity. Those who mistake the Vatican organization for the Church instead of some other sort of establishment within some mere portion of which a portion of the Church "subsists" must necessarily deny that the characteristics and marks of the Church can be found anywhere today. Such ones must conclude that either the Holy Spirit has abandoned the Church, leaving it to error, or alternatively that the Church was never meant to exist for all time, or at least meant to disappear for some (hopefully brief) season before the world ends, or else they must redefine those characteristics and marks so as to deprive them of their original meaning and power as defined by the Church. "Infallibility" of the sort where "he is always right, except when he is wrong" is very useless and unreliable, as is "infallibility" of the sort where "he is always right, even when he is wrong."

Since the Vatican organization lacks the characteristics and marks of the Church, as demonstrated in Chapter 4, many Catholics have fallen rather neatly into four categories: Some came to believe that the Holy Spirit has abandoned the Church, perhaps in connection with some special period in prophecy associated with the presence of the Antichrist.

Others came to believe that those characteristics and marks were mere accidents, in that a church (such a belief structure makes the concept of a "The Church" impossible) might only happen to have these characteristics and marks if things happen to be going well at the moment. When the Vatican organization lost those characteristics and marks of the Church, these people took recourse to the Protestants, the East Orthodox, or the Old Catholics.

Others, taking the heretical home-aloner position, are claiming that this current cycle of ecclesiastical crucifixion cannot be complete until every last shred of legitimacy and validity has been eradicated. When Jesus was yet to be crucified, He spoke in no uncertain terms that He would be laying down His life and taking it up again. All the disciples needed to have done then was to have listened to Him. For the present situation, what He has said is that He shall be with us always. This implies that there shall always be an "us" for Him to be "with" "always." The Resurrection of the Church is not that it somehow magically reappears after some complete eradication but rather the Resurrection Life of Christ miraculously sustaining Her, in all of Her plenitude of teaching and ruling authority and sacramental power, throughout even such dark periods as ours that I have documented, though to a vast majority it must seem truly like a resurrection.

Still others have redefined and reinterpreted those characteristics and marks in such a fashion as to eliminate any supernatural aspect to them, so as to make it sound as if the Vatican organization might still possess them. The ways in which they rationalize this are so silly and bizarre that they would be laughable if only there weren't souls out there who take those new definitions and interpretations seriously. All four of these courses are mistaken.

When the disciples came to the tomb to finish the burial of Christ, they did not find His body there, only an angel who told them, "He is not here; He is risen! See for yourselves the place they have laid Him. He has gone ahead of you into Galilee." The Body of Christ is no longer held within that modern whitewashed sepulcher, the Novus Ordo Vatican apparatus, or "Church of the People of God." The typical Novus Ordo parish, with its false new "sacraments" and teachings, is an empty tomb without even the corpse of Christ for company. He is not to be found there at all. Instead, He has risen! He has gone to Galilee. Let us focus on the meaning of that.

In the days of the original Resurrection of Christ, the center of the Faith instituted by God was in Jerusalem. It had long been so with the Temple, the High Priests, and the Kings of the nation of Israel. Under the Christian covenant it continued there in Jerusalem until Peter took his See to Antioch, and then later to Rome. What was Galilee in all of this? What is it now? A boondock! A backwater! Nowhere important! Jerusalem had abandoned the Faith, with the Jews cooperating with the pagan Romans to crucify their Messiah, and even Peter himself running for his life, and denying his Lord. What was in Galilee? Simple, pious folk who only knew to obey God and precious little else.

What Galilee was then, Ecône is today! And not only Ecône, but Winona, Weissbad, Spokane, St. Mary's, Campos, Oyster Bay Cove, Fribourg, Acapulco, Omaha, Post Falls, Denton, and indeed every other bastion of the Traditional Catholic Faith! These places, where the true Faith is taught, and where the true sacraments of the Church are dispensed precisely where and how Christ Himself wants, are where Christ Himself resides and lives in the sacraments and activities of the Church.

For I offer here a fifth alternative to the four mistaken alternatives given above: The ontological identity with the Mystical Body of Christ has been formally and relinquished by the Vatican organization, and with that, the characteristics and marks of the Church remain only with all faithful priests and bishops who hold and teach the Catholic Faith and dispense Catholic sacraments, regardless of what relationship they might have, if any, to the Vatican organization. I claim here and now that the characteristics and marks of the Roman Catholic Church exist and live, in their fullness, throughout the entire Traditional Catholic community!

Let us see how the Traditional Catholic community proves itself to be truly and completely identical to the Roman Catholic Church:

ONE: Oneness may seem to be the most difficult mark of the Church to see in this troubled time. Yet the Oneness is clearly and strongly there for those who take the time to look carefully. Remember that Unity has never meant that all Catholics must be "buddy-buddy" pals with one another. Indeed, there have been many bitter controversies throughout the history of the Church, even between saints. Did the Oneness of the Church vanish during the First Great Western Schism in the fourteenth century when there were two, and then three claimants to the Papacy? No! How was the Oneness expressed then? In their unity of Faith, Morals, and Worship. They all believed the same things; they all lived by the same rules; they all worshipped the same God using the same Mass. They all believed that the entire Church was meant to have only one visible head, the Pope. As Msgr. G. Van Noort writes of the episode in Church history, "hierarchical unity was only materially, not formally, interrupted." So it is today.

All Catholics were trying to be obedient to the Supreme Pontiff. The only problem was one of "Which one is he?" Even then, there were not two or three Catholic Churches since they all came from one and returned to one. More importantly, those, such as St. Vincent Ferrer who mistakenly adhered to the wrong "pope," were not outside the Church. So it is today. Those who reach the wrong conclusion regarding whoever is then the current leadership of the Vatican organization or any other claimant are also not outside the Church. Indeed, the situation back then very closely approximates what we have today. During that First Great Western Schism, when a pope in Rome would die, their priests said Mass as if they had no pope, even though a "pope" was reigning in Avignon, and likewise when a "pope" in Avignon would die, their priests said their Mass as if they had no pope, even though a pope was reigning in Rome. How very like today, when some priests say Mass in union with whomever they count as pope, while others say Mass as if there is no pope.

Fortunately, it is Christ Himself, in His Mystical Body, who is taking action, not us ordinary believers. We may not know all of what He is planning, and how it will all work out, but we make ourselves a part of His plans by adhering to our Traditional Catholic Faith, Sacraments, and congregations. All Traditional Catholics today are similarly united in their Faith, their Morals, and their Latin or other authentic traditional worship. Priests from the FSSP, the ICR, the SSPX, the SSPV, the priests loosely associated with Bp. Dolan, the CMRI, the Sociedad Sacerdotal Trento ("Trento Priests"), and all other "independent" Catholic priests, all teach from the same Catechism, and say the same Mass.

One important thing which all sides of the current question have in common is the Universal and Historical Magisterium of the Church on which the various disputants all base their arguments. Such a shared basis for belief is essential since without it, their arguments would become like the example where one person states, "The Bible teaches that…" and the other person yells back, "It doesn't matter what the Bible says…" and after that it would degenerate into a shouting match. But not here! No matter how much Traditional Catholic priests may disagree with each other as to how to interpret or respond to the present crisis, they all agree with each other that the Magisterium of the Church must be consulted as authoritative, and agreed upon everything that the Church has authoritatively taught. It is there that they all look for principles and precedents.

Above and beyond that, all Traditional Catholics are bound together in a unity of purpose, namely the restoration of the Catholic Church. What divides them is not the goal, but simply the means to attain that same goal. Those differences are merely the result of different ways the current crisis has been understood as we have come to do by using our limited human minds. We presently lack a formal and official approved answer to these questions which is morally binding on all.

Certainly, there is at work the Satanic strategy of "divide and conquer" which definitely has a role in these painful divisions, but one must remember that what Satan (or wicked men) mean for evil, God means for good. The present crisis places the faithful remnant in a quite bizarre and unusual position. The Church needs to be several things all at once, some of which are mutually exclusive and seemingly contradictory.

On the one hand, in the hopes of rehabilitating the Vatican establishment, there have been those who have encouraged any inclination on their part to allow or better yet promote the true Catholic Faith. **Since the current Vatican leadership refuses to show the nobility of the true shepherd by teaching the truth in season and out of season, then let them at least show the practical wisdom of the hireling by teaching the true religion which they already find much more profitable!** Two competing factions cooperate in an attempt to bring this about: the "Motu Proprio" priests (those using the "extraordinary form usage") and their regular parishioners who serve as a beachhead of real Catholicism on the shores of the Vatican organization, and the SSPX which serves as the surgeon who by virtue of not being quite within their "jurisdiction" yet faithful in support of them is therefore in a position to perform "open-heart surgery" on the latter as an unconscious patient. It is widely conceded by those who enjoy the Extraordinary form that had not the SSPX done its job, there would be no Latin Rite Catholic worship anywhere within the Vatican organization.

But on the other hand, the Church must be prepared to continue in the virtually certain event that the Vatican establishment never gets rehabilitated, and for that reason there are sedevacantists and the SSPX. They both serve the function of providing self-enclosed bastions of the Catholic Faith which are totally beyond the reach of the apostate leaders of the Vatican society. These bastions serve as a "base of operations" from which the Church raises Her children strong in the Faith and also launches Her missionary forays. They also provide the Catholic Faith and Sacraments which would otherwise be totally unavailable to those living in unfriendly dioceses which have no "extraordinary form." The sedevacantists also accept the duty of investigating the nature of the present crisis from the standpoint of Catholic Theology, doctrine and dogma.

Where the Novus Ordo religion is filled with people who feel free to disobey their pope, even in the event they were to get anyone clearly and universally recognizable as such, Traditional Catholics all recognize their Catholic duty of being obedient to the Supreme Pontiff, and that, in these times, is what constitutes one's attachment to the Barque of Peter. On that, one can solidly base the claim that unity can and must only continue to grow and be strengthened among the Traditional Catholics as the various pockets of the Church, left behind in the backwash of that catastrophic tidal wave, Vatican II, find each other and overcome their fear and distrust of each other. Once the flood waters recede, it can only become clear that what now seem to be scattered islands are in fact the tops of mountains on the same continent.

As one can therefore readily see, God has permitted these seeming divisions in order to place his servants in a rich variety of strategic positions so as to bring about the termination of the present crisis. Already, despite the occasional "potshots" one traditional group may yet hurl at another, there is present a very strong bond of unity among all Catholics of the Traditional Catholic community, especially lay Catholics of the various groups who have cooperated in such things as Rosary marches and Anti-Abortion demonstrations. There is a tremendous crescendo of cries for the traditional priests and bishops to stop their bickering. It is my deepest and fondest hope that even this volume may, in its own small way, contribute towards this unity by providing its readers with a sense of proportion, perspective, and the true and hope-giving Vision of "Where the Church Is" today.

HOLY: Holiness is the one Mark which many find to be the very easiest to see. The traditional priest approaches the altar of God with a reverential awe appropriate for so great and humbling of an honor. He wears the liturgical vestments which all priests used to wear, and he goes on to preach a homily so good and solid and based on the readings of the day and all of the church's teaching that were such homilies to be gathered up into a book and published in the "good old days" before Vatican II, such a book would have had absolutely no trouble getting an imprimatur. The holy example of the priests and bishops of the Traditional Catholic community is amply echoed in the lives of their parishioners, in whose lives all of Catholic Morals are lived and taken seriously.

Where else can one enter a Catholic church half an hour before Mass, and find the church to be completely silent and the pews full of families—men, women, little children, teenagers—all at prayer, and a line outside the confessional? Where else can one so favorably compare the excellent conduct and piety of the young men honored to serve at the Holy Altar against that of so many of their schoolmates and neighbors? Where else can you see a generation of young people so set apart as the Traditional Catholic youth of today, not in gangs, not living sinful lives, but trying their best to please God? Where else do you see marriages that hold together despite sickness and hard times, marriages that are truly sacramental, in which mutual sacrifice and not concupiscence is the goal? Where else do you find those who, having been away from the Church for nearly 50 years, on discovering that the Traditional Faith is still practiced, come back, make a good Confession and return to the practice of the Faith as known in their youth? Within and throughout the Traditional Catholic community, **and nowhere else**! It is only there that these holy and exclusively Catholic fruits of the Spirit are to be found, and that in rich abundance!

CATHOLIC: Traditional Catholics are universally all supremely confident that their belief is meant for all persons. When the temporal order is done away with, all persons will believe what all those of the Traditional Catholic community believe now, and it is their desire that all persons come into that knowledge now while they still have a chance to act on it. It is interesting to note that all of the schismatic dissenting groups have some secular power which created them. The Eastern heretical groups such as the Arians or the Monophysites, all became heretical under the pressure of secular rulers who wanted the Church to serve their state. When the Eastern Orthodox broke off, it was under the same sort of pressure, but exerted by secular rulers whose theology was closer to orthodox. The Protestant breakaway was sponsored by the Germans, certain Swiss leaders, and the King of England. The Dutch originally persuaded the Old Catholics to break off and center their new church in Utrecht. The Chinese government, under "Chairman Mao," created the Patriotic Chinese Church. Certain German, American, and Dutch bishops, priests, and conciliar "periti," under the influence of their Protestant, Modernist, Sillonist, Communist, and Americanist puppet-masters invented the Novus Ordo religion and their new mode of worship.

In contrast to that the Traditional Catholic community exists without even so much as the sympathy of any secular power. The traditional priests and bishops and those religious and lay who follow them have all taken their stand solely and strictly on theological and doctrinal grounds. Bishop Kelly did not become a sedevacantist because the Governor of New York wants to start a "Church of New York!" and neither, for that matter, did Bishop Vezelis. On the other hand, Paul VI did install another Archbishop in Hungary while the proper Archbishop, Mindszenty, was still there in prison, so as to set up a Hungarian vernacular-mass-saying national Church which, unlike anyone set up by any reliable pope, had no trouble playing ball with the Communist government.

The Traditional Catholic community adheres to the entirety of the Magisterium of the Church, as taught throughout all ages and in all places. That certain recent Vatican proclamations may be ignored or denied is strictly based on the fact that those particular recent proclamations are strictly disciplinary, and furthermore flatly contradict Faith or Morals, as promulgated by all reliable popes who have ever discussed the issues ever since the particular questions came up. Where the current Vatican leadership speaks in union with the Magisterium of the Church, the Traditional Catholics offer no criticism, but also refuse to quote the recent proclamations in an authoritative manner. Why quote Paul VI's rare fine words in *Humanae Vitae* when we already have the same moral teaching far better expressed (both more precisely and more tastefully) in Pius XI's *Casti Connubii*?

The atmosphere at any Traditional Catholic chapel, parish, or Mass center is so markedly Catholic that those who remember the Pre-Vatican II Catholic Church all unanimously recognize that uniquely Catholic atmosphere the minute they walk through the door. All the details are there, right down to blessed Holy water fonts and chapel veils for the ladies. Yes, there have been imitation Catholic Churches, but their non-Catholic status shows itself readily enough, as in the case of a "High" Episcopal Church where the worship was nearly Catholic, far more so in fact than in Fr. Bozo's nearby parish, and yet posters could be found in the back recommending the services of Planned Parenthood to pregnant girls. Those kinds of betrayals of the Faith are never found anywhere within the Traditional Catholic community! The traditionalists are Catholic through and through.

Catholic is what those in the Traditional Catholic community call themselves, because Catholic is what they are. Not just any Catholics, like Old Catholics or Anglo Catholics, for example, but **Roman** Catholics, **Homoousian** Catholics (as distinct from the Arian heretics, or Homoiousian Catholics, a case in which a single Greek iota made such a difference). Where necessary, we make such a distinction today, **Traditional** Catholics (and no less "Roman" and "Homoouisian" as well), so as not to be mistaken for the Modernist pseudo-Catholics of the Novus Ordo Church of the People of God. If Traditional Catholics are sometimes somewhat cautious in their use of the word Roman, or Rome, it is simply because they adhere only to Eternal Rome, the Rome of Peter himself, of Clement, Leo, Gregory, and Pius, but not to at least the dysfunctional aspects of the Modernist infested "Rome" of the current Vatican establishment. Traditional Catholics seek to bring all persons into the Church, Jew or Gentile, rich or poor, black or white, male or female, with no distinction. Only the Traditional Catholic community truly qualifies as Apostolic.

APOSTOLIC: If ever there was a Church which is primarily concerned with maintaining the apostolic faith, a valid apostolic succession, and the apostolic mission to reach the entire world, it is the Traditional Catholic community! The whole point of all the consecrations of bishops described in this work is the continuation of the Apostolic Succession. If the present crisis continues much longer (and as the Eastern Rites also get destroyed) the next reliable Pope, as Bishop of Rome, will have to trace his Episcopacy to Thục, Lefebvre, de Castro Mayer, Mendez, and/or any similar Eastern Rite bishop as may yet arise, in order to be a truly Apostolic and validly consecrated bishop.

It is the Traditional Catholic community alone whose existence is smoothly continuous with the Catholic Church throughout the ages. There are some Traditional Catholics who have gone clear through this crisis without ever missing the Traditional, or "Tridentine" Mass, never attending a Novus Ordo "Mass" or having anything to do with the synthetic new Church of the People of God. In many cases their stand cost them a great deal of persecution, loss of friends, and severe inconveniences as they often traveled long distances or even took inferior employment in far off towns in order to continue their Catholic worship in what few places it was offered during the lean times.

The people in the Traditional Catholic community all believe that the Catholic religion is the truth and will not tolerate any treatment of other religions as though they were just as true. You will never see Traditional Catholics going to Assisi to worship demons with or alongside those who don't know God (1 Corinthians 10:14–22) no matter what manner of idolatry the Vatican leadership is willing to indulge in. You will never see Traditional Catholics going along with the recent directive not to evangelize the schismatic East Orthodox.

THE CHURCH IS THE MYSTICAL BODY OF CHRIST: The Mystical Body of Christ has in common with the actual physical Body of Christ the characteristic of the resurrection life. It cannot be killed. It may be injured at times, but it always grows back while the parts cut off simply wither up and die unless rejoined. The different parts of the Traditional Catholic community work together to bring back the full size and structure of the Roman Catholic Church which Vatican II has destroyed. This is true membership, as different parts of a body of various shapes, sizes, and functions are truly members of the same body. Their coordination may not be consciously planned by those involved (in most cases), but neither is it accidental. It is God Himself who has shaped the pieces of the Church so that they can one day be fitted together like pieces of a jigsaw puzzle. Man divides, but God unites, and it is God's own authority and power which is at work among the Traditional Catholic community.

THE POPE IS INFALLIBLE: Only traditionalist Catholics and conservative Novus Ordo believers believe in this principle. And conservative Novus Ordo believers have unwittingly redefined Papal Infallibility to a definition which is at odds with the definition given within the Magisterium of the Church. Like Traditional Catholics, conservative Novus Ordo believers believe that Papal Infallibility means that the teachings of the Pope (on Faith and Morals) are authoritative and must be obeyed. So far, so good.

But Papal Infallibility does not merely mean that the Pope speaks authoritatively; there is another aspect to Infallibility. What the Pope teaches on Faith and Morals is not merely to be believed and obeyed, it is **Truth**! There is an objective reality "out there" which all Infallible Papal pronouncements are in full accord with, an Eternal and Unchanging Reality. A conservative Novus Ordo believer must believe that there is no such objective reality. To them, as long as Pope X teaches that **A** is true, then **A** is true and must be believed, but when Pope Y comes along and teaches that **A** is false, then **A** is false and must be denied. There is no further reality for the conservative than the current "Pope's words."

A Traditional Catholic believes that if Pope X teaches that **A** is true, then not only is **A** true and must be believed, but that there is an Eternal Reality "out there" for which (or Whom) **A** is really true. It is therefore not the prerogative of any later pope to come along and teach that **A** is false. If Pope Y comes along and teaches that **A** is false, then either Pope Y is speaking only as a private theologian and has not engaged his Infallibility or else "Pope Y" is not really a pope. That is in line with the definition of the Universal and Historic Magisterium of the Church. Traditional Catholics are therefore the only Catholics who really believe in Papal Infallibility as taught by the Church explicitly at Vatican I and implicitly throughout Her history.

Even the absence of a pope does not change this principle. The Charism of Infallibility continues to rest among the universal and ordinary magisterium of the Church's real and truly faithful bishops, those adhering to Sacred Tradition, whenever the Church is between popes. The Charism is fully manifested in a bishop who is given Peter's monarchical authority and jurisdiction over the entire Church and who understands and accepts the responsibilities of that office.

THE CHURCH IS INDEFECTIBLE: The faithfulness and stability of the Traditional Catholic parishes, especially when set in stark contrast with the inherent instability and constant upset of the Novus Ordo Church of the People of God, is one of the most stunning and dramatic evidences that the Traditional Catholic community is the lawful recipient of this Charism of the Church. Despite seeming dangers from every side, the Traditional Catholic community has marched directly ahead in a straight line while the Novus Ordo paddles around erratically with one oar out of the water. The Roman Schism of 1964 so very directly parallels the English Schism of 1534. In each case one side marched straight ahead unchanged (even though a lot smaller) while the other side descended rather rapidly into changed and abused liturgies, frequently invalid sacraments, religious error, confusion, and apathy.

Even the questionable, or even asinine, behavior of a few Traditional Catholic leaders, has not in any way weakened the doctrine or morals of the Traditional Catholic community. It has enjoyed the same Divine Intervention of Indefectibility which protected the Church during the times of such wicked and corrupt popes as Alexander VI. Mark my words: The sedevacantists will **NOT** become addicted to not having a pope, nor will they be taken over by extremist political or religious factions; the "Motu Proprio extraordinary form" crowd will **NOT** be sucked into the Modernism of the Modernist Church; the SSPX **WILL** maintain its delicate balance between the "Motu" crowd and the sedevacantists until there is a clear que for anything different to happen; all three groups, and any other Traditional Catholic, **WILL** maintain their total adherence to the entirety of the Universal and Historic Magisterium of the Church.

THE GATES OF HELL SHALL NOT PREVAIL AGAINST IT: Clearly, the Power of God (and nothing else) has sustained the Traditional Catholic community through this difficult and confusing time. The small size it seems to have is mostly illusory. As in the time of Prophet Elisha, the mountains are full of horses and chariots of fire, and as in the time of Judge Gideon it may seem that the Lord's army is reduced to 300 men, but those 300 men are all the Lord needs to defeat the Devil's 10,000 men. The Traditional Catholic community may seem small in numbers, but it is largely composed of spiritual warriors very much like Gideon's 300 men, and the Heavenly help the Church always receives rests upon it.

In so many ways, the Devil has tried to shut it down, with false suspensions and false excommunications, with confusion, division and discord, and even with a few personally corrupt leaders. The only result is that all parts of the Traditional Catholic community grow stronger in numbers and holiness. We are in the fire of persecution being refined like choice gold! The attacks of the Devil continue, and so does the Church. Despite the enormous blow the Church has received at Vatican II the Gates of Hell have not prevailed against it, nor can they ever!

There is a good reason that "Traditional" can be quite properly added to distinguish real Catholics from the false kind, precisely as did "Homoousian," and "Roman," which we traditionalists one and all are as well. And that is the manner in which Sacred Tradition is so intrinsic and interwoven into each of the Marks and attributes of the Church. Fr. Anthony F. Alexander wrote in his book, *College Apologetics*, "Nowhere in the Gospels is it recorded that Christ gave his Church a technical or proper name. And it is easy to surmise why He did not. There is nothing about a proper name which prevents its from being appropriated or copied by a false church. Instead of a name He gave it four marks. *A mark is defined as a fixed characteristic given by Christ to His Church as a whole to serve as a means of easily distinguishing it from unauthorized agencies which attempt to usurp its prerogatives.*" (*italics* his) Let us examine the role of Tradition in each mark:

Tradition means the passing on to future generations exactly that which was passed on to us, no addition, subtraction, or modification. The mark of Unity cannot exist unless the full content of the faith "delivered once for all time to the saints" (Jude 3) is indeed so passed along from the very beginning. If today's faith be different than that originally delivered to the saints, then there is no Unity between then and now. The doctrine as originally given by Christ who is perfectly holy is therefore also perfectly holy; any deviation, any failure to uphold it in its entirety, is to defile it with imperfection, unholiness. To be Catholic, the one Church, with one belief, has to be all over the world, and all through time from the Apostolic beginning through now, and clear to the end; to have different beliefs in different places or times, to deprive any place or time of the one original and Traditional Gospel, is to lack Catholicity. And one cannot be Apostolic unless one holds to the doctrine of the Apostles, a doctrine is either kept or deserted, but never legitimately modified.

The same goes for the other attributes and characteristics of the Church. To be about pushing any doctrine other than that originally given is to be adulterated in purpose, a defected church. For a Pope to lead the Church into any other doctrine is to prove fallible and defeat the purpose of Peter's authority; such is guaranteed not to happen; wherever it does happen, there antecedently and a priori is no real papal action. And how can a system of doctrines be authoritative if they are ever in a state of flux? An agency might well have the power to enforce a uniformity of thought about whatever they choose among a given group of people, but what is the point of learning a thing to be "true" when one can already sense that another day it may well be "false"? How can the Mystical Body of Christ enforce anything but the doctrine of Christ? And how can any agency that enforces anything other than the doctrine of Christ be His Mystical Body? Tradition runs throughout the warp and woof of all the marks and attributes of the Church. One cannot bear any of the marks and attributes of the Church unless the fullness of the Tradition is also upheld. It is therefore reasonable and proper that that

Church, once differentiated from false churches by being additionally described as "Homoousian" and "Roman" in various past epochs should be similarly described as "Traditional" today.

The survival of the Church as the Traditional Catholic community is a great and ongoing miracle which all can see for themselves. The marks and characteristics of the Church are Divine realities which **CANNOT** be destroyed or removed from the earth. By all human standards and knowledge, the humble sheep of the fold should have all been successfully deceived into embracing the Novus Ordo religion. They weren't! The principle at work is the one preached by Christ when He said that "the Sheep know His voice **AND ANOTHER THEY WILL NOT FOLLOW**!" There is no precedent quite comparable to the existence of the Traditional Catholic community even after so great a Death (or Separation) of the Church as that mandated at Vatican II, except for the resurrection of Jesus Christ Himself after His Death on the Cross mandated by Pontius Pilate.

So there you have it all before you, dear reader: Conclusive proof that "**The Traditional Catholic community IS the Visible Unity of the Roman Catholic Church today!**" I have made my proof. I have shown precisely where the formerly hierarchical personnel of the once-Catholic Vatican organization states that is has formally relinquished its exclusive claim of jurisdiction over the Roman Catholic Church, and to that extent, their hold on the historic Sees of the Church. I have documented the logical, and (at least in retrospect) predictable consequences of that schismatic event. I have demonstrated by formal test the fact that the Traditional Catholic community has all four Marks of the Church, and indeed all other characteristics of the Church as well, in rich abundance running throughout its entire length, depth, breadth, width, and extent, and furthermore that the Novus Ordo Church of the People of God entirely **lacks** all of those Marks and characteristics.

I have chronicled the Resurrection of the Roman Catholic Church in the form of the Traditional Catholic community, holding back nothing, not even the faults and foibles of the sometimes weak and imperfect men who have led the Church through this crisis. **The burden of proof now rests on those who would attempt to prove me wrong**, who would attempt to prove that the Novus Ordo religion, or any other group that is not obviously Catholic as verifiable through their open and public profession of the Catholic Faith in their creeds, their liturgy, or their doctrines, could somehow still be the religion of Popes Peter, Linus, Anacletus, Clement, Leo the Great, Gregory the Great, Pius V, Pius X, and all other reliable popes clear up to Pius XII!

It is reasonable to expect that there will be some who will attempt a refutation of this book. I divide such attempts into two basic categories: honest, and dishonest. As to those which are dishonest, their "case" will invariably depend on misquoting their sources, quoting them out of context, and/or faulty logic. While such an attempt may be convincing to the uninformed, the person making it shall lose all credibility as I demonstrate their culpable misrepresentation of their sources, their deceptive pseudo-logic, or else the uncatholicity of their basic outlook or premises. On the other hand, I most earnestly hope that many will attempt an honest refutation, for it is in their failure to do so that they shall have a "Road to Damascus" experience and it is from the ranks of such the Apostle Paul himself was drawn.

Many of the more likely and predictable objections some might raise are discussed and addressed in Appendix B of this book, Questions and Objections. This includes not only questions from the Novus Ordo standpoint, but also some questions those of certain Traditional Catholic orders might have regarding the legitimacy of certain other Catholic orders. Those with more detailed theological or ecclesiological questions should read my *Sede Vacante!* volumes, for further information. Finally, I am available for public lectures or debates on the topic of this book.

With all that said, the sincere Catholic should now know what to do. There is no salvation outside the Church. The exact boundaries of the Church have been defined in detail here. The sincere Catholic should readily enough be able to ascertain whether they are in the Church or outside it. Is your Mass a Tridentine Mass? Alternatively, are you an Eastern Rite Catholic attending your Eastern Rite, and furthermore one which you can discern to be faithful from your long familiarity with your particular Rite, together with a lack of any compromise with the Novus Ordo or East Orthodox? If neither of those is true, you are outside the Church. That does not mean that you are a bad person, but merely that you have been in the same boat as all those other fine people who attend non-Catholic Churches. If you are in the Novus Ordo, God will forgive your having been led away from Him by the fallen Vatican organization, but now that you know the truth God will hold you responsible for what you do with that knowledge. You must follow the traditional integrity wherever it leads, without qualification, with no reservations whatsoever. What it leads to is that which is truly the visible unity of the Traditional Roman Catholic Church, outside of which there is no salvation.

The next logical step at this point is to locate a Traditional Catholic Mass near you and begin attending it. If you are uncertain of the location of any Tridentine Masses near you, one of the references I strongly recommend is a directory which gives the location of all such Masses within the United States and Canada, the *Official Catholic Directory of Traditional Latin Masses* (See Bibliography or On-Line). This same reference also contains the names and addresses of persons to contact for Mass sites in other countries. If none of these are available and if you also know of no other "independent" Catholic priest or bishop near you, there may also be the last recourse of an Eastern Catholic Church, if there exists one nearby which can be seen to be faithful. Plus, traditional priests may visit locations not listed in the directory, on a less frequent basis, and they would also gladly provide you with the Mass and Sacraments, even if only once every several months on a weekday and in a private home.

Chances are you will be surprised just how near you might find Catholic worship. You may even find you have the option of two or more different Mass locations from two or more different orders! If on a given Sunday there is a traditional Mass you can travel to in one hour or less (regardless of whether you are so poor that the hour would be spent walking, or so rich that the hour would be spent flying in your own private jet), attendance at such a Mass is necessary and sufficient to fulfill your Sunday obligation. If no such worship can be found within an hour's traveling distance, you are excused from the Sunday obligation, the presence of Novus Ordo or other schismatic worship notwithstanding.

However, many Traditional Catholics receive much Grace from traveling a greater distance even though they know their situation excuses them from their Sunday obligation. Some travel as much as six or seven hours each way on a Sunday just to go to Mass. Others consult their traditional Roman Missal for the prayers and readings of the Mass of the day and pray them in spiritual union with all Catholic Masses they know are taking place in other parts of the world. Most Roman Missals include a prayer for those who are not in a position to receive communion at that time. That prayer, prayed at the appropriate point of one's such prayers offered privately in union with the Church and the Mass, is what is called a spiritual communion, the closest one can come to receiving Communion when physically separated from it.

However, I would doubt the sincerity of anyone who simply used the points brought out here as an excuse to avoid going to Church. Such ones might as well at least continue their attendance at a non-Catholic church, if only from the standpoint of maintaining the useful habit of going to Church every Sunday, but that is not Church and cannot fulfill a Sunday obligation. Indeed, one's presence at such a service must be entirely passive, if to be tolerable at all. Those who live between one and two hours away from the nearest Catholic worship ought to consider going twice a month or so, and those with Catholic worship not too much beyond that should go about once a month, or at least several times a year.

Attendance at your first Catholic Mass in a long time (if ever) will be a deeply moving experience. For those old enough to remember the Church from before "the changes" it will be a trip down memory lane as they say to themselves, "Oh that's right! I forgot about that. Yes, now I remember that, and that, and that! I never thought I'd see this again; it's just like coming home." For those who never saw the real Catholic Church before, it is a time of wonder, as the beauty and majesty of the ritual touches their spirit in places they never knew they had. It is like a person having lived all their life in darkness suddenly being put in the light for the first time and only then discovering they have eyes. Attendance at single Mass offered at any Traditional Catholic chapel will do more to demonstrate the truth of its being an official parish of the Catholic Church far better than all this book and all the words and logic anyone and everyone could offer. If it is really all that remote from your location, **make the pilgrimage**.

Having once again (or now for the first time) seen such beauty, one of the most frustrating and painful aspects of returning to the true Catholic Faith is inviting all of your Novus Ordo friends and family, and then having them decide not to come, or even having to leave them behind. Certainly, you should invite them along. Sometimes, they might object because the Novus Ordo villains have been lying to them about the traditional Faith. I have even seen some refuse to attend a Modernist-approved Mass for fear of being "schismatic."

If their refusal to come is this fear and not merely a matter of convenience for their schedule or some other practical reason, may I make a suggestion? Give the person a copy of this book. (If this book is too large, one of the smaller recommended books listed in the Bibliography, such as the *Open Letter to Confused Catholics, The Ottaviani intervention: Short Critical Study of the New Order of Mass*, or *The Problems With the New Mass* may be used instead.) As you give them the book tell them, "Take this book and read it. I don't care whether you agree with it or not, just make sure you read it so that you can discuss its contents intelligently. Then come back to me and we'll talk about it."

That is also a good approach to use if harassed by a Novus Ordo "priest" or "bishop" or another leader who wants to pressure you into remaining in the false Novus Ordo religion. In that case I would strictly recommend only this book. Chances are they won't read it. Instinctively they sense that with this knowledge comes responsibilities and sacrifices they are not willing to make. However, so long as they don't read it you have the perfect rejoinder to their harassment: "Have you read that book yet?"

While attendance at the Mass is obligatory, or at least a valuable and Grace-filled experience for those who go even though excused from it on account of distance, a faithful Catholic should also belong to a Traditional Catholic parish. Other than the few hundred Modernist-approved "extraordinary form" parishes – as distinct from merely a Mass – (and maybe some few among the Eastern Rite parishes, proceed with caution), all other Catholic parishes (and the best and safest) are those operated by priests of the Church who are outside the Vatican organization. Oftentimes, Modernist-approved "extraordinary form" Masses which might be perfectly fine in and of themselves, are part and parcel of a Novus Ordo parish. While a Catholic might reasonably attend such a Mass to fulfill a Sunday obligation if traveling or particularly convenient, their membership can only properly belong to a Catholic parish. If one lives a long distance from the nearest Catholic parish, but can find a Mass location without a parish a great deal closer, they should become and count themselves as a member of that far away parish and try to go there at least once or twice a year, and probably more frequently than that if possible. On other Sundays and Holy Days of obligation it is enough to attend the closer Mass. Do not be concerned with whether or not a Traditional Catholic parish community identifies itself as a parish; so long as it functions like one, full and integral, it should be so counted.

One's primary financial support of the Church should go to their traditional Catholic parish. Any other Mass location they attend should be nominally supported with a few dollars dropped anonymously in the collection. By giving to these "independent" priests, that is how one gives to the Church, which is the lawful recipient of all tithes and offerings. Of course, such giving does not relieve anyone of their other charitable obligations or commitments. Many traditional priests operate on a shoestring budget, often traveling hundreds or even thousands of miles each Sunday to say Mass in widely separated regions. You should regard these parish priests as your regular confessors, and it is to them you would have to go for the other priestly functions such as Baptisms, Confirmations, weddings, funerals, and for burial. Giving to them is a tremendous honor, as it is giving to the Church in Her most desperate hour of need. It is like buying a stock when nobody wants it, and which you can utterly know will rise to the top, the spiritual equivalent of buying IBM stock in 1938.

While a Traditional Catholic should properly belong to one parish, I strongly encourage lay Catholics to pay occasional visits to parishes and Mass centers other than their own, especially while traveling. Though priests are often bound by their parish and other commitments to serve in their particular order, laypeople are morally free visit all different kinds of Traditional Catholic parishes and Mass centers, clear from sedevacantist through SSPX and other "independent" priests to "extraordinary form" and back again. There is no more vivid method of seeing the fundamental Oneness of the Church than to view firsthand the startling uniformity of worship, belief, and *atmosphere*, that exists throughout all varieties of traditional parishes and Mass centers. Such consistency and unity is like a breath of fresh air after seeing so many Novus Ordo "parishes," each of which is practically a "religion unto itself."

It is true you may often encounter traditional priests who will talk as if their order is the only truly safe or "Catholic" order. The proper response is to keep in mind that when they do this they are talking not from certain knowledge (which will not exist until the next reliable pope pronounces on the question) but from his own opinion or that of his bishop or priestly order. The uncertainties of the present crisis coupled with his responsibility as a pastor of souls may also cause him to feel that he is not competent to recommend any congregation but his own. All I can ask is that, while taking such particular statements with a grain of salt, do try to be patient with him on this. If he can't handle your occasional attendance at another Catholic Mass, there is no need to rub his nose in the fact that you have attended it and found it to be utterly Catholic.

When you go to the Traditional Catholic Mass, please keep in mind that a Catholic Church is a quiet and decorous place. The Blessed Sacrament is there, and respect should always be shown. Talking, other than to join in the prayer of the rosary or other vocal prayers recited before and after the Mass, or to whisper such things as "Excuse me," or "Mind if I sit here?" is simply not done. There will usually be some other area nearby, outside, where the parishioners can gather and talk before and after the Mass.

Traditional Catholics dress modestly and conservatively to Church, not at all like Novus Ordo People of God. Women's skirts or dresses should be long enough to cover the knees while seated, and their dress should not show any shoulders, backs, or fronts. Slacks and pantsuits (and jeans) are not appropriate clothing for women in Church. Men's clothing should also be nice and modest, along the lines of suit and tie, although the suit (or coat) and tie themselves may be omitted in hot weather. Shorts, tank-tops, athletic clothing, and jeans are not appropriate clothing for men in Church. Those living in other cultures with different standards of Sunday dress should adapt that advice accordingly. Women are advised to wear a hat or bring a mantilla to cover their heads, although many Traditional Catholic parishes will have mantillas available by the door which the women may borrow for Mass. I say this, not to establish some sort of dress code for Catholics, but so that first-time visitors would not have to endure the embarrassment of showing up inappropriately dressed.

Finally, I cannot recommend strongly enough the cultivation of a strong and vibrant interior life with God. Having been thus fed with the Heavenly Food of the true Mass and true Blessed Sacrament and the true Catholic teaching and the interior life, let us convey to others what God has given us, for to whom much is given, much is to be asked (and by whom much will be achieved). This conveying shows itself (and can be aided by) a serious living out of the Gospel, forgiving those who wrong us, helping those in need, being honest and having integrity in all that we do, and showing proper respect for the dead and for all clerics. It also involves being activists in our local communities for all that is good, the true Faith, the true Sacraments, the true Church, support of upright laws, and the common interests of all. It is a life of self-sacrifice as we give of ourselves freely and without requiring anything back. Find and read the good Catholic books, such as *The Imitation of Christ* or *The Spiritual Combat* and many other saintly writings and be immersed in them. If possible, teach the Faith at any opportunity, even in the Novus Ordo classes if possible. For did not the Apostle Paul preach even in the Jewish synagogues? Enough practical advice! You should learn the rest from the Traditional Catholic priest who is your spiritual director and do take the time to get to know your fellow parishioners, lay, nun, and monk alike. Welcome home, we're glad to have you!

The present crisis cannot endure forever. In 1950, everyone more or less thought that the Church would just continue to grow and expand along a roughly mathematical curve as it had for the previous several centuries. Who could have predicted that in a mere thirty years the Church would be almost three orders of magnitude smaller and virtually entirely exiled from Her visible and historic facilities? The Church has a nature, expressed in Her four marks and other characteristics, which demonstrate themselves clearly enough at all times, even when one or another is specifically injured by particular temporal events. This era of no popes other than those (?) who resemble shifting sand much more than the Rock of Peter will have to pass one day. Every crisis eventually ends; there will be once again a reliable pope seated on the throne of Peter. There will be a vast and widespread triumph of the Immaculate Heart.

An awareness of that fact should help to put many things into perspective. At present we lack any pope who is willing to address these complex issues. The only persons who at least nominally claimed that title since Vatican II have been constantly off somewhere trying to make push the Communist agenda or some other equally non-Petrine activity. There they go, one after the other, so busy gallivanting all over the landscape trying to fix (break?) the whole wide world, while their own backyard is in such a mess. The Church is confused. Should they even be counted as popes at all? Which annulments can we trust? Which explanation of events will be accepted by the Church as the true one? What goes for those who mistakenly went along with false alternative explanations? What will the restoration of all things in Christ actually look like once it happens?

Since we can know as an article of Faith that there **WILL** once again be a reliable and indisputably Catholic Successor of Peter who will solve these problems, it makes sense to view the current situation as if in retrospect from the vantage point of such a time in the future. Although we don't know the exact time or manner in which this will happen, there appear to be two main forms it could take. The first and by far most likely is that the Traditional Catholic community (real Catholic Church) recognizes itself and its full scope and domain, and its leaders all cooperate to organize a conclave to elect a Pope and the election takes place. The other main scenario formerly considered reasonably possible, but fading fast, would be the resuscitation of the Vatican organization. If by some major divine miracle, a major intervention effectively overriding the hearts of about a billion people to such a degree as God has never been done before, all to make those who wrongly called themselves Catholics to decide to become Catholics for real, the repentant Vatican leadership and its functionaries would have a rather clear and simple path to follow from there. It would have to begin with the abrogation/repudiation of Vatican II in its entirety.

Once we have a Pope, the next concern is what he should or might do. We are not in total darkness as to the way out of the present ills. Let us take it from this standpoint: There will one day be a reliable pope; when that happens, what things **OUGHT** he do? Certainly, a saintly pope (a truly saintly pope would really be needed since an unsaintly pope might never be able to rally true Catholics around him despite his orthodoxy) can be expected to do as he ought, or at least something very close to it. So what is it exactly that he ought to do?

First, he must formalize in detail the revocation of the entire Second Vatican Council. The first portion, from its beginning clear through *Lumen Gentium*, must be abrogated. Everything after that, being the Council of a schismatic Church opposed to Rome and the Pope must be formally disowned and condemned, as were the Fourth Council of Ephesus and the Council of Pistoria. That includes not only that Council itself, but everything that follows which is in any way based upon it, or even what merely pretends to be based upon it. Perhaps that may sound rather extreme. After all are there not many good things said in Vatican II? But why go through sifting for them? As the saying goes, "You can even dig some diamonds out of Rock and Roll," but why bother? Centuries hence, when worldly fashions of thought all lean in a different direction and those who could actually wax enthusiastic about the whole Vatican II direction have all gone the way of the dodo, perhaps some academic types might want to peruse the condemned documents of shining examples of twentieth century stupidity. Doubtless the documents and all the disorder they brought about will make a fit subject for many dissertations.

The revocation of the Council carries with it the acknowledgment that while it was Law, the Catholic Church could only subsist partially within parts of the Vatican organization and that those portions of the Church subsisting outside the Vatican organization were also authoritative, juridical, and canonical. The revocation of the Council also means the end of diversity within the Western Rite; there cannot be both Tridentine ("extraordinary form") and Novus Ordo worship in union with Rome, only Tridentine alone, but more about that concern a bit further down.

The next duty of the next reliable pope is to attend to the problems of the members of the hierarchy. The traditional bishops and priests who have functioned outside the Vatican organization need to be given dioceses and parishes with conventional territorial jurisdiction or comparable. Indeed, because of their heroic stand for the Faith, I know of no more trustworthy persons to put in all the important posts, those of Archbishops, Cardinals, Deans of Seminaries, and Curial officials. The clergymen who have been within the Vatican organization are a different story. Many of them are quite rebellious and will not obey the Pope; they must be put out. Many more never knew the true Faith or Sacraments and will have to be retrained. Many of those and others besides who are at risk of having not been validly ordained or consecrated will need to be given unquestionably valid conditional ordinations or consecrations. Those who accept to be retrained and ordained or consecrated, whether conditionally or unconditionally, as appropriate, can then be officially added to the hierarchy.

His next order of business is to put the Rites of the Church in order. The bogus Novus Ordo "Rites" must be canceled. Ideally the work of the subversive Liturgical Movement should be entirely rooted out, restoring the Liturgy, other Sacraments, and Breviary to the forms they each had in the 1940's and 1950's before anything stemming from the Liturgical Movement had been done to them. At the very least he must go back to the 1962 Liturgical books, nothing later. All bishops and priests must follow what he is mandating here or else they are suspended or excommunicated, as appropriate for each case.

The fourth duty of the next reliable pope is to convene a council for the purpose of setting straight the issues raised by the crisis recently terminated. In particular, were the doubtful popes really popes? Which group is to be regarded as having followed the correct path? What Dogmas in ecclesiology can be formally defined, as discovered throughout this crisis, so as to prevent something similar from happening again? Such a Council needs to pronounce the answers to these questions with the Supreme and extraordinary (and infallible) Magisterium of the Church. Whatever approach is retroactively decided to have been in any sense the "correct" one (and there might well be reasons to delay that action for a generation or so), mercy must be shown towards all who have striven to keep the Faith, by whatever approach was available to them.

The fifth and last duty of the Pope in restoring the Church is a change in policy towards any nation which would like to be a Catholic nation. The Pope must be once again willing to have truly Catholic nations which base their laws on the teachings of the Church. It may be some time before that happens and the reliable pope who has the pleasure of signing a Concordat with the first Catholic nation could very likely come several popes after the pope who performs the first four duties.

Some may find it hard to believe that there may ever be a reliable pope ever again. They look at the present Vatican organization and see almost no one there willing and able to become, elect, or even obey a true pope. There is only the deepening of their new perverse direction, a clear commitment to lock themselves into their errors and heresies. Then when they turn to look to the sedevacantists, what are they doing to get a pope? How many cardinals do they have? However, there are several ways to get a pope.

The manner in which the Church may obtain its next reliable pope would be affected by whether the Vatican leaders repent. There do remain among them those who sense that Vatican II has been an unqualified disaster. Many more others have already seen that the only parts of their organization which are growing are the traditional parts, and at least ordinarily the ability to demonstrate growth in one's diocese or religious order while everyone else's only experiences shrinkage would certainly a good recipe for advancement. Not only are such ones such a tiny minority, but their success is such a rebuke to those incompetents they work alongside, that Vatican politics is totally against them. Their growth has often come at the shrinkage of those other areas where there is no "extraordinary form" and the people migrate to where there is, and then from there to other real Catholics outside their organization. Either the Traditional Catholic theology regains the upper hand in the Vatican organization, or else that society must continue its descent into error. Here is how things must look if Traditional Catholic theology were to regain the upper hand within the Vatican organization:

Their leader ("pope" or whatever he may be properly called at this time) decides to end the current crisis. He mandates a return to Catholic worship and sacraments as described above. Next, he meets with the traditional bishops who have functioned outside the Vatican organization and explains what he intends to do and secures their cooperation in returning the Vatican organization to the Catholic Faith. He must give them reason to believe that he is really going to function as a real Pope, such that they may accept him as such, at least conditionally. In the event there is doubt concerning the validity of his episcopal consecration he has himself conditionally consecrated by them. If he is to be Bishop of Rome, he must first of all be bishop. He then returns to Rome and revokes Vatican II as described above, the traditional clergy then acclaim him as Pope, and the rest follows logically. The crisis is over.

The one great advantage such a scenario has is that it would be relatively swift and painless. It could happen tomorrow. Francis could wake up tomorrow having come to his senses and decide to start cleaning up his own backyard by proceeding as I have just described and undoing the colossal damage he has worked, and then he would have to turn to the traditional bishops for a valid episcopal consecration. And if not him, then a successor could do that same thing. There is of course the far more likely scenario, namely that the Vatican organization will continue its journey into heresy.

In that event there will come a point at which "God's Spirit shall no longer strive with Man," at which something will happen making it so that the Catholic Church will no longer subsist within it. Once all Catholic worship ceases within it by having the "extraordinary form" revoked and completely done away with, and the Eastern Liturgies also all destroyed or surrendered to the schismatic East Orthodox, once it is clear that there is no longer any overlap between the Catholic Church and the Vatican organization, that is the cue for **ALL** Traditional Catholics to become sedevacantists. All real Catholics among those who used the Modernist's "extraordinary form" will have to move over to the traditional priests and bishops operating outside the Vatican organization, or else they are not Catholics and were never really with us. The SSPX's mission to rehabilitate the Vatican organization will have terminated at that point, a heroic failure. All such sedevacantist Catholic bishops whether newly so or not, will get together so as to elect a man to lead them. The man may either be one of them or any priest, monk, or even layman whom they would all trust to lead the Church in the Petrine office, and who is willing to be consecrated a bishop and accept such a duty. In accepting the office, he must also, in the same breath and sentence, revoke Vatican II, at least generally in a manner consistent with that given above.

That man, so elected, is the next reliable pope. His duties are from that point on pretty much the same as those of a pope elected by the Vatican "cardinals." In such a case, the Vatican organization will have voluntarily relinquished all claims to having any portion of the Roman Catholic Church within it, and thereby also all claims they have to the Vatican properties. The Catholic Church would continue to grow and the "People of God" would continue to shrink and shrivel until the point shall come that we can afford either the legal or military horsepower required to drive out whatever usurpers and squatters will have been inhabiting Vatican City by then!

Divine Judgment may also come in a different form. There is a real possibility that the Vatican will one day be occupied neither by Catholics nor Novus Ordo believers, but by others (Muslims, most likely) who could sack Rome and either execute, confine, or exile the then current Vatican leadership. It is a well-known and frequently observed principle that a spiritual fall eventually leads to a military fall as well. God works that way. Such an upcoming and terrible destruction of the Vatican and all its treasures and slaughter of its personnel would at least have the good effect of rendering moot all questions which set Catholic against Catholic (such as who is legitimate and who is not, and whether or not there is a pope), and serve to make full unity among the Catholics far more attainable. There would no longer be the various camps of the Church as there are today, but merely Catholics all of whom know the Faith and desire to restore the "One Shepherd" in order that all may be in one flock, and the Mark of unity could shine again in its full glory, no longer materially interrupted. It is even possible that once the archives fall into the hands of the Muslims, they might hand them to their scholars and learn that the Christian Church has **not** in any way concealed any prophecies of their prophet, but rather that there was never any talk of anyone coming after Jesus. Perhaps it is finally time for us to see why the heresy of Islam (unlike all others) has only continued to grow and expand despite its nearly 1500 years, and with almost no converts to Christianity.

We Traditional Catholics belong to Rome, the Eternal Rome of Peter, of Leo, of Gregory, and of Pius. Just as being a Father's child implies that the Father is also that child's Father, our belonging to Rome also implies that Rome belongs to us. Those who inhabit Vatican City now must either repent and return to the Faith or face an eviction mandated by none less than the Almighty Himself! Vatican City was deeded to the Roman Catholic Church which exists now exclusively as the Traditional Catholic community. Its history, heritage, and treasures are ours, **NOT** Fr. Bozo's.

Always remember: We are always morally free to do the right thing; We have the peace that comes from knowing.

On Easter of 1982, the twenty-five priests of the Diocese of Campos, Brazil (five of whom were Monsignori) jointly signed the following Profession of the Catholic Faith in the Face of Present Errors, to which all contemporary Traditional Catholics of all stripes can adhere to without reservation:

> WE BELIEVE FIRMLY in all that our Holy Mother, the Catholic, Apostolic and Roman Church, believes and teaches, and in this Faith we wish to live and die, since only in the Church is God honored and salvation found.
>
> We believe that Jesus Christ has founded one only Church, the Catholic hierarchical Church, whose chief pastors are the Pope and the Bishops in union with the Pope, and whose aim is to ensure that

the man of all ages reach the salvation obtained for us by Our Lord Jesus Christ, together with all the benefits that radiate from it, and to promote on earth the Reign of Our Lord (Mt. 28:19–20). To do this, the Church does not preach a "new doctrine" but, with the aid of the Divine Holy Spirit, faithfully explains the Deposit of Faith received from the Apostles and religiously preserved by her (*Vatican Council I*).

We profess communion with the See of Peter, in the legitimate successor to which we recognize the primacy and government over the Universal Church, pastors and faithful, and nothing in this world would separate us from the Rock on which Jesus Christ has founded His Church. We firmly believe in papal infallibility as Vatican Council I has defined it. We respect the power of the Holy Father, which is supreme even if not absolute or without limits. It is subordinate to and cannot contradict Sacred Scripture, Traditions and the definitions already proclaimed by the Church in her constant Magisterium. Moreover this power cannot be arbitrary or despotic, such as to impose unconditional obedience or absolve subjects of personal responsibility. We owe unconditional and unlimited obedience only to God.

We are Catholic, Apostolic, Roman and shall be so till death, with the grace of God, and no power or authority will drive us from Holy Church.

We profess the Catholic Faith in an integral and total manner, as it was always professed and transmitted by the Church, by the Sovereign Pontiff, by Councils, and therefore in loyal and perfect continuity and consistency, without excluding a single article of faith.

We reject and anathematize, with the same firmness, all that has been refuted and condemned by Holy Church.

Together with all the Popes, we condemn heresy and all that can favor it: we particularly condemn Protestantism, liberalism, spiritism, naturalism, rationalism, and modernism, in all forms and variants whatever, just as the Popes have done.

We reject equally, together with the Popes and in the same way they have done, all the consequences of these errors.

For these reasons we condemn the present heresy that takes the name of "Progressivism," an improper name besides, since it is nothing but the repetition of errors long ago condemned by Holy Church.

We accept therefore the entire applicability of the words of St. Paul: "Even if we ourselves or an angel from heaven preach to you a gospel different from that which we have preached, let him be anathema" (*Gal 1:8*).

Thus, the tendencies of whatever sort emanating even from persons of authority, whenever they be contrary to traditional Catholic doctrine as it has always been taught, giving free rein to the errors already condemned by the Popes and by the Councils, demand from our conscience a formal rejection.

We affirm in consequence that, whatever contradiction becomes manifest between what is taught today and what Tradition teaches, there is a duty to follow what was *always* taught *by all and in every part* of the Church, because only this is truly and properly Catholic.

FOR ALL THESE REASONS, and to be consistent with what the Church our Mother has always taught us in her constant Magisterium, with the Faith of our baptism, of our confirmation, of our first Communion and of our priesthood, in order not to be perjurers and contradict what we have always believed—for all these reasons—WE REJECT:

THE NEW MASS, whether in Latin or in the vernacular, since "it represents, both as a whole and in its details, a striking departure from the Catholic theology of the Holy Mass as it was formulated in Session XXIII of the Council of Trent" (*Letter of Cardinals Ottaviani and Bacci to Paul VI, 5/10/69*). In fact the new Order of Mass obscures the expressions intended to underline the Eucharistic dogmas, bringing the Mass close to the Protestant supper and not accentuating the clear profession of the Catholic Faith.

THE NEW MORAL THEOLOGY—subjective, opportunist, contaminated by permissive liberalism, in which little or nothing any longer constitutes a sin.

THE PROFANATION OF THE CHURCH by dress always considered immodest, just as by certain kinds of music and instruments already rejected by the Church for her places of worship.

THE NEW THEOLOGY of modernism that founds the whole Catholic religion on the evolution of the religious instinct of the first Christian communities.

THE NEW CATECHISMS, purveyors more or less covertly of the modernist errors and not infrequently vehicles of subversive political doctrines.

THE THEOLOGY OF LIBERATION, based on a new interpretation of the Gospel, completely opposed to the teachings always held by the Church, and calculated to favor Marxist machinations.

THE TREND TOWARDS SOCIALISM AND COMMUNISM that manifestly contradicts the whole social doctrine of the Church and scorns the excommunications of Pius XII directed against those who collaborate in any way with Communism.

THE SECULARIZATION OF THE CLERGY, causing grave scandal to the faith and the inevitable impoverishment of Christian life.

THE CONFORMING TO THE SPIRIT OF THE WORLD on the part of the clergy and faithful in complete opposition to the teaching of Our Lord and to the spirit of mortification, teaching and example of Jesus Crucified.

THE REFORM OF THE SEMINARIES, carried out in line with these new tendencies.

THE OBSESSIVE CONCERN FOR HUMAN PROGRESS that leads to disregard for the specific purpose of the Church which is the salvation of souls.

THE DILUTION of true spirituality in a vague religious sentimentalism.

THE ECUMENISM that makes the Faith grow cold and makes us forget our Catholic identity, seeking to negate the antagonism between light and darkness, between Christ and Belial (cf. II Cor. 6:14–18), and leads to a panchristianity "a most grave error and capable of destroying the foundations of the Catholic Faith at their base" (Encyclical, *Mortalium Animos* of Pius XI).

RELIGIOUS LIBERTY, understood in the sense of an equalization of rights between truth and error, giving supremacy to a supposed subjective right of man, to the prejudice of the absolute right of Truth, of Good, of God, and as a consequence laicizing the State, rendered agnostic towards the true Religion.

THE HORIZONTALISM OF THE RELIGION OF MAN, that concretizes what St. Pius X calls the "monstrous and detestable iniquity proper to the times in which we are living, through which man substitutes himself for God" Enc. *Supremi Apostolatus*).

DEMOCRATIZATION OF THE CHURCH by means of a collegial government in opposition to the hierarchical and monarchical constitution given to her by Our Lord.

LAICIZATION OF SOCIETY, that brings to life once more the cry of the Jews at the death of Jesus refusing to accept His social Kingship and not even recognizing in Him the Supreme Legislator of human society.

Thus, in the name of Faith, with tranquil conscience, we reject all those who introduce the "smoke of Satan" into the Church and apply themselves to her self-destruction (*cf. Paul VI, allo. of 7/12/68*).

WHAT WE ARE FOR

We love, praise and adopt all the traditional practices, uses and customs of Holy Church that have contributed, and do contribute, so much to the sanctification of the faithful, among them, for example:

AURICULAR CONFESSION;

ESTEEM FOR THE CONTEMPLATIVE LIFE that ought to have pre-eminence over the active life;

USE OF THE CASSOCK AND HABIT by priests and religious, that marks so well their separation from the spirit of the world and keeps consciences alert to the spirit of Christ;

VENERATION OF IMAGES and relics of the saints;

COMMUNION RECEIVED ON THE TONGUE AND KNEELING, in token of respect and adoration;

PENANCES AND MORTIFICATIONS, internal and external;

THE ORNAMENTATION AND MAGNIFICENCE of the Churches which contribute so much to the splendor of worship;

THE SOLEMNITY AND POMP OF THEIR CEREMONIES, that so impress and move the good people and stimulates their devotion;

LATIN IN THE LITURGY, a factor of the unity and universality of the Church, a precious casket befitting the sacredness expressed by the liturgical prayers;

GREGORIAN CHANT, which has nourished piety for so many centuries.

In particular, we wish to profess and spread a deep and burning devotion to the Mother of God, in whose Immaculate Conception, perpetual virginity and universal mediation we believe. We maintain that such practices of devotion, principally the holy Rosary, have a special efficacy for the sanctification of souls and triumph of Holy Mother Church.

We profess a convinced acceptance and a sincere love of holy priestly celibacy, one of the sources of pastoral zeal that constitutes a firm response to the hedonism in which the neo-pagan society of our day is immersed.

Our position is not one of rebellion, nor of disobedience, or contestation, but of fidelity. It is a question of loyalty to the Faith of our baptism, to our priesthood, to our legitimate superiors and to the faithful.

We do not judge the consciences of others, since this judgment belongs to God, we only claim the right and exercise the duty of every Catholic. We confront what is taught today with what was always taught; we hold on to what is of the Church and refuse what serves only for her self-destruction, whoever may promote it knowingly or unknowingly.

We are not against progress if it represents an organic development of Revelation; we are against that false "progress" that is not consistent with Tradition but is in discontinuity with it.

It is not a matter of indiscriminate attachment to the past, but of clinging to the Faith which does not pass away.

We believe in the permanence of the doctrine traditionally taught by the Church and in the objective sense of the formulas that express the dogma and truth that Holy Church teaches.

We believe that the truths of Faith remain absolutely independent of the ways of thinking and living of men, since Truth comes from God through the Church and her Tradition and does not arise from the instinct and religious feeling of the people.

We have absolute certainty that out position is legitimate, not by virtue of our arguments and ideas, but because we take our stand on that which the Church herself has taught us and in such a way that we could make our own the words of St. Augustine when he exclaimed that, if what we believe were an error, then God Himself would have deceived us.

We have firm hope that within a short time the Church will have overcome the current crisis she is living through, and, dissipating the darkness of heresy will return to shine out as always, as a glorious beacon to the nations: "The gates of Hell shall not prevail against it" (*Matt. 16:18*).

We love from the bottom of our hearts our Mother, the Holy Church, Catholic, Apostolic, Roman. For her we wish to give our lives if it is necessary.

EPILOGUE

♦

A PARABLE FOR THE CHURCH TODAY

A tree in the forest grew great and majestic above all others, so that the birds of the air came and nested in its branches (Matthew 13:32), but then one day the call came down, "Chop down the tree and use the wood for lumber." And so it was done. Many looking on mourned for the tree and followed it to the lumberyard where it was carved up into lumber.

In time, the lumber came to be used in the construction of a billboard, on which was advertised various brands of cigarettes, airlines, and telephone companies, etc. The many who were saddened by this were truly appalled that their beloved tree should be put to such ignoble uses.

Loyally, they stuck by the billboard made from the tree because they believed that the tree could never die, even though the lumber in the billboard showed every sign of being dead.

Yet the tree was still very much alive, not in the part that had been cut off, but in the stump left behind in the forest. Small branches had sprouted out of the sides of the stump, just under the place it had been cut down. These small branches sprouted green shoots and leaves and grew by leaps and bounds.

Because the stump had been so large, some of the branches grew quite some distance from each other. Because of that, some people mistook these branches for distinct little trees with no relationship to each other or to the original tree, but anyone having even a little bit of curiosity could easily look and see that the leaves were the same as the leaves of the original tree, and if only they dug down just a little bit, they could see where the stump was and where each of them was attached to it, growing from it.

At first, only a scattered few dared to leave the billboard to return to the stump. Many at the billboard denigrated those who left it to return to the stump for deserting the tree, but as the health and vigor of the tree at the stump grew, and as more people became aware of that, the small trickle of persons transferring from the billboard to the stump grew into a steady stream and finally into an avalanche as everyone eventually came to realize that the life of the tree is in its roots, and not in the branches (nor even the trunk) which have been cut off.

"For Saint Augustine answers precisely: 'The branch lopped off has the shape of the vine; but what avails the form if it has not the **root**?'" (*Mirari Vos*, Paragraph 14). Needless to say, in time the branches grew into an even bigger tree, more glorious than the original had been, and all the stronger for what it had been through.

AFTERWORD

This Revised and corrected edition serves not only to provide historical updates to the book as have occurred since the publication of the previous edition in 2004 and correct a smattering of typographical errors, or update for clarification some of the terms and expressions, but most of all to make emends to some whom I have accidently wronged in the previous editions. I can only assure the reader that this was entirely unintentional, but justice requires that I nevertheless right the wrongs done, that the good names of certain individuals not be besmirched any longer, at least not by my pen, so my conscience may be clear.

There are two basic categories wherein correction is needed, and to that extent this edition is to be regarded as a retraction and correction of those contents of my earlier editions. The first of these categories concern particular individuals, key figures, whose good memory I defaced in the telling of the history of the Church's struggle to preserver Herself and Her faith in these times. There are also those I have exalted somewhat excessively and who need to be more realistically appraised, appreciated for the good they have done, but also shown to be limited and even unreliable at times as visible and widely known representatives of the Catholic Faith in this dire struggle. The other category pertains to the canonical and juridical aspects of the Church of which I had merely "picked up on" from others. Within each category I will explain what, who, and why, not to excuse my previous failures (for if that could be done there would be no need for this correction/retraction), but to demonstrate how, as a diligent and careful researcher, mistakes were made. At least I can take some solace in that the mistakes made are not my own, but of the sources upon which I depended, and at least in the absence of better information since obtained, I did my best to mitigate the bad reports and false narratives of others, but without real information these mitigations were of quite limited value, and perhaps even counterproductive at times.

The august person of Archbishop Pierre Martin Ngô Đình Thục first and foremost requires this corrective. What little positive information as I had at the time had come, and rather indirectly at that, from Doctors Kurt Hiller and Eberhard Heller, who, though they had worked with Thục during that period in 1981 to early 1982 in the selection of the three Catholic clergy he raised to the episcopacy, were unable to speak for the time before their presence with Thục, and were in any case very selectively quoted by other sources quite committed to destroying Thục's reputation. I could see the slant and endeavor to compensate for it to some degree but that was all. I could not discover very much useful in the way of facts with which to balance the seriously vile and calumnious narrative of his life, as told by even his friends let alone his many enemies. But I had no access to his autobiography (which covers everything up to 1975, shortly before the Palmar de Troya trip that so changed his life), nor whatever other memoirs he would have written to cover the period of time since then, no detailed biographical document at all, except for the Rev. Terence R. Fulham's *Corona Spinarum*, Fr. Cekada's attempt to defend the validity of the Thục consecrations, various essays pro and con regarding Thục's mental state, and other such material, all the rest of which was utterly hostile to the saintly Archbishop.

And I had one other hard fact, namely the dates and names and notes about how personally ugly the people were that he consecrated, beginning with those at Palmar de Troya, and ending with the various "Old Catholics." Facts like these, out of context, can sound truly sinister (or scatterbrained, mindless, or bribed, depending upon the narrative), but to illustrate: What would one say of a man who shot and killed 63 people? Here is the list of those he shot, and what dates and times he shot them, and there are their gravesites. And all of that really is true; a narrative can easily be put to those hard facts fit to make him out to be a truly horrible mass murderer! But then you learn that he was a soldier in a just war, and as a combatant in that war he shot and killed 63 enemy combatants, all in the line of duty, so no mass murderer at all but merely a soldier, and one to be commended at that, fit to be given a medal. Sometimes context is everything, and I had no context. Even his friends seemed unable to provide it.

Without knowing anything of Thục's side of the story, what he was trying to do and why, his reasons, his goals, and most of all his extreme clarity of thought by which he could see his clear and present duty to continue the Church as he did, it is easy as duck soup to spin the worst possible scenarios. Why did he consecrate those five persons for the Palmar de Troya? Why did he consecrate those "Old Catholics" (or whatever they were)? Was he so "out of it" he didn't know or care what he was doing? Was he maliciously trying to "screw things up" for the Church in what few little ways he could? Was he crassly selling these consecrations for money? Even one of the Doctors who worked with him had ventured something roughly and crudely along the lines of that last as something of a guess to account for it. But that was just him guessing; he was not speaking from actual direct personal knowledge (since all of these other consecrations came either before or after Thục's association with him and the other Doctor).

So with only this vile patchwork of calumnious narrative to work from, all I could do in my original edition was stress the extreme tragedy of his past life, the murder by the Communists of almost every family member, the betrayal of Paul VI who threw in with Communist usurpers in the role of the Vietnamese bishops to take his place and office and others like them, the loneliness and devastation he lived with as a permanent exile from his homeland or even from anyone who could remember him or appreciate him or even speak his native language, and so forth. In the middle of all the controversy all there seemed to be was a lonely and shell-shocked old man asking, "Why does everyone want to be a bishop?" I cannot help but cringe with embarrassment at how I pretty much had to take this "You'll have to excuse my friend; he hasn't got all of his marbles" approach to "defending" or "explaining" him. That may well have been worse than outright attacking him. And all for nothing, for the real story is far more interesting.

In the years since originally writing this *Resurrection* book, several things seriously punctured that narrative. The very first hint I came across was the question, how is it that, when Fr. Revas came to Abp. Lefebvre in 1975, Lefebvre, in turning him away managed to think of Thục, out of all of his fellow bishops he could have chosen from? Of all the bishops he could have named, how many (besides Thục) would have been willing to even just consider performing an episcopal consecration, let alone actually do one? How did he know? Obviously, he had been in some sort of contact with Thục in recent times (another source I have quite recently discovered mentioned that Thục had been to Ecône and even given talks there, and it is also stated that Thục had known Fr. Revas, a Canon Law professor from Ecône from their having been both present there), and apparently the two must have discussed the prospect of performing such consecrations, which Lefebvre was nowhere near ready for yet, but Thục was. This in turn indicates a clear and expressed intention on Thục's part to consecrate real and traditional bishops to continue the Church, which demonstrates what a lie it is to say or imply that Thục was merely willing to do whatever someone wanted of him, no matter how outrageous or illegal. Do you begin to see a crack in the vicious old narratives? Somehow, Thục had conveyed to Lefebvre his intention/desire to continue the Church by consecrating bishops.

Finding information from his autobiography, and also his episcopal successor Bp. Vezelis (who writes, "Once a lie has done its work, truth must labor hard and long to dislodge it. Lies fly as if on wings; and truth comes slowly limping far behind.") there was much more to his past in Vietnam, his ancestors and family, his acquaintance with Fr. Revas and others connected with the consecrations of the final part of his life. The next big break was discovering real and good hard facts about the Palmar de Troya group. Since the late 1970's or so, they have become not only strange but also secretive, closed in upon themselves, ingrown, and utterly incapable of being contacted for any reason. As a result, I too knew little, and again much of it was flawed or incomplete information. In their case however, I seem to have nothing to retract merely because I scarcely even mention them and do not attempt any detailed description of any sort. I had however assumed that their disreputableness (so evident in the later years to what few people could even find them) had been a going concern even when Abp. Thục was paying his visit thereto, and this, it turns out, had been nearly entirely incorrect. Though some "seeds of destruction" could have been ferreted out with some considerable due diligence at the time (and of course all private revelations or apparitions are to be considered suspect until approved by the Church), by all evidences they seemed quite respectable at the time, even able to attract at least one instructor from Ecône (Fr. Revas) whom Lefebvre had continued to respect enough to see him (even if only to send him elsewhere) rather than not.

The next major break however was the combination of two events, one being my own comprehensive doctrinal review of the Church's present circumstance, as recently published in my *Sede Vacante!* volumes (of which more is said further down) and my meeting up with a former follower of the Palmar de Troya group who helped me discover their book, *Palmar de Troya: The Light for the Church and for the World,* which contains Abp. Thục's own explanation of his consecrations at Palmar de Troya. Had I not done the comprehensive doctrinal study, the document Thục wrote by which he explained himself and his action might well have been practically incomprehensible, as doubtless most of what few who even encountered it at all might well have found it to be. But in the light of that study, what it reveals is a stunningly clear grasp of the true status of the Church, and of his role in continuing it, and even specifically the fact that ("thanks" to Paul VI who promulgated Vatican II, including *Lumen Gentium*) "no minister, even an admittedly Catholic one or not, can be excluded from performing [all ministerial] functions within anyone's parish or diocesan territory, even altogether without permission or consent," thus canonically justifying his performance of the consecrations despite the attempt of the bishop (?) of Seville to exclude him.

If Thục had attempted to explain it all to anyone (as one might normally expect someone taking so drastic a step to do) before performing it, it might have taken more than the rest of his life merely to figure out how to explain it, how to prove it, and even then whoever he explains it all to would still have the ability, as many sadly do, to willfully refuse to acknowledge when a thing has been proved, and then merely, stupidly, tell him not to do what he knew he needed to do. That would have forced him into a deadly inaction or else to go against advice sought and given. How long might it have taken for the stupidity of any immobilizing advice to have been seen for what it is?

He knew in his own mind what he was doing, why it was not only right but obligatory before God, and given that he doubted he would have the time to explain it all to anyone's satisfaction, all he could do was proceed with the consecrations and document it with merely sufficient hints by which the Church would be able to figure it out again. After all, when something true is discovered, that means that it can be discovered again, and he was trusting in Providence that the Church would, at least eventually, rediscover what he knew, and then be able to recognize the soundness and fitness of his actions. He knew then and there that he was attempting to convey not only a sacramentally valid episcopacy, but also the canonical mission of the Church, a true hierarchical status. If only the consecrands put forth to him by those who should have been trustworthy had been fit subjects to receive the canonical mission, he would have succeeded then and there.

A keen mind, fit to pierce through and make sufficient sense of a complex situation, comprehended by few, and to see through it all what action to take, despite what persecutions would come his way, is not to be mistaken for someone confused, disloyal, easily manipulated, or even mercenary. Knowing all this, one begins to look even at his other mistakes, the supposed "Old Catholics" (none of whom actually were, but merely that they had, in a somewhat similar-minded desperation, turned to such a succession merely for the valid power of the sacraments) were turning to him, not so much out of doubt as to the validity of their Orders, but desiring the Canonical Mission and a legitimate place in the Church. But none of that was written down anywhere (that survives, anyway) and practically nothing is known of the facts of these consecrations. For one thing, a consecration on a given date mentions nothing of the months of careful investigation, research, prayer and meditation, soul-searching, and consideration, discussion and negotiation, and finally spiritual preparation that preceded it.

But the first of such, Jean Laborie, had lived at or near the same French town (Toulon) as Thục had settled in; as fellow clerics they visited each other often and knew each other well. And (before his questionable involvements) Jean Laborie had held and earned the respect of no less than Cardinal Alfredo Ottaviani for his doctrinal knowledge. One does not see from the simple list of names and dates of consecrations how many hoops each of the few who passed muster had leapt through in "proving" themselves fit, how many more had tried and failed. At a later point one gets a glimpse of Abp. Thục not yet satisfied with the bona fides of no less than the Rev. Otto Katzer, a noted and Doctorate-holding theologian, who had already met the tough standards of (then three professors, Lauth, Hiller, and Heller), only later was he finally convinced and ready to consecrate him, but then Otto Katzer died before it could happen and another, Fr. Guérard des Lauriers, was chosen in his place and had to go through all the same hoops before being consecrated.

And as the seeming other complaints admit of proper explanations, they demonstrate nothing supportive of the vicious attacks made on his character: the bizarre suggestions ventured at Vatican II had been solicited of those present and feeling obliged to participate, he came up with some of his own, but obviously and realistically, would never have countenanced seeing any of those suggestions being put into practice, his complaint about who was not invited to Vatican II was based on a seemingly racial unfairness in which white European "Christians" of various sects were being invited as observers while Asiatic persons (generally of Asiatic religions) were not; in early 1981 he was granted permission to hear confessions in a Novus Ordo cathedral, a service he particularly desired to render to his fellow Vietnamese living in the area, and for their sake and to be able to help them he accepted associations he otherwise wanted nothing to do with, and became one frequently approached for that sacrament. However, being found there he was once pressed into serving (as like an Altar server, specifically the ceremonial role of "Deacon," not the celebrant) in a Novus Ordo service, which for him was, of itself and regardless of his coerced participation, the mere simulation of a sacrament, and which afterwards made him feel soiled and which he actively sought to disavow and repent of. Nominally, the position that enabled him to be the confessor could require such participation of him at any time, but only that one time was it unavoidable. Not being the celebrant, he was not in any way responsible for any lack of validity of the supposed sacrament.

Even his lack of any strong stand about the "Pope" question, until 1982 anyway, only reflects how his focus was not so much on that rather narrow question, but on the far broader question regarding the Church as a whole society. That is the same as my approach by emphasizing the status of the Church as a whole society rather than focusing so exclusively on the "Pope" question. Even many sedevacantists miss the boat on this by picturing it all being a heretic masquerading as Pope leading the real Church into heresy and it just follows him. But doctrinally that is impossible, just as a heretic cannot be Pope, neither can a "church" that follows a heretic into heresy be the real Catholic Church. Next to that great issue of the whole Church, the mere "Pope" question fades into near insignificance. Many sedevacantists seem to see it as being like realizing that the steering wheel has been stolen. But seeing the fullness of the truth is like realizing that the whole rest of the car has been stolen with it.

And so it goes with every vacuous accusation made of Abp. Thục to besmirch the good name of a truly holy and saintly prelate of Holy Mother

Church, and furthermore he who first truly grasped the scope and scale of the problem. While so many others hemmed and hawed on the sidelines, wondering what was going on and hoping that things would just get better again of their own accord, Thục was out there doing what every faithful bishop of the Church should have been doing. That he had to act alone is to be taken as a truly sharp rebuke to all of those who did not join him in doing what he did, who did not work with him to continue the Church, to provide the checks and balances that could have spared him many of the cases that did not turn out well. And as to those mistakes, he applied every precautionary measure possible within his meager and unsupported means, and all that anyone can rightly learn from any of his mistakes is that even the greatest precautionary measures are no guarantee of success.

There is a world of difference between looking for gossip versus looking for the truth, and all of those who have taken scandal at Abp. Thục can rightly be accused of pharisaical scandal and gossip, for he has given no actual cause of scandal whatsoever throughout his entire career. And I can only apologize for my accidental part in having contributed to that gossip by passing it along unknowingly, apologize to the readers who have been misinformed, and to the memory and soul of Abp. Thục who has been so egregiously wronged.

My mistake extends a bit beyond Abp. Thục himself personally, since some of the activities of the earliest Thục bishops, most notably Musey and Vezelis, also seemed incomprehensible to me. I did not realize, as I do now, that the proper action for bishops to take, in the absence of all but the barest handful of all others, is what they did, namely to form "mega-dioceses" and agree among themselves who is to be in charge of which and where their boundaries shall be. Ordinarily a bishop would never be set over more than one diocese, or one single assignment of whatever nature. By that same policy a given bishop would never be the Abbot or Superior General of a religious order or community and also the bishop of a conventional diocese at the same time. However, extreme reductions in the number of bishops and Faithful can and do justify exceptions to that policy. That so many other otherwise faithful bishops coming after them failed to understand and recognize this correct path the early Thục bishops had taken is not merely their own individual faults, but also the fault of the general run of faithful Catholics (including myself) who took a purely pharisaical scandal at their action, rendering it impracticable and pressing the real hierarchy of the Church into an unusual and largely unprecedented form: Catholic flocks based on particular individual societies rather than territorial dioceses. Even in this, the Divine Constitution of the Church is retained: At the top, the Pope (even if the office be vacant), under him all bishops, each set over some particular and definable flock of Catholics: priests, consecrated religious, and laity, thus leading all into the Gospel all around the world.

Next, there is the person of Francis Schuckardt who has also been similarly wronged by me. While I have always acknowledged the value of his earlier work with the Fatima Crusade from its origins in the 1960's clear until at least fairly well into the 1970's, I must do so even more, and also of the things that went wrong towards the end of the 1970's and into the 1980's, the exaggerations and misrepresentations I had accidently passed along, portraying him as a grasping greedy power-seeker, openly decadent in his own life and degrading to those who followed him, all of which were unjust accusations, the claims of those with an axe to grind against him and what he achieved, an assault upon the memory of a man who, though possessing some real failings, accomplished much for the Kingdom of God; people may fail us ultimately, but God does not.

For this I have one person to thank, Bishop Joseph Marie, one of two bishops that Bp. Francis Schuckardt consecrated in 2006 shortly before his death on November 5, and the only member of Bp. Schuckardt's group to go public with their cause, and the closest firsthand account of the events relevant to him and his departure from Spokane.

Now on balance, I must bring down much of my former enthusiasm for Michael Davies; the praise I had for him was undeniably excessive. Whatever traditional leanings and sympathy he evidently has, it has since become clear to me, especially as documented in the book *Michael Davies – an Evaluation* by John S. Daly – though I grant that book to have been overly harsh at certain points – that Michael Davies, whatever legitimate praise he might still deserve, was no great scholastic genius, nor even perfectly honest in his scholarship at all times. How easy it is to be taken in by pseudo-scholarship; due diligence must always be applied, even when the points seem so reasonable, so obvious, so barely in need of proof as to make citations seem superfluous, one must nevertheless look them up. His continual reliance upon "theologians," nearly all of whom go unnamed, and what few as are named, are just names to me, such as Van der Ploeg, their works unknown, should have been a warning flag.

I also note that at one point in my previous editions I named a "Bp. Thomas Fouhy" as being among the legitimate bishops serving in the Church. However, through no fault of his own (other than perhaps a lack of intensive due diligence) he failed to realize that the man who "consecrated" him, one Jean Gérard de la Passion Antoine Laurent Charles Roux, had quite falsely claimed to have been consecrated by Abp. Thục, but in fact was not, and who has proven to be an extremely unscrupulous person, merely taking advantage of the crisis. This is not to denigrate the teachings and writings and actions of "Bishop" Thomas Fouhy, but it does mean that every episcopal sacramental action of his is null and void.

Finally, there are the juridical/canonical issues which were understandably quite fuzzy and unclear in my original edition. The only serious books I had read on such topics were *The Catholic Controversy* by St. Francis de Sales, which has a surprisingly large amount of information on the visibility and nature of the Church, Heribert Jone's *Moral Theology*, and an assortment of standard catechisms. While I knew there must also exist solid pre-Vatican books on Theology and Canon Law, I had no idea what authors or titles to search for, nor expectation that any such would be in English instead of Latin or some other language. In the years since 2004 I have acquired such volumes, studied them intensely, and come to both firm up and refine my understanding of events as I suspected as far back as 1995 when, already exposed to the fundamental differences between the authentic Catholicism of nearly twenty centuries and what was being served up as a "new Catholicism" by the Novus Ordo church of the People of God, and puzzled by that discrepancy as were virtually all faithful Catholics back then, I made my first real breakthrough in understanding the true nature of our present ecclesial circumstance.

Most notable was that it was not clear to me that the Church must have an official existence as such, namely that supplied jurisdiction for individual acts, while potentially useful for some actions of this or that particular individual ministry, cannot comprise the whole basis for the whole Church to exist. Somehow, I had actually thought that with nothing but epikeia and ecclesia supplet the Church could nevertheless carry forth. I also had a vague notion of "delegated jurisdiction" which bore extremely little relation to delegated jurisdiction as described in Canon Law.

My far more detailed understanding of things began with my obtaining the three *Dogmatic Theology* volumes by Msgr. G. Van Noort, followed by E. Sylvester Berry's *Church of Christ* and the translations of St. Robert Bellarmine's works by Ryan Grant, especially those upon which St. Francis de Sales had based his work that I had read so many years before. All of this, and several years of controversy as seen (and occasionally participated in by me) on the various Traditional Catholic blogs and forums culminated in my summarizing my findings in my two *Sede Vacante!* volumes. I will not here attempt to repeat what those volumes demonstrate and prove in great detail but suffice to say that what they prove serves as the theological, ecclesiological, and canonical basis for the viewpoint of this work. While the basic view taken in the original book holds perfectly true and valid, by far all the more so now that I have studied a good selection of the standard recognized pre-Vatican II texts of the Church on these subjects, there were nevertheless a number of minor points in which my general "layman's" description of things could be misleading, even outright incorrect.

Starting with my take back then on "delegation" or even "delegated authority" as mentioned in passing a time or two therein. There does exist an extremely broad and general sense in which it can be properly said that all authority in the Church is "delegated," since it is always a matter of Catholics who are not only members of the Church, but leaders in at least some sense, who choose the leaders in and of the Church for all future ages. In this sense, Jesus Christ delegated His juridical authority (and even miraculous) authority to the Apostles, the Apostles then delegated His juridical authority to their succession, and so on. The Pope is elected by some manner of leaders of the Church, as represented by certain clerical persons in and near Rome (the "Romans"), later replaced by an explicit Cardinalate which served in that role clear to about the middle of the twentieth Century, and the Pope is the final arbiter of who has what authority in the Church, whether his will be directly and personally involved, or tacitly and legally involved in cases where access to the Pope is impossible but decisions required. The key thing is that no one arrogates authority to themselves in the Church, and nor does anyone gain a position owing to their popularity or learning or wealth, nor from the King's favor; they must either be elected (as is the Pope, and perhaps in some cases of the leader of a religious house), or else appointed – "delegated" in my loose use of language in the previous editions – by a superior or at least as designated by some superior-imposed law, superiors and electors being in all cases those in and of the Church and not of any other society, not of the secular government, nor of any economic concern.

But Canonists speak specifically of three basic categories of jurisdiction, one which is ordinary or regular (that of the Pope over the whole Church, and of any legitimate bishop over a diocese, religious order or congregation, military or other ordinariate, and so forth, the second of which is what they properly mean by "delegated" inasmuch as someone with ordinary/regular jurisdiction may "delegate" some facet of their authority to another person for whatever reason, and the third of which is commonly called "supplied" wherein the person does not have the jurisdiction either by ordinary/regular appointment/election nor by delegation, but by circumstances of a complex situation or legal status, or else by "common error," the honest mistake of any random individual availing himself of the powers of a priest who may not have the powers assumed by the individual. It is best to restrict the use of such phrases to their proper context.

In particular, I had supposed that the Law of the new "Constitution" for the Church (*Lumen Gentium*) had granted ("delegated") jurisdiction to any and every Catholic cleric. In point of fact, this is not entirely my own mere layman's fancy, but actually a concept seriously ventured by no less than Msgr. Charles Journet on pages 506-509 of his book, *The Church of the Word Incarnate*, who theorized such a "blanket jurisdiction" as being tacitly extended to schismatic, but sacramentally valid, clerics – all without their having been designated to any territories or other roles by the Pope, and in fact even in opposition (or at least rivalry) to the Pope – could nevertheless be real and have real salvific effects. But for many good reasons I refuse to base the legitimacy of the Church's present officers (traditional clergy) on such a claim for a host of reasons given and discussed in pages 47-50 of *Sede Vacante!* Volume Two. I therefore withdraw and remove any hint of that thought for the present edition.

The next main defect is most evident in the previous editions in that the Vatican organization is repeatedly spoken of as the "Vatican institution," failing to bring out that it is actually we traditionalists, the Church, who are the original institution of the Son of God, the Mystical Body of Christ, and whatever of an "institutional" aspect of the heretics represents a new and separate "institution," a peculiar one at that, on the part of the Vatican heretics.

Unfortunately, that plays directly into the hands of villainous persons, agents provocateurs to be precise, of the Vatican organization who deliberately labor to sow discord and disunity within what they know as well as I to be the real Church by systematically destroying any notion of authority. Authority has the power to step in and force disagreeing individuals to set aside their disagreements until said same authority can resolve the conflict and require them to work together for the good end of helping the Church return to Her former strength and recognition. If authority is doubted, it has no such power and every disagreement further atomizes the Church. This also enables them to set up the Novus Ordo apparatus as being "the only institutional game in town," and Catholics are nothing if not institutional. This one ploy alone has brought many Catholics out of the Church and into the Novus Ordo religion.

Most Traditional Catholics have at least some vague intuition that we Traditional Catholics, alone but all together, comprise the real Catholic Church, the real and true inheritor of that grand legacy of Catholic history and doctrine and morals and saints and miracles etc. But then when authority is spoken of as something that doesn't exist at all, that lie divides us by depriving us of any mechanism of resolving conflicts or of taking concerted action, and then leaves the Catholic with no apparent institutional place to turn but the heretics. But the real Church is also an institution, in fact the Divine institution which alone can possess real authority in all doctrinal, moral, liturgical, and spiritual things, and this implies the existence of authority within our own ranks. And who would that be but the usual Church officers, the traditionalist clergy, as we have them now, such as they are. A pope would be at the head, assuming someone existed of such description and willing to so function; under that the bishops of the Church, those whom I have identified herein and who can be shown to be truly Apostolic, and under them the priests of the Church, and then comes the consecrated religious and we laity. This chain of command has always existed and necessarily must and does exist among us real Catholics even today, despite the naysayers who deny the existence of all authority, as proven in *Sede Vacante!* Volume One, pages 26-33.

In parallel to that, I seemed to have supposed that mere supplied jurisdiction and epikeia would be "enough" (perhaps backed up with the *Lumen Gentium* "delegated jurisdiction") to sustain the Church, as that was so often so spoken of by others whom I assumed understood these things better than I. It was in learning about the nature of the authority of the Church that I realized that the Church positively, dogmatically, cannot be reduced to persons all functioning under the provisions of supplied jurisdiction due to common error, that so many clerics, unable or unwilling to explore their true canonical status, seem to fall back upon as if that were enough to justify their ministries. It IS enough to justify this or that individual ministry, especially for someone who has no canonical mission but only a valid power of Orders, a correct doctrine, and a real desire to help the Catholic Church, but if all ministries around the world be so reduced to such functioning then that amounts to there being no actual hierarchical Church anymore, a dogmatic impossibility.

On a minor note, I now refrain from speaking of Traditional Catholics as a "movement" but instead as a "community." This is because a "movement" often and most appropriately refers to a political group, and often a subversive one. We were never merely some "movement" within the Church or wherever, but the Church itself. I can only hope that the use of the word "community" does not lull us into inactivity; our duty remains to evangelize the whole world. At least the word "movement" suggested action, which itself was a good thing (and why I formerly used that word).

Finally, in this revision I have carefully refrained from calling the various leaders, topmost and localized, a "hierarchy," since specifically, when theologians speak of hierarchy they refer solely and specifically to that hierarchy which is set up by and comprising the sum total leadership of the Church. All else, though it may take on a "hierarchical-like" pyramidal organizational form, is best and most properly not so spoken of. I get this from Journet (Ibid. page 501):

> The theologians alone can give the word "hierarchy" its highest meaning. They do not use it indiscriminately for any subordination of powers whatever. The hierarchy, in its most general sense, is in their eyes, a sacred principate (*sacer principatus*), comprising several

co-ordinated degrees (*gradus et ordo*), and endowed with power and knowledge (*potestas et scientia*) to lead a multitude (*actio inducens ad finem*) to union with and likeness to God (*finis intentus*). The hierarchy is thus defined by its essential principles (*ordo, scientia, actio*) and by its end (*ad Deum unitas et similitudo*). But – as St. Thomas notes – its most formal definition is that taken from the end: union and conformity with God. That is why there is no hierarchy among the devils…"

For that reason, none but the Traditionalist clergy alone are ever spoken of as being "hierarchical" within this book, for to no others does such a description apply, at least in its "highest meaning" as given by Journet, and that is the meaning consistently used herein.

APPENDIX A

THE POPE CONDEMNS VATICAN II

We Catholics do not have to wait for some future reliable pope to come along and formally condemn the heretical implications and results of the Second Vatican Council. It so happens that many reliable popes have already spoken out quite clearly against the New (Novus Ordo) Religion even before its official creation at Vatican II. There is nothing miraculous about that. The Novus Ordo religion existed at least as far back as the French Revolution, both hiding inside the Church (in the twentieth century as the "Liturgical Movement" once driven underground by Pope Saint Pius X), and operating more openly outside the Church (primarily as the Old Catholics and Protestants). It was only with the Second Vatican Council that official recognition and approval was given to a religion which had been systematically condemned by every reliable pope who ever faced any aspect of it. Let us hear a very small sampling of what the reliable popes have had to say about the Novus Ordo religion.

The most obvious aspect of the New Religion is the introduction of a new form, or order, of the Mass, coupled with a total rejection of the Tridentine Mass. Hear what Session Seven of the council of Trent (confirmed by Pope Paul III) has to say about that:

> If anyone says that the received and approved rites of the Catholic Church, accustomed to be used in the administration of the sacraments, may be despised or omitted by the ministers without sin and at their pleasure, or may be changed by any pastor of the churches to other new ones, let him be *anathema*. (*Canons on the Sacraments In General*, Canon 13).

The phrase "any pastor," having no qualifiers, necessarily includes even the Supreme Pastor, namely the Pope. Yet Paul VI both despised the approved rites of the Catholic Church by virtually forbidding from 1971 onward their practice as "accustomed to be used in the administration of the sacraments" since long before the Council of Trent confirmed them, and also he changed all sacramental rites to new ones of doubtful validity. Paul VI therefore comes under the anathema of Paul III and the Holy Council of Trent.

Another phrase of interest in that quote is the reference to all rites "received and approved." Even if the new "mass" is regarded as being "approved" because Paul VI and those after him liked it, it still doesn't come under this protection on account of the fact that it was never "received," only "invented." The Tridentine Mass was received by the early Church from Christ and the Apostles as a part of Revelation, but the Novus Ordo Missae was invented by Bugnini and the various Protestants whose work he plagiarized and was guided by.

Let us now see what Pope Saint Pius V had to say about the Roman Missal as promulgated in his day:

> In perpetuity We grant and permit that they may by all means use this Missal in singing or reciting Mass in any church whatsoever without any scruple of conscience, without incurring any penalties, sentences, or censures; in order that they may be able to do this and be able to use this Missal freely and lawfully, We by virtue of Our Apostolic Office, and by virtue of the present document, We grant and permit this forever. No one may be required to offer Holy Mass in another way than has been determined by Us; no one, neither Pastors, Administrators, Canons, Chaplains, and other secular priests of whatever Order; and We likewise determine and declare that no one be compelled or pressured by anyone to change this Missal, or that this letter should ever be recalled or its effectiveness restrained but that it may always stand firm and strong in all its vigor.
>
> …No one is allowed to go contrary to this letter which expresses Our permission, statute, regulation, mandate, precept, grant, Indult, declaration, or will and Our decree and prohibition; no one is allowed to act against it with rashness or temerity. But if anyone would presume to attempt this let him know that he will incur the wrath of Almighty God and of Saints Peter and Paul, His Apostles (*Quo Primum Decree*).

Although he talks of reform in other parts of this decree, that reform was a mere smoothing out of local variations. In such-and-such a diocese, an extra prayer might be said at the end of the Mass; in so-and-so's parish, an extra chant might be sung before the reading of the Gospel; in another place some small rubric might be added or even deleted at some point in their Mass. The main thing Pope St. Pius V wanted to put a stop to was any tendency on the part of priests to compromise their liturgy in order to escape Protestant persecution.

Certain priests in hostile lands were already starting to make various liturgical concessions to the surrounding Protestant culture. These minor variations were so subtle that few parishioners even noticed them at all, yet the damage they were causing to the Faith and the Mass was quite serious. What the reform of Pope Pius V did was enforce a uniform consensus of what the Mass had always been. Paul VI rashly and temerariously acted against this decree and thus incurs the wrath of Almighty God and His Apostles, Peter and Paul. May God have mercy on his soul?

Pope Saint Pius V also attempted to safeguard the words by which Jesus Christ confected the sacrament by saying:

> Defects may arise in respect of the formula, if anything is wanting to complete the actual words of consecration. The words of consecration, which are the formative principle of this sacrament, are as follows: **For this is My Body** and: **For this is the Chalice of My Blood, of the new and everlasting Testament, the Mystery of Faith, which shall be shed for you and for many unto the remission of sins.** If any omission or alteration is made in the formula of consecration of the Body and Blood, involving a change in meaning, the consecration is invalid. An addition made without altering the meaning does not invalidate the consecration but the celebrant commits a mortal sin (*De Defectibus Decree*).

At best, the formula promulgated under Paul VI could be considered ambiguous, leaving one to wonder whether the priest in using the new formula means to say what the old formula said or not. Clearly, the priest who uses the Novus Ordo Missae formula sins mortally, and even the validity of the Mass he says is open to doubt. Witness also what the dogmatic Catechism of the Council of Trent (also approved by Pope Saint Pius V) has to say about the use of the Novus Ordo forms as long promulgated in the vernacular versions:

> The additional words *for you and for many*, are taken, some from Matthew, some from Luke, but were joined together by the Catholic Church under the guidance of the Spirit of God. They serve to declare the fruit and advantage of His Passion. For if we look to its value, we must confess that the Redeemer shed His blood for the salvation of all; but if we look to the fruit which mankind have received from it, we shall easily find that it pertains not unto all, but to many of the human race. When therefore (our Lord) said: *For you*, He meant either those who were present, or those chosen from among the Jewish people, such as were, with the exception of Judas, the disciples with whom He was

speaking. When He added, *And for many*, He wished to be understood to mean the remainder of the elect from among the Jews or Gentiles.

With reason, therefore, were the words *for all* not used, as in this place the fruits of the Passion are alone spoken of, and to the elect only did His Passion bring the fruit of salvation. And this is the purport of the Apostle when he says: *Christ was offered once to exhaust the sins of many*; and also of the words of our Lord in John: *I pray for them; I pray not for the world, but for them whom thou hast given me, because they are thine.* (*Catechism of the Council of Trent*, pages 227–228, TAN Books edition).

Pope Pius XII had a lot to say about certain attempts being made in the Liturgical Movement to enforce or promulgate false, primitivistic forms of the Mass:

> It is neither wise nor laudable to reduce everything to antiquity by every possible device. Thus, to cite some instances, one would be straying from the straight path were he to wish the altar restored to its primitive table-form; were he to want black excluded as a color for the liturgical vestments; were he to forbid the use of sacred images and statues in the Churches; were he to order the crucifix so designed that the Divine Redeemer's Body shows no trace of His cruel sufferings; lastly were he to disdain and reject polyphonic music or singing in parts, even where it conforms to regulations issued by the Holy See.
>
> …Just as obviously unwise and mistaken is the zeal of one who in matters liturgical, would go back to the rites and usage of antiquity, discarding the new patterns introduced by disposition of Divine Providence to meet the changes of circumstance and situation.
>
> This way of acting bids fair to revive the exaggerated and senseless antiquarianism to which the illegal Council of Pistoria gave rise. It likewise attempts to reinstate a series of errors which were responsible for the calling of that meeting as well as those resulting from it, with grievous harm to souls, and which the Church, the ever watchful guardian of the "deposit of faith" committed to her charge by her Divine Founder, had every right and reason to condemn. For perverse designs and ventures of this sort tend to paralyze and weaken that process of sanctification by which the sacred Liturgy directs the sons of adoption to their Heavenly Father for their soul's salvation (*Mediator Dei*, paragraphs 62, 63, 64).

Keep in mind that the "usage of antiquity" of which he speaks is merely the hypothetical forms advocated by the Liturgical Movement, while the "new patterns introduced by disposition of Divine Providence" are those which were in use in his day (1947) and which had been used at least as far back as one finds any detailed description of Christian worship. One sees here the exact forms of the Novus Ordo Missae (the altar "restored" to a primitive table-form, black excluded as a color for the liturgical vestments, destruction and forbidding of the use of sacred images and statues in the Churches, the crucifix so designed that the Divine Redeemer's Body shows no trace of His cruel sufferings) being condemned as being of grievous harm to souls.

The notion that any liturgical prayers could be arbitrarily changed as to content or meaning is indirectly condemned by Pope Pius XII in *Mediator Dei* when he wrote:

> The entire Liturgy…has the Catholic faith for its content, inasmuch as it bears public witness to the faith of the Church.
>
> For this reason, whenever there was question of defining a truth revealed by God, the Sovereign Pontiff and the Councils in their recourse to the "theological sources," as they were called, have not seldom drawn many an argument from this sacred science of the Liturgy. For an example in point, Our Predecessor of immortal memory, Pius IX, so argued when he proclaimed the Immaculate Conception of the Virgin Mary. Similarly during the discussion of a doubtful or controversial truth, the Church and the Holy Fathers have not failed to look to the age-old and age-honored sacred rites for enlightenment. Hence the well-known and venerable maxim: "*Legem credendi lex statuat supplicandi*"—let the rule for prayer determine the rule of belief (*Mediator Dei*, paragraphs 47, 48).

What a grievous risk to Catholic teaching it would be if the liturgical prayers on which the indisputably infallible teaching of the Immaculate Conception are based on were to be changeable willy-nilly by any Church authority! Church doctors, theologians, and even popes have often quoted Liturgical prayers, basing their arguments on the exact turn of phrase they use, just as one might quote Scripture. Clearly, the one can be no more open to change than the other! And how can the "public witness to the faith" be so substantially altered without professing another, different, faith?

What about the "reconciliation service" which is increasingly taking the place of the Sacrament of Penance? While one may indeed find some devotional value or even remission of venial sins in such public prayers as the Confiteor recited at the beginning of the Mass (and maybe even, in its modified Novus Ordo form, at the "reconciliation service"), that can be no substitute for individual Absolution from a priest. Pope Pius XII wrote of this:

> The same result follows from the opinions of those who assert that little importance should be given to the frequent confession of venial sins. Far more important, they say, is that general confession which the Spouse of Christ, surrounded by her children in the Lord, makes each day by the mouth of the priest as he approaches the altar of God. As you well know, Venerable Brethren, it is true that venial sins may be expiated in many ways which are to be highly commended. But to ensure more rapid progress day by day in the path be highly commended. But to ensure more rapid progress day by day in the path off virtue, We will that the pious practice of frequent confession, which was introduced into the Church by the inspiration of the Holy Spirit, should be earnestly advocated. By it, genuine self-knowledge is increased, Christian humility grows, bad habits are corrected, spiritual neglect and tepidity are resisted, the conscience is purified, and grace is increased in virtue of the Sacrament itself. Let those, therefore, among the younger clergy who make light of or lessen esteem for frequent confession realize that what they are doing is alien to the Spirit of Christ and disastrous for the Mystical Body of our Savior. (*Mystici Corporis*, Paragraph 88).

For another example, let us look at the Sacrament of Extreme Unction. The Catechism of the Council of Trent teaches the following regarding the form of that Sacrament:

> The form of the Sacrament is the word and solemn prayer which the priest uses at each anointing: *By this Holy Unction may God pardon thee whatever sins thou hast committed by the evil use of sight, smell, or touch.*
> That this is the true form of this Sacrament we learn from these words of St. James: *Let them pray over him…and the prayer of faith shall save the sick man.* Hence we can see that the form is to be applied by way of prayer. The Apostle does not say of what particular words that prayer is to consist; but this form has been handed down to us by the faithful tradition of the Fathers, so that all the Churches retain the form observed by the Church of Rome, the mother and mistress of all Churches. (*Catechism of the Council of Trent*, pages 309, TAN Books edition).

Regarding the notion that the form could ever be changed, Session Fourteen of the council of Trent (confirmed by Pope Julius III) says:

> If anyone says that the rite and usage of extreme unction which the holy Roman Church observes is at variance with the statement of the blessed

> Apostle James, and is therefore to be changed and may without sin be despised by Christians, let him be anathema. (*Canons Concerning The Sacrament of Extreme Unction, Canon 3*).

And yet the Novus Ordo religion despises that rite by deleting the form and changing it beyond recognition into a mythical "Sacrament of the Anointing of the Sick," even omitting from its "form" any reference to the forgiveness of sins.

Pope Pius XII also had something to say about the form required to perform a valid consecration of a bishop:

> Regarding the matter and the form used in the conferring of each of the Orders, We, by the Apostolic Authority, ordain and decree the following:…in the ordination or consecration of a bishop the matter is the imposition of the hands which is done by the bishop consecrator. The form consists in the words of the preface, of which the following are essential and therefore required for validity: Fulfill in Thy priest the completion of Thy ministry, and adorned in the ornaments of all glorification sanctify him with the moisture of heavenly unguent. (*Sacramentum Ordinis*).

What right has any man to attempt to change or improve upon what the Pope has ordained and decreed?

Enough on the Sacraments. Anyone reading the documents of Vatican II would have to admit that they perfectly fit the description of books written by the Modernist heretics given by Pope Saint Pius X:

> In their writings and addresses they seem not infrequently to advocate doctrines which are contrary one to the other, so that one would be disposed to regard their attitude as double and doubtful. But this is done deliberately and advisedly, and the reason of it is to be found in their opinion as to the mutual separation of science and faith. Thus in their books one finds some things which might well be approved by a Catholic, but on turning over the page one is confronted by other things which might well have been dictated by a rationalist. (*Pascendi Dominici Gregis*, paragraph 18, The Methods of Modernists).

Is that not a sufficiently perfect description of the documents of Vatican II so as to raise suspicion as to the intents of their authors? A close study of his encyclical *Pascendi* would reveal many similarities between the Modernists whose religion Pope Saint Pius X defines as "the synthesis of all heresies," (*Pascendi Dominici*

Gregis, Paragraph 39) and the religion created and instituted as a result of Vatican II. For example:

> It remains for Us now to say a few words about the Modernist as reformer. From all that has preceded, it is abundantly clear how great and how eager is the passion of such men for innovation. In all Catholicism there is absolutely nothing on which it does not fasten. They wish philosophy to be reformed…Regarding worship, they say, the number of external devotions is to be reduced, and steps must be taken to prevent their further increase, though, indeed, some of the admirers of symbolism are disposed to be more indulgent on this head.
>
> They cry out that ecclesiastical government requires to be reformed in all its branches, but especially in its disciplinary and dogmatic departments. They insist that both outwardly and inwardly it must be brought into harmony with the modern conscience, which now wholly tends towards democracy; a share in ecclesiastical government should therefore be given to the lower ranks of the clergy, and even to the laity, and authority, which is much too concentrated, should be decentralized. The Roman Congregations, and especially the *Index* and the *Holy Office*, must be likewise modified. The ecclesiastical authority must alter its line of conduct in the social and political world; while keeping outside political organizations, it must adapt itself to them, in order to penetrate them with its spirit.
>
> With regard to morals, they adopt the principle of the Americanists, that the active virtues are more important than the passive, and are to be encouraged in practice. They ask that the clergy should return to their primitive humility and poverty, and that in their ideas and action they should admit the principles of Modernism; and there are some who, gladly listening to the teaching of their Protestant masters, would desire the suppression of the celibacy of the clergy. What is there left in the Church which is not to be reformed by them and according to their principles? (*Pascendi Dominici Gregis*, paragraph 38, The Modernist as reformer).

Is that not the same as the dilution of authority mandated by Vatican II? "Popes" must share their authority with a "College of Bishops"; bishops no longer run their dioceses but must follow guidelines voted on in "Bishop's Congresses," parish priests must allow their parishes to be run by "parish councils" run by the laity. And whatever happened to the *Index* and the *Holy Office*?

The present confusion results from this decentralization of authority mandated at Vatican II, exactly what Pope Saint Pius X said was what the Modernists want. Let us see what else he had to say about Congresses of Priests and Bishops:

We have already mentioned congresses and public gatherings as among the means used by Modernists to propagate and defend their opinions. In the future, Bishops shall not permit congresses of priests except on very rare occasions. When they do permit them it shall only be on condition that matters appertaining to the Bishops or the Apostolic See be not treated in them, and that no resolutions or petitions be allowed that would imply a usurpation of sacred authority, and that absolutely nothing be said in them which savors of Modernism, presbyterianism, or laicism. At congresses of this kind, which can only be held after permission in writing has been obtained in due time and for each case, it shall not be lawful for priests of other dioceses to be present without the written permission of their Ordinary. Further, no priest must lose sight of the solemn recommendation of Leo XIII: "Let priests hold as sacred the authority of their pastors, let them take it for certain that the sacerdotal ministry, if not exercised under the guidance of the Bishops, can never be either holy, or very fruitful, or worthy of respect." (*Pascendi Dominici Gregis*, Paragraph 54, Congresses).

And see the rest of it! Philosophy is changed. External devotions are cut way back. Americanist principles of religious indifference are everywhere taught and imposed. The ceremonial pomp and circumstance of the clergy is stripped down to poor essentials (if even that much), and pressure is being put on them to cease being celibate. It is as if vandals have taken over, stripping the Church buildings themselves of the beauty they once had, as if they adhere to the notion condemned by Pius IX that:

> The Church has not the innate and legitimate right of acquiring and holding property. (*Syllabus of Errors*, Condemned Proposition 26).

In the realm of philosophy, St. Thomas Aquinas, the Angelic Doctor, and patron Saint of Scholasticism has always been totally upheld by all reliable popes since his appearance. Pope Pius IX condemned those who were wanting to push aside the Angelic Doctor and his Scholasticism:

> The method and principles according to which the ancient scholastic Doctors cultivated Theology are in no way suited to the necessities of our times and to the progress of the sciences. (*Syllabus of Errors*, Condemned Proposition 13).

And again, Pope Pius XII said:

> If one considers all this well, he will easily see why the Church demands that future priests be instructed in philosophy "according to the

method, doctrine, and principles of the Angelic doctor," (C. I. C., can. 1366, 2.) since, as we well know from the experience of centuries, the method of Aquinas is singularly preeminent both for teaching students and for bringing truth to light; his doctrine is on harmony with divine revelation, and is most effective both for safeguarding the foundation of the faith, and for reaping, safely and usefully, the fruits of sound progress. (A. A. S. Vol. 38, 1946, p. 387.)

    How deplorable it is then that this philosophy, received and honored by the Church, is scorned by some, who shamelessly call it outmoded in form and rationalistic, as they say, in its method of thought. They say that this philosophy upholds the erroneous notion that there can be a metaphysic that is absolutely true; whereas in fact, they say, reality, especially transcendent reality, cannot better be expressed than by disparate teachings, which mutually complete each other, although they are in a way mutually opposed. Our traditional philosophy, then, with its clear exposition and solution of questions, its accurate definitions of terms, its clear-cut distinctions, can be, they concede, useful as a preparation for scholastic theology, a preparation quite in accord with medieval mentality; but this philosophy hardly offers a method of philosophizing suited to the needs of our modern culture. They allege, finally, that our perennial philosophy is only a philosophy of immutable essences, while the contemporary mind must look to the existence of things and to life, which is ever in flux. While scorning our philosophy, they extol other philosophies of all kinds, ancient and modern, oriental and occidental, by which they seem to imply that any kind of philosophy or theory, with a few additions and corrections if need be, can be reconciled with Catholic dogma. No Catholic can doubt how false this is, especially where there is a question of those fictitious theories they call immanentism, or idealism, or materialism, whether historic or dialectic, or even existentialism, whether atheistic or simply the type that denies the validity of reason in the field of metaphysics. (*Humani Generis*, Paragraphs 31,32).

And yet today, where alone is Thomistic theology taught, but in the traditional seminaries? Throwing aside the Angelic Doctor is barely the surface of the problem. At its base, the Novus Ordo religion rests upon a total denial of Divine Revelation. In the Novus Ordo religion, "truths" of religion owe their existence to the thinking processes of mankind, and God is merely a concept which evolves with the needs of man, a notion Pius IX condemned:

> All the truths of religion have their origin in the innate vigor of the human reason: hence it follows that reason is the sovereign guide by

> which man can and ought to attain to the knowledge of all truths of every kind. (*Syllabus of Errors*, Condemned Proposition 4).

And again:

> There exists no Supreme Being, perfect in His Wisdom and in His Providence and distinct from the universe. God is identical with nature and consequently subject to change. God is evolving in man and in the world, and all things are God and have the very substance of God. (*Syllabus of Errors*, Condemned Proposition 1).

As a result, they feel free to tamper with the sources of religious dogma, namely Divine Revelation. Pope Pius IX condemned the Novus Ordo teaching that:

> Divine Revelation is imperfect and consequently subject to a continual and indefinite progress which corresponds to the progress of human reason. (*Syllabus of Errors*, Condemned Proposition 5).

And yet one sees that notion taught again and again in their treatment of Sacred Scripture, especially in the seminaries where the professors convey an attitude to their students, which was condemned by Pope Saint Pius X, to the effect that:

> They display excessive simplicity or ignorance who believe that God is really the author of the Sacred Scriptures. (*Lamentabili Sane*, Condemned Proposition 9).

In the Novus Ordo seminaries, the condemned (by Pius X) claim is repeatedly taught as truth that:

> Until the time the canon was defined and constituted, the Gospels were increased by additions and corrections. Therefore there remained in them only a faint and uncertain trace of the doctrine of Christ. (*Lamentabili Sane*, Condemned Proposition 15).

If the Bible were really written that way, and the Sacred Liturgy as well, why not change it willy-nilly to fit the "needs" of modern man? The Novus Ordo religion falsely accuses the early Church of doing what they themselves obviously feel free to do, an accusation already condemned by Pope Saint Pius X:

> In many narrations the Evangelists recorded, not so much things that are true, as things which, even though false, they judged to be

more profitable for their readers. (*Lamentabili Sane*, Condemned Proposition 14).

It is one thing to say that the early Church chose from among the Apostolic writings those which were most edifying for use in the Bible and the Sacred Liturgy, but quite another to claim that those Apostolic writings were written and modified at will by any Church authorities for such a purpose. To them, Sacred Scripture and Sacred Liturgy are nothing but tools of propaganda.

With nothing solid to stand on, even their basic dogmatic beliefs are subject to change, and therefore reduction. Pope Pius XII had the following to say about those in the Church who were already fomenting for the Novus Ordo religion:

> In theology some want to reduce to a minimum the meaning of dogmas; and to free dogma itself from terminology long established in the Church and from philosophical concepts held by Catholic teachers,…They cherish the hope that when dogma is stripped of the elements which they hold to be extrinsic to divine revelation it will compare advantageously with the dogmatic opinions of those who are separated from the unity of the Church and that in this way they will gradually arrive at a mutual assimilation of Catholic dogma with the tenets of the dissidents.
>
> Moreover they assert that when Catholic doctrine has been reduced to this condition, a way will be found to satisfy modern needs, that will permit of dogma being expressed by the concepts of modern philosophy, whether of immanentism or idealism or existentialism or any other system. Some more audacious affirm that this can and must be done, because they hold that the mysteries of faith are never expressed by truly adequate concepts but only by approximate and ever changeable notions, in which the truth is to some extent expressed, but is necessarily distorted. Wherefore they do not consider it absurd, but altogether necessary, that theology should substitute new concepts in place of the old ones in keeping with the various philosophies which in the course of time it uses as its instruments, so that it should give human expression to divine truths in various ways which are even somewhat opposed, but still equivalent, as they say. They add that the history of dogmas consists in the reporting of the various forms in which revealed truth has been clothed, forms that have succeeded one another in accordance with the different teachings and opinions that have arisen over the course of the centuries.
>
> It is evident from what We have already said, that such tentatives not only lead to what they call dogmatic relativism, but that they actually contain it. The contempt of doctrine commonly taught and

of the terms in which it is expressed strongly favor it. (*Humani Generis*, Paragraphs 14–16).

Little did he realize that the very persons he was speaking against would soon take over what many would mistake for the Church, and change it drastically. Pope Pius XI condemned the notion that the Church could ever be changed, when he wrote:

> Hence, not only must the Church still exist today and continue always to exist, but it must ever be exactly the same as it was in the days of the Apostles. Otherwise we must say—which God forbid—that Christ has failed in His purpose, or that He erred when He asserted of His Church that the gates of hell should never prevail against it (Mt. 16:18) (*Mortalium Animos*, Paragraph 6).

Looking at the establishment ruled from the Vatican today, it obviously has been changed drastically. However, the Traditional Catholic community continues to be exactly the same as it was in the days of the Apostles, and is therefore the true fulfillment of Christ's prophecies and the lawful object of His purposes.

Some years later on, Pope Pius XI warned about those who were changing the religious conceptions in a similar manner to what the Novus Ordo religion would do:

> You must be especially alert, Venerable Brethren, when fundamental religious conceptions are robbed of their intrinsic content and made to mean something else in a profane sense. (*On the Church In Germany*, Paragraph 23).

Despite the Pope's teaching, the functionaries of the Novus Ordo leadership no longer feel answerable to the teachings contained in Divine Revelation, or established as dogmas, but feel free to do as they please, even to the point of attempting to change the organic constitution of the Church with their "Dogmatic Constitution on the Church," *Lumen Gentium*. Pope Saint Pius X has already condemned the very thought years in advance:

> The organic constitution of the Church is not immutable. Like human society, Christian society is subject to a perpetual evolution. (*Lamentabili Sane*, Condemned Proposition 53).

The newly reconstituted Novus Ordo Church of the People of God has been totally redirected towards questionable social and political ends. They have supplanted the determined body of Christian doctrine applicable to all times and all men as they exploit the Church as merely a religious movement to be adapted to different times and places as suits their nefarious purposes, precisely what Pope Saint Pius X condemned:

> Christ did not teach a determined body of doctrine applicable to all times and all men, but rather inaugurated a religious movement adapted or to be adapted to different times and places. (*Lamentabili Sane*, Condemned Proposition 59).

During the reign of Pope Pius X, a certain faction in the French Church known as the Sillon was already developing the Novus Ordo religion, and the Pope expressed his concerns about it:

> We fear that worse is to come: the end result of this developing promiscuousness, the beneficiary of this cosmopolitan social action, can only be a Democracy which will be neither Catholic, nor Protestant, nor Jewish. It will be a religion (for Sillonism, so the leaders have said, is a religion) more universal than the Catholic Church, uniting all men to become brothers and comrades at last in the 'Kingdom of God', "We do not work for the Church, we work for mankind." (*Our Apostolic Mandate*, Paragraph 39).

Unfortunately, the Sillonists simply expanded and became a Socialist faction existing in so many more countries that Pope Pius XI had to say of them:

> Accordingly, Venerable Brethren, you can well understand with what great sorrow We observe that not a few of Our sons, in certain regions especially, although We cannot be convinced that they have given up the true faith and right will, have deserted the camp of the Church and gone over to the ranks of Socialism, some to glory openly in the name of socialist and to profess socialist doctrines, others through thoughtlessness or even, almost against their wills to join associations which are socialist by profession or in fact. (*Quadragesimo Anno*, Paragraph 123).

How very like in the Novus Ordo, especially with its "Liberation Theology," to desert the camp of the Church for a bankrupt political philosophy. And many even have been dragged along somewhat against their wills, as for example, the conservative Novus Ordo believers who secretly would rather be Catholics, but who mistakenly feel that they have to follow their Non-Catholic diocesan "bishops," even into religious error.

Nor can one properly say that the Post-Vatican II Church has found a viable compromise between Socialism or Communism, and the true Faith. Of such an absurd notion, Pope Pius XI said:

> Yet let no one think that all the socialist groups or factions that are not communist have, without exception, recovered their senses to this extent either in fact or in name. For the most part they do not reject the class struggle or the abolition of ownership, but only in some degree modify them. Now if these false principles are modified and to some extent erased from the program, the question arises, or rather is raised without warrant by some, whether the principles of Christian truth cannot perhaps be also modified to some degree and be tempered so as to meet Socialism half-way and, as it were, by a middle course, come to agreement with it. There are some allured by the foolish hope that socialists in this way will be drawn to us. A vain hope! Those who want to be apostles among socialists ought to profess Christian truth whole and entire, openly and sincerely, and not connive at error in any way. If they truly wish to be heralds of the Gospel, let them above all strive to show to socialists that socialist claims, so far as they are just, are far more strongly supported by the principles of Christian faith and much more effectively promoted through the power of Christian charity.
>
> But what if Socialism has really been so tempered and modified as to the class struggle and private ownership that there is in it no longer anything to be censured on these points? Has it thereby renounced its contradictory nature to the Christian religion?...We make this pronouncement: Whether considered as a doctrine, or an historical fact, or a movement, Socialism, if it remains truly Socialism, even after it has yielded to truth and justice on the points we have mentioned, cannot be reconciled with the teachings of the Catholic Church because its concept of society itself is utterly foreign to Christian truth. (*Quadragesimo Anno*, Paragraphs 116, 117).

And what was the ultimate goal of the Sillonists and those Socialists and Communists who followed in their footsteps? To get progressively higher and higher Vatican authorities to defect from the Faith, until even the Pope should become reconciled with the modernist errors, a notion which Pius IX condemned:

> The Roman Pontiff can and ought to be reconciled and come to terms with Progress, Liberalism and modern Culture (or Civilization). (*Syllabus of Errors*, Condemned Proposition 80).

Alas for today's Catholics, John XXIII, Paul VI, and all Vatican leaders since have been completely reconciled and brought to terms with "Progress, Liberalism and modern Culture (or Civilization)." As the Pope feared, the Sillonists finally got their way.

No real defense for all of this manipulation and continual upset has ever been put forth, for it cannot be justified. It is a grave and ongoing injustice which the current leaders and functionaries of the Vatican establishment indulge in for no other reason than the fact that they can, as if "Might makes Right," or:

> Right consists in the material fact: all the duties of men are a word devoid of meaning and all human happenings have force of right. (*Syllabus of Errors*, Condemned Proposition 59).

Now let us take a look at what the popes have said about other religions, ecumenism with these other religions, and the duties of the State with regards true and false religions. I start here with Pope Gregory XVI:

> We now come to another and most fruitful cause of the evils which at present afflict the Church and which We so bitterly deplore; We mean indifferentism, or that fatal opinion everywhere diffused by the craft of the wicked, that men can by the profession of any faith obtain the eternal salvation of their souls, provided their life conforms to justice and probity. But in a question so clear and evident it will undoubtedly be easy for Us to pluck up from amid the people confided to your care so pernicious an error. The apostle warns us of it: "One God, one faith, one baptism." Let them tremble then who imagine that every creed leads by an easy path to the port of felicity; and reflect seriously on the testimony of our Savior Himself, that those are against Christ who are not with Christ, and that they miserably scatter by the fact that they gather not with Him, and that consequently they will perish eternally without any doubt, if they do not hold to the Catholic Faith, and preserve it entire and without alteration. (*Mirari Vos*, paragraph 13).

This condemnation of indifferentism, or the treating of all religions, true and false, as if they were all of equal value, continued with Pope Pius IX who condemned the following heresies:

> Every man is free to embrace and profess that religion which, guided by the light of reason, he shall have come to consider as true. (*Syllabus of Errors*, Condemned Proposition 15).

> Men can find the way of eternal salvation and reach eternal salvation in any form of religious worship. (*Syllabus of Errors*, Condemned Proposition 16).
>
> Protestantism is nothing else than a different form of the same True Christian Religion, and in it one can be as pleasing to God as in the Catholic Church. (*Syllabus of Errors*, Condemned Proposition 18).

Pope Pius X, in condemning Sillonism, also faulted Sillonism for committing the very same error, when he said:

> What are we to think of this appeal to all heterodox, and to all the unbelievers, to prove the excellence of their convictions in the social sphere in a sort of apologetic contest? Has not this contest lasted for nineteen centuries in conditions less dangerous for the faith of Catholics? And was it not all to the credit of the Catholic Church? What are we to think of this respect for all errors, and of this strange invitation made by a Catholic to all the dissidents to strengthen their convictions through study so that they may have more and more abundant sources of fresh forces? What are we to think of an association in which all religions and even Free-Thought may express themselves openly and in complete freedom? For the Sillonists who, in public lectures and elsewhere, proudly proclaim their personal faith, certainly do not intend to silence others, nor do they intend to prevent a Protestant from asserting his Protestantism, and the skeptic from affirming his skepticism. Finally, what are we to think of a Catholic who, on entering his study group, leaves his Catholicism outside the door so as not to alarm his comrades? [Therefore] the social action of the Sillon is no longer Catholic. (*Our Apostolic Mandate*, Paragraphs 37, 38).

Pope Pius XI continued this teaching against putting religious error on the level with truth in his Encyclical *Mortalium Animos*, which certainly bears reading straight through:

> Assured that there exist few men who are entirely devoid of the religious sense, they seem to ground on this belief a hope that all nations, while differing indeed in religious matters, may yet without great difficulty be brought to fraternal agreement on certain points of doctrine which will form a common basis of the spiritual life. With this object, congresses, meetings, and addresses are arranged, attended by a large concourse of hearers, where all without distinction, unbelievers of every kind as well as Christians, even those who unhappily have

> rejected Christ and denied His divine nature or mission, are invited to join in the discussion. Now, such efforts can meet with no kind of approval among Catholics. They presuppose the erroneous view that all religions are more or less good and praiseworthy, inasmuch as all give expression, under various forms, to that innate sense which leads men to God and to the obedient acknowledgment of His rule. Those who hold such a view are not only in error; they distort the true idea of religion, and thus reject it, falling gradually into naturalism and atheism. To favor this opinion, therefore, and to encourage such undertakings is tantamount to abandoning the religion revealed by God. (*Mortalium Animos*, Paragraph 2).

See how friendly meetings with unbelievers of every sort are not for Catholics to participate in. Yet Vatican II itself turned that principle on its head with all its talk of "dialogue." Pope Pius XI continues:

> There are actually some, though few, who grant to the Roman Pontiff a primacy of honor and even a certain power or jurisdiction; this, however, they consider to arise not from divine law but merely from the consent of the faithful. Others, again, even go so far as to desire the Pontiff himself to preside over their mixed assemblies. For the rest, while you may hear many non-Catholics loudly preaching brotherly communion in Jesus Christ, yet not one will you find to whom it ever occurs with devout submission to obey the Vicar of Jesus Christ in his capacity of teacher or ruler. Meanwhile they assert their readiness to treat with the Church of Rome, but on equal terms, as equals with an equal. But even if they could so treat, there seems little doubt that they would do so only on condition that no pact into which they might enter should compel them to retract those opinions which still keep them outside the fold of Christ.
>
> This being so, it is clear that the Apostolic See can by no means take part in these assemblies, nor is it in any way lawful for Catholics to give such enterprises their encouragement or support If they did so, they would be giving countenance to a false Christianity quite alien to the one Church of Christ. (*Mortalium Animos*, Paragraphs 8, 9).

As one can see, the idea of having the Supreme Pontiff lead or participate in such meetings is unthinkable, so how can one reconcile that with John Paul II's behavior at Assisi and many other occasions? Incidentally, Pope Pius IX also condemned the notion that the Church is at fault for the divisions of Christianity, such as between the Catholic Church and the schismatic East Orthodox:

> The exorbitant pretension of the Roman Pontiffs contributed to the division of the Church into Eastern and Western. (*Syllabus of Errors*, Condemned Proposition 38).

And yet there went John Paul II apologizing to the whole world on behalf of "the Church" for the alleged "sins" of the Church, such as causing division with the schismatic East Orthodox, the treatment of Galileo, or the persecution of the Jews at the hands of the Nazis. Never mind the fact that Galileo was well treated and only his theological conclusions were condemned, not his scientific theories, nor that Pope Pius XII sacrificed much to rescue as many Jews from Nazi tyranny as possible. The teaching of the Church against ecumenism and treating all religions alike whether true or false continues with Pope Pius XII:

> Another danger is perceived which is all the more serious because it is more concealed beneath the mask of virtue. There are many who, deploring disagreement among men and intellectual confusion, through an imprudent zeal for souls, are urged by a great and ardent desire to do away with the barrier that divides good and honest men; these advocate an "eirenism" according to which, by setting aside the questions which divide men, they aim not only at joining forces to repel the attacks of atheism, but also at reconciling things opposed to one another in the field of dogma. And as in former times some questioned whether traditional apologetics of the Church did not constitute an obstacle rather than a help to the winning of souls for Christ, so today some are presumptive enough to question seriously whether theology and theological methods, such as with the approval of ecclesiastical authority are found in our schools, should not only be perfected, but also completely reformed, in order to promote the more efficacious propagation of the kingdom of Christ everywhere throughout the world among men of every culture and religious opinion (*Humani Generis*, paragraph 11).

It can therefore be safely concluded that the notion that all religions can be treated the same, or be valid means of salvation, or that dialogue and mixed assemblies between Catholic and unbeliever are a good thing can be regarded as totally condemned for all time. That is the constant teaching of the Church as demonstrated over an entire century. The same principle applies regarding the activity of secular States, namely that the secular State has a moral duty to the Church. Let us see some more notions condemned by Pope Pius IX:

> The Ecclesiastical Power must not exercise its authority without permission and assent of the civil government. (*Syllabus of Errors*, Condemned Proposition 20).
>
> Nay, more, even in clerical seminaries, the method to be adopted in studies is subject to the control of the civil authority. (*Syllabus of Errors*, Condemned Proposition 46).
>
> The laws of morality do not require the divine sanction and there is absolutely no need that human laws should be in conformity with natural law or receive from God the power of obliging in conscience. (*Syllabus of Errors*, Condemned Proposition 56).
>
> In (or by) natural law, the bond of matrimony is not indissoluble, and in various cases a divorce in the strict sense of the term can be sanctioned by the civil authority. (*Syllabus of Errors*, Condemned Proposition 67).
>
> At the present day it is no longer advantageous that the Catholic religion should be considered as the only religion of the State to the exclusion of all other forms of worship. (*Syllabus of Errors*, Condemned Proposition 77).
>
> Accordingly, it is a matter for commendation that, in certain Catholic countries, the law has provided that foreigners who come to live there enjoy the public exercise of their particular forms of religious worship. (*Syllabus of Errors*, Condemned Proposition 78).

Pope Leo XIII had the same thing to say about the relation between Church and State, and the necessity of the State to profess one religion, the true one:

> This [mistaken] kind of liberty, if considered in relation to the State, clearly implies that there is no reason why the State should offer any homage to God, or should desire any public recognition of Him; that no one form of worship is to be preferred to another, but that all stand on an equal footing, no account being taken of the religion of the people, even if they profess the Catholic faith. But, to justify this, it must needs be taken as true that the State has no duties towards God, or that such duties, if they exist, can be abandoned with impunity, both of which assertions are manifestly false. For it cannot be doubted but that, by the will of God, men are united in civil society; whether its component parts be considered; or its form, which implies authority; or the object of its existence; or the abundance of the vast services which it renders to man. God it is who has made man for society, and has placed him in the company of others like himself, so that what was wanting to his nature, and beyond his attainment if left to his own resources, he might obtain by association with others. Wherefore, civil society must acknowledge

God as its Founder and Parent, and must obey and reverence His power and authority. Justice therefore forbids, and reason itself forbids, the State to be godless; or to adopt a line of action which would end in godlessness—namely, to treat the various religions (as they call them) alike, and to bestow upon them promiscuously equal rights and privileges. Since, then, the profession of one religion is necessary in the State, that religion must be professed which alone is true, and which can be recognized without difficulty, especially in the Catholic States, because the marks of truth are, as it were, engraven upon it. This religion, therefore, the rulers of the State must preserve and protect, if they would provide—as they should do—with prudence and usefulness for the good of the community. For public authority exists for the welfare of those whom it governs; and, although its proximate end is to lead men to the prosperity found in this life, yet, in so doing, it ought not diminish, but rather to increase, man's capability of attaining to the supreme good in which his everlasting happiness consists: which can never be attained if religion be disregarded. (*Libertas Praestantissimum*, Paragraph 21).

And again:

> His empire [Christ's] includes not only Catholic nations, not only baptized persons who, though of right belonging to the Church, have been led astray by error, or have been cut off from her by schism, but also all those who are outside the Christian faith; so that truly the whole of mankind is subject to the power of Jesus Christ. (*Annum Sacrum*, Paragraph 3)

Pope Pius XI clearly agreed with his predecessors Pope Pius IX and Pope Leo XIII when he wrote that:

> If We ordain that the whole Catholic world shall revere Christ as King, We shall minister to the need of the present day, and at the same time provide an excellent remedy for the plague which now infects society. We refer to the plague of secularism, its errors and impious activities. This evil spirit, as you are well aware, Venerable Brethren, has not come into being in one day; it has long lurked beneath the surface. The empire of Christ over all nations was rejected. The right which the Church has from Christ Himself, to teach mankind, to make laws, to govern peoples in all that pertains to their eternal salvation, that right was denied. Then gradually the religion of Christ came to be likened to false religions and to be placed ignominiously on the same level with them. It was then put under the power of the State and tolerated

> more or less at the whim of princes and rulers. Some men went further, and wished to set up in the place of God's religion a natural religion consisting in some instinctive affection of the heart. There were even some nations who thought they could dispense with God, and that their religion should consist in impiety and the neglect of God. The rebellion of individuals and of nations against the authority of Christ has produced deplorable effects. We lamented these in the Encyclical *Ubi arcano*; We lament them to-day: the seeds of discord sown far and wide; those bitter enmities and rivalries between nations, which still hinder so much the cause of peace; that insatiable greed which is so often hidden under a pretense of public spirit and patriotism, and gives rise to so many private quarrels; a blind and immoderate selfishness, making men seek nothing but their own comfort and advantage, and measure everything by these; no peace in the home, because men have forgotten, or neglect their duty; the unity and stability of the family undermined; society, in a word, shaken to its foundations and on the way to ruin. (*Quas Primas*, Paragraph 24).

Pope Pius XI displayed a tremendous amount of depth and wisdom when he wrote about nations making their religion impiety and neglect of God, in other words, making Atheism a religion. Atheism is a religion, even if most people don't know that, but here we see that the Pope has so declared it. He also shows here the root source of the present social disorder. It is inevitable that nations which put all religions on the same level are all shaken to their foundations. Pope Pius XII also speaks of the duties of nations toward the true religion:

> But on the other hand, to tear the law of nations from its anchor in Divine law, to base it on the autonomous will of States, is to dethrone that very law and deprive it of its noblest and strongest qualities. Thus it would stand abandoned to the fatal drive of private interest and collective selfishness exclusively intent on the assertion of its own rights and ignoring those of others. (*Summi Pontificatus*, Paragraph 76).

I know this may sound somewhat unpatriotic, but it is the truth: Americanism is a heresy. In particular, the separation, or "wall," between Church and State is not only wrong, but complete nonsense. Every nation necessarily draws its ethos from its actual religion, be that Catholicism, Protestantism, Islam, Judaism, Atheism, or anything else. See what notion Pope Pius IX condemned:

> The Church should be separated from the State and the State from the Church. (*Syllabus of Errors*, Condemned Proposition 55).

Another aspect of the Americanist heresy is the notion that people should be free to say or print anything that strikes their fancy, without any reality check. Pope Leo XIII here points out quite clearly and firmly the folly of that position:

> We must now consider briefly *liberty of speech*, and liberty of the press. It is hardly necessary to say that there can be no such right as this, if it be not used in moderation, and if it pass beyond the bounds and end of all true liberty. For right is a moral power which—as We have before said and must again and again repeat—it is absurd to suppose that nature has accorded indifferently to truth and falsehood, to justice and injustice. Men have a right freely and prudently to propagate throughout the State what things soever are true and honorable, so that as many as possible may possess them; but lying opinions, than which no mental plague is greater, and vices which corrupt the heart and moral life, should be diligently repressed by public authority, lest they insidiously work the ruin of the State. The excesses of an unbridled intellect, which unfailingly end in the oppression of the untutored multitude, are no less rightly controlled by the authority of the law than are the injuries inflicted by violence upon the weak. And this all the more surely, because by far the greater part of the community is either absolutely unable, or able only with great difficulty, to escape from the illusions and deceitful subtleties, especially such as flatter the passions. If unbridled license of speech and of writing be granted to all, nothing will remain sacred and inviolate; even the highest and truest mandates of natures, justly held to be the common and noblest heritage of the human race, will not be spared. Thus, truth being gradually obscured by darkness, pernicious and manifold error, as too often happens, will easily prevail. Thus, too, license will gain what liberty loses; for liberty will ever be more free and secure in proportion as license is kept in fuller restraint. In regard, however, to all matters of opinion which God leaves to man's free discussion, full liberty of thought and of speech is naturally within the right of every one; for such liberty never leads men to suppress the truth, but often to discover it and make it known. (*Libertas Praestantissimum*, Paragraph 23).

Once again, the constant teaching of the Church is seen saying that secular nations have no right to attach themselves to false religions nor to treat all religions as equals. How does one reconcile that teaching with the activities of Paul VI and John Paul II in their signing of Concordats with the various nations, even the nations already Catholic, to the effect that all religions must be treated equally, and that religious information in the public forum is no longer to be checked for accuracy the way that public commercial information (advertising) is checked? As a result of what those two did, no Catholic nation remains.

It is bad enough that nations have been doing this on their own, but it is unthinkable to these faithful and reliable popes that the Church could ever direct nations to commit such a sin. Truly, almost every effect and consequence of Vatican II has been already systematically condemned by these indisputably reliable popes, and who are we to disagree with the Pope?

Furthermore, one cannot simply write these things off, saying "Well, none of those statements were spoken with the extraordinary and infallible authority of *ex cathedra* statements; therefore we can just ignore the published opinions of those popes." See what those same popes have to say about that opinion, starting with a notion which Pope Pius IX saw fit to condemn:

> The obligation strictly incumbent on Catholic teachers and writers is limited to those points which have been defined by the infallible judgment of the Church as dogmas of faith to be believed by all. (*Syllabus of Errors*, Condemned Proposition 22).

And yet again, by Pope Pius XII:

> Nor must it be thought that what is expounded in Encyclical Letters does not of itself demand consent, since in writing such Letters the Popes do not exercise the supreme power of their Teaching Authority. For these matters are taught with the ordinary teaching authority, of which it is true to say: "He who heareth you, heareth Me"; (Lk. 10:16) and generally what is expounded and inculcated in Encyclical Letters already for other reasons appertains to Catholic doctrine. But if the Supreme Pontiffs in their official documents purposely pass judgment on a matter up to that time under dispute, it is obvious that that matter, according to the mind and will of the same Pontiffs, cannot be any longer considered a question open to discussion among theologians. (*Humani Generis*, Paragraph 20).

Even with all that the reliable popes have taught, the problems had become so great that Pope Saint Pius X instituted the following Anti-Modernist Oath to be taken by every new priest who is being ordained, a law which Paul VI rescinded:

> I, [Name], firmly embrace and accept each and every definition that has been set forth and declared by the unerring teaching authority of the Church, especially those principal truths which are directly opposed to the errors of this day.

And first of all, I profess that God, the origin and end of all things, can be known with certainty by the natural light of reason from the created world, that is, from the visible works of creation, as a cause from its effects, and that, therefore, his existence can also be demonstrated.

Secondly, I accept and acknowledge the external proofs of revelation, that is, divine acts and especially miracles and prophecies as the surest signs of the divine origin of the Christian religion and I hold that these same proofs are well adapted to the understanding of all eras and all men, even of this time.

Thirdly, I believe with equally firm faith that the Church, the guardian and teacher of the revealed word, was personally instituted by the real and historical Christ when he lived among us, and that the Church was built upon Peter, the prince of the apostolic hierarchy, and his successors for the duration of time.

Fourthly, I sincerely hold that the doctrine of faith was handed down to us from the apostles through the orthodox Fathers in exactly the same meaning and always in the same purport. Therefore, I entirely reject the heretical misrepresentation that dogmas evolve and change from one meaning to another different from the one which the Church held previously. I also condemn every error according to which, in place of the divine deposit which has been given to the spouse of Christ to be carefully guarded by her, there is put a philosophical figment or product of a human conscience that has gradually been developed by human effort and will continue to develop indefinitely.

Fifthly, I hold with certainty and sincerely confess that faith is not a blind sentiment of religion welling up from the depths of the subconscious under the impulse of the heart and the motion of a will trained to morality; but faith is a genuine assent of the intellect to truth received by hearing from an external source. By this assent, because of the authority of the supremely truthful God, we believe to be true that which has been revealed and attested to by a personal God, our creator and Lord.

Furthermore, with due reverence, I submit and adhere with my whole heart to the condemnations, declarations, and all the prescripts contained in the encyclical *Pascendi* and in the decree *Lamentabili*, especially those concerning what is known as the history of dogmas.

I also reject the error of those who say that the faith held by the Church can contradict history, and that Catholic dogmas, in the sense in which they are now understood, are irreconcilable with a more realistic view of the origins of the Christian religion.

I also condemn and reject the opinion of those who say that a well-educated Christian assumes a dual personality-that of a believer and at the same time of a historian, as if it were permissible for a historian to hold things that contradict the faith of the believer, or to establish premises which, provided there be no direct denial of dogmas, would lead to the conclusion that dogmas are either false or doubtful.

Likewise, I reject that method of judging and interpreting Sacred Scripture which, departing from the tradition of the Church, the analogy of faith, and the norms of the Apostolic See, embraces the misrepresentations of the rationalists and with no prudence or restraint adopts textual criticism as the one and supreme norm.

Furthermore, I reject the opinion of those who hold that a professor lecturing or writing on a historico-theological subject should first put aside any preconceived opinion about the supernatural origin of Catholic tradition or about the divine promise of help to preserve all revealed truth forever; and that they should then interpret the writings of each of the Fathers solely by scientific principles, excluding all sacred authority, and with the same liberty of judgment that is common in the investigation of all ordinary historical documents.

Finally, I declare that I am completely opposed to the error of the modernists who hold that there is nothing divine in sacred tradition; or what is far worse, say that there is, but in a pantheistic sense, with the result that there would remain nothing but this plain simple fact—one to be put on a par with the ordinary facts of history—the fact, namely, that a group of men by their own labor, skill, and talent have continued through subsequent ages a school begun by Christ and his apostles.

I promise that I shall keep all these articles faithfully, entirely, and sincerely, and guard them inviolate, in no way deviating from them in teaching or in any way in word or in writing. Thus I promise, this I swear, so help me God, and these holy Gospels of God which I touch with my hand.

When Abp. Lefebvre finally got his audience with Paul VI shortly after the Mass at Lille, Paul VI accused Lefebvre of having his seminarians take an Oath against the Pope. The charge was false of course, but there may be a grain of truth to it after all, in the form of the Anti-Modernist Oath instituted by Pope Saint Pius X, which Abp. Lefebvre (as have all bishops who are truly Catholic) continued to insist all his seminarians take. If Paul VI were to have seen himself as Modernism personified, then he could quite reasonably interpret this Oath as being against himself personally. Such an admission would of course be a truly damning indictment of Paul VI and the position he took.

So there you have it, the teaching of pope after pope regarding the state of the modern Church. That is the teaching of all the popes, and therefore the teaching of any pope now or to come, insofar as that pope can be properly said to be reliably functioning as a truly Catholic pope. Given the large number of popes who have repeatedly reaffirmed the principle of having no share with those who walk in darkness or seek equality under law for truth and error, it should be quite clear that we are not dealing here with some single isolated statement from some single isolated pope, but a systematic doctrine most emphatically taught in the strongest possible terms by each and every reliable pope to whom the problem has been posed. The above papal quotes therefore represent the true and eternal teachings and principles of the Church, and no one, not even a pope, can deviate from the above without separating himself from the Barque of Peter.

"The Pope," whoever he is, or may yet come to be, must only affirm, enforce, and develop the teachings of the past reliable popes. No attempt to develop the teachings of the previous reliable popes could ever validly consist of, involve, nor entail in any way, the diminution, attenuation, or negation of anything taught by the previous reliable popes, including the above. And yet Vatican II seems to have somehow swept all of that away.

Allow me to conclude this section with the teaching of the First Vatican Council, as confirmed by Pope Pius IX:

> And since, by the divine right of Apostolic primacy, one Roman pontiff is placed over the universal Church, We further teach and declare that he is the supreme judge of the faithful, and that in all causes the decision of which belongs to the Church recourse may be had to his tribunal, but that none may re-open the judgment of the Apostolic See, than whose authority there is no greater, nor can any lawfully review his judgment. Wherefore they err from the right path of truth who assert that it is lawful to appeal from the judgments of the Roman pontiffs to an ecumenical council, as to an authority higher than that of the Roman pontiff (*Vatican Council I, Session 4, Chapter 3*).

Those who have implemented the New Religion have all done precisely what was condemned by the First Vatican Council and Pope Pius IX, namely "appeal from the judgments of the Roman pontiffs [as sampled here in this chapter] to an ecumenical council [Vatican II], as to an authority higher than that of the Roman pontiff." It is as if Vatican I were a prophecy of Vatican II, and a warning against it!

Those who embrace the heresy of Collegialism may fondly imagine that a pope may have more authority when backed by an ecumenical council than when he acts alone, but as we see, both the First Vatican Council and Pope Pius IX have already condemned than notion. In the opinion of this writer, Rome has already spoken against the New Religion; the Novus Ordo cause is finished.

APPENDIX B

QUESTIONS AND OBJECTIONS

GENERAL QUESTINIONS ABOUT TRADITIONALISM

Why Be a Traditional Catholic?

A Traditional Catholic is one who believes that Jesus Christ created a Church, and that the Roman Catholic Church is that Church. That being the case, the Traditional Catholic is one who endeavors to serve God in the manner that God directs rather than one's own preference, or the preference of those who are hostile to the Faith. A Traditional Catholic is simply a Catholic, as all serious and devout Catholics were only sixty years ago. In the more recent decades, many have slurred and distorted the commonly understood meaning of "Catholic" into something quite alien to what that word has meant for the past almost twenty centuries. By calling ourselves "Traditional Catholics" we identify ourselves as Catholics in the sense understood over most of that historical period rather than in the slurred and distorted sense many have mistakenly come to think it means today. By being Traditional Catholics, we are connected to God's own Church, and through that to God Himself. No one else has that kind of unity with God.

Are not the newer Catechisms easier to read and follow?

The problem with the new Catechisms is that they are tainted with the follies of our modern age, with Communism, Humanism, Atheism, pluralism, multiculturalism, collectivism, and so forth. As a result, many even basic Catholic doctrines get short shrift, if treated of at all, and frequently in vague terms that may mislead or else oversimplify to the point of distortion. Traditional Catholics have no patience for such blather, fluff, and psychobabble; they want the **straight stuff** as God gave it, which they can commit to memory and to heart and believe and live by.

No one likes having to "unlearn" something they previously made the mistake of learning and taking to heart. And no one likes to waste time learning something they already sense they must one day unlearn. While the older Catechisms immediately begin laying the groundwork, first with standard prayers we truly ought to know, and then by teaching us what we are to believe (the Creed), what we are to do (the Commandments), and how we are to obtain God's Grace (the Sacraments), and finally, in some cases (such as the Course in Religion by Fr. John Laux M. A., published by TAN), how we can know and prove to ourselves and to others that what we have been taught is really the Truth of God (Apologetics), the new "Catechisms" blather on and on about circles of family and community, political correctness, the environment (!?!), and how much better off we are today now that Vatican II has swept away all those strict and meaningless old rules from the Dark Ages.

Why do you want to go back to what was? Can't you just accept the fact that things have changed now, and are going to be different from now on?

This is not a question of going "back." There have always been Catholics holding to the traditional Latin Mass and Sacraments and teaching and Church throughout this entire period of mass defection. That Church is eternal and cannot be destroyed and has already begun a tremendous comeback which shall continue to expand. These Catholics have already marched forward into the new millennium precisely as Catholics were meant to march forward. Those who have gone over to the new ways have not matured their Faith but abandoned it. Everyone is obligated before God to return to the Faith (or keep it), and whoever fails to do so shall have much to answer for.

Why should we be so concerned about such things as the way the Mass or sacraments are said, when so many more important things such as abortion or starvation require more urgent attention?

The Biblical precedent applicable here is Israel's war against Amalek, as told in Exodus 17:8–16. An interesting point about this war is that when Moses held up his hands, Israel prevailed, but when Moses let down his hands, Amalek prevailed. When Moses got too tired to hold his hands up, they sat him down and two other men held his hands up. It may seem a bit odd that God should place such significance on what a man does with his hands, and that almost all alone at the top of a hill instead of down fighting on the battlefield, but there it is. Moses' raising of his hands represented his efforts as leader of the Israelites to push back the Amalekites, while his letting his hands fall meant his relaxing at his post.

Traditional Catholics today serve in that same capacity as the two men who held up Moses' hands. It is Traditional Catholics who keep the Pope's teaching alive, by living it to the fullest and so teaching others. It is the growing strength of the Traditional Catholic community which is turning back the tide of modernism, liberalism, hedonism, and utter dissipation which became so fashionable during the same period the Church was being destroyed.

People who fight abortion or hunger etc. do perform a useful function. Theirs is the function of those on the battlefield fighting the war while Traditional Catholics hold up the hands of Moses by keeping their Tridentine Masses and other traditions.

Vatican organization leaders often make such a big fuss about being ecumenical and friendly with other churches. Why can't they be so friendly to Traditional Catholics?

The real problem is that the Vatican establishment, by going Novus Ordo, has simply entered the same darkness that all other schismatic bodies have entered. Their ecumenism with Protestants etc. is perfectly easy to explain, merely the fellowship which darkness has with darkness. Their inability to deal in a similar "ecumenical" fashion with Traditional Catholics (other than the few within their questionable "care") is a direct result of the fundamental (and eternal) schism that exists between light and darkness. Only true light can fellowship with true light, hence it is only a matter of time before the SSPX and the other Traditional Catholic bodies and independent priests must eventually sort out their differences, and recombine as the one unified Roman Catholic Church. True, many Motu Proprio Catholics are also a true light, but theirs is a true light held hostage by the surrounding darkness of the Vatican establishment. It is only a matter of time before the captive light of the Motu Proprio Catholics must either escape, convert the surrounding darkness back into light (by restoring the totality of Catholic tradition to the totality of the Vatican establishment, a daunting task indeed!) be extinguished, or be joined to the rest of the Church outside the Vatican organization, having then fully ceased to allow any portion of the real Catholic Church to subsist within any portion of their apostate society.

Why do you feel free to disregard the contentions the home-aloners make regarding the lack of valid and/or licit orders among the various groups that comprise the Traditional Catholic community?

First of all, allow me to make an important distinction. There are many Catholics who pray the Mass at home, going to no traditional Mass (let alone any non-Catholic worship service). There are some who, due to geographical distance, ill health, or other valid concerns, are unable to attend at any of the thousands of valid and licit traditional Latin (or any Eastern Rite) Masses. However, there are others who deny the validity or licitness of the Catholic bishops, priests, or orders who have maintained the Traditional Catholic Faith and Sacraments and Church. The first sort are not "home-aloners" in the heretical or dissident sense at all, and as such merely doing what they can.

It is the other group with which I take issue, since they are, in fact, heretics. There is one indisputable Catholic teaching which these people have ignored or flagrantly disregarded, always in practice and sometimes in theory as well: The relative priorities of the various types of Law and Teaching which exist. At the absolute top is Divine Positive Law, and in a similar category would go all laws and teachings which are Divinely revealed as being from God himself. In a qualitatively lower rank goes "Ecclesiastical Law," namely those laws and disciplines which the Church imposes. Chief among the first would be the Ten Commandments of God; chief among the second would be the Six Commandments of the Church. In a qualitatively lower rank still would go all manner of secular laws, of nations, provinces, cities, families, and the various by-laws of businesses and social clubs.

At the top level, God's Laws, Teachings, and Divine Revelation, which all Catholics are morally bound to believe, is the teaching that the Church is Apostolic (the fourth Mark, after One, Holy, and Catholic). Part of what that means is a valid and licit Apostolic Succession, namely that there must always exist valid and licit bishops of the Roman Catholic Church. At a lower level, namely that of Ecclesiastical Law, comes all of the finer points of what ordinarily makes for a licit bishop of the Church. In particular, though the job of appointing and accepting each bishop resides with the legal will of the Papacy by Divine Law, the manner in which the Papacy applies this Divine Law is rightfully determined by Ecclesiastical Law, whether to allow the specifics of the choice to be delegated to named individuals or even tacitly allowed to devolve to yet other approved individuals. As such, it is merely a piece of Ecclesiastical Law that would impede against Patriarchs and others set over particular Rites, or failing that, other approved bishops, from continuing the Apostolic Succession during a prolonged papal vacancy. Where Ecclesiastical Law makes Divine Law impossible to carry out, it is the Ecclesiastical Law, and not the Divine Law, which gives way by becoming ipso facto null and void at that point.

The heretical home-aloners have reversed that by putting Ecclesiastical Law above Divine Law. It is that inversion itself which is heretical. They believe that a mere disciplinary law, imposed as late as 1951 by Pope Pius XII, can actually prohibit God from carrying out His promises to continue the Church! Now think it through: You have a Divine law which states that each episcopal consecration must be approved by the legal will of the Pope, and an Ecclesiastical Law which limits that "legal will" to the direct and personal intervention of a living Pope. Now assuming (as most of these folks in fact do) that we have had no pope since 1958, it is only a matter of time before the very last bishop approved by Pope Pius XII dies off, due to old age (as I write this only one remains alive, and he has retired from all active service). All other bishops are illicit, and many are invalid as well. Once that last one dies off, the valid and licit Apostolic Succession is permanently and irrevocably removed from the face of the earth! The Church ceases to be Apostolic and can never be so again and God has failed us all! We might as well make a religion of playing golf every Sunday morning…

For the heretical home-aloner, all bishops in the Vatican organization are heretics, or at least have been party to heresy which amounts to the same thing, and all bishops outside the Vatican organization are illicit due to the lack of any Papal mandate or place within the Vatican organization. That means (if one follows their position to its logical conclusion) that the Church has already vanished. It's all over!

Granted, most home-aloners don't actually claim that the Church has vanished, even though that is the logical implication of their beliefs about it. Many of them wiggle out of that necessary and inescapable logical deduction by positing that maybe, somewhere, somehow, perhaps in some far off country, maybe in Siberia, who knows(?), owing to some technicality resulting in some delayed reaction, there might still exist some faithful bishop of the Church, lawfully consecrated, and (due to his confinement and isolation) unaware of any of the heretical direction which the Vatican organization has taken over the last sixty years or so, in other words pure, faithful, and licit. While I concede the remote hypothetical possibility of there being such a bishop, maybe as many as half a dozen or so, as of this writing, their numbers must be vanishingly small and falling precipitously. Such conditions of confinement and isolation (and doubtless torture as well) do not engender longevity, let alone sanity. Such elderly and physically weak persons could hardly be expected to make any escape, and so long as they continue faithful, they can forget about ever being released. And the fact remains that so long as such remain perfectly hidden, their ability to rule any portion of the Church is already nil. Authority, if it be real, must be asserted and visible, and that in turn makes it knowable and known. A potential authority is of no use to the Church unless it is realized, at which point it becomes known.

We already have before us a case of a bishop who really was confined in China since 1955, a Bishop Ignatius 龚品梅 (Kung, Pin-Mei), to be precise. At great length, the Chinese communist government finally released him. After years of badgering, pressuring, and torturing him, the schismatics finally got to him, and he lost either his Faith, or his sanity. I think he still possessed these as late as 1985 (which is why he was not released then) when he was allowed to eat at a large and crowded table opposite some "Apostolic visitors" and broke out in song to commemorate his continued fidelity to Peter. There is no evidence that he had any idea at that time what the "Peters" had been doing or saying over the 30 years of his confinement. However, when he got out (in 1989), it was with perverse relish and sickening glee that this bishop celebrated the heretical Novus Ordo Missae, the "Mass" of the Chinese Patriotic Church and of all other (non-Eastern-Rite) schismatics. They released him because he finally caved and came to be in full, public union with the Chinese Patriotic Church, and for that reason, John Paul II made him a "Cardinal." He then solicited funds, ostensibly so as to fight the Chinese Patriotic Church, but actually his "Pope" has already granted recognition to the Chinese Patriotic Church and was actually in warm and fuzzy ecumenical union with them. This, to me, is fraudulent. Such a horrific example does not bode well for the possibility of there being some other as yet unreleased bishop who is still faithful and sane. As for Archbishop 龚 (Kung) himself, one can only hope and pray that the peace and quiet of his closing days allowed for him to regain something of his sanity or integrity. Evidence that this might indeed have happened is that after he died (March 12, 2000), his funeral Mass, by his own request, was in the Tridentine form. Alas, he died before ever gaining such strength or integrity to continue the Church as the home-aloners would have expected of such a bishop.

When the last of such faithful bishops dies off, that idea must die. If it has not happened already, it is bound to happen quite soon. And the crisis will still be with us. Worse still, even if, by some incredible miracle, one such bishop should finally arrive out of some prison after decades of confinement, isolation, and torture, still faithful and sane, how will we know for sure that the man we have is really the actual bishop and not some impostor? What records can prove that he is really the same man? I doubt that it can truly be valid or licit to recognize as a bishop a man whose identity is impossible to verify.

The same can be said regarding a number of other "secret" promises, dispensations, and so forth which several otherwise fine and good priests of the traditional community have had recourse. I have already come across the following examples: "My bishop was given special permission by the Pope (Pius) to consecrate bishops without a papal mandate," "My priest was made irremovable

Pastor (or given permanent faculties) in his parish in the good old days," "Our parish was incorporated in the good old days to belong to a group of laymen, and to grant temporary faculties to any visiting priest, for all time," "My priest was given a special dispensation by the Pope to retain his faculties no matter where he goes or where he says Mass." Doubtless, there are others as well. Of all such things I must say this: All such "promises," "dispensations," and "installments" are, for all practical and reasonable purposes, to be regarded as entirely bogus. Even if some were real, they all might as well be treated as bogus, or at least ineffectual since otherwise that would limit the Church to whoever has such a "something special" at some one isolated and unknown location. Anything of that sort will die off soon, if not already, and the sustenance of the Church cannot depend upon anything so ephemeral. However, even in such a case, the authentic documentation needed to prove it is no doubt deeply hidden in some recess of the Vatican archives, where no one will ever find it. We have reached a point at which it would be vastly easier to create an utterly convincing forgery than it would be to track down an authentic original, even if it really did exist.

All such claims made by the heretical home-aloners are therefore dead ends, and furthermore inimical to the serenity of God's saints. What is needed is an obvious, visible, and accessible official document which enables jurisdiction to be granted to Catholic bishops consecrated without any Modernist approval. I maintain that *Lumen Gentium* is such a document. It is obvious; it is confirmed at the time by at least visible and apparent Pope and Council; it is readily available to everyone for checking and verification. It is visible, and therefore provides a clear legal basis for Catholic bishops consecrated since November 21, 1964 to be truly Catholic and canonical despite their non-membership in the Vatican organization, as discussed in Chapter Three of this book. At another point, that same document explicitly states that bishops consecrate bishops in that convey the Apostolic and canonical mission by virtue of the consecration itself.

So, to sum up, there is the basic fact that the Divine necessity of continuing the Church renders (at least for the duration of this current crisis) the mere Ecclesiastical law imposed by Pope Pius XII, which gives way to the extreme necessity of the Church's own existence even if my theory regarding *Lumen Gentium* were somehow mistaken, and which in *Lumen Gentium*, was effectively abrogated, and thus granting the ability to hold real and visible jurisdiction to all such traditional bishops and priests. Finally, there is the basic existence of the Church. Either the Traditional Catholic community as I have written about **is** the Roman Catholic Church, or else there is no such church at all. And it is sheer nonsense to claim membership in a church which one denies the existence of.

LITURGICAL QUESTIONS

Why can't the Novus Ordo be considered acceptable if done without any abuses?

Novus Ordo intrinsically engenders abuses. It is just like fornication (which it is spiritual fornication, since it being false worship, is therefore worshipping a false god). Just as fornication engenders contraception, unwanted pregnancies, abortion, transmission of sexual diseases, and even perversions, the Novus Ordo ceremony engenders such abuses as Communion in the hand, Altar girls, invalid breads, irreverence, and even such sick and goofy things as clowns, donkeys, hand puppets, animal Communions, animal sacrifices, and so forth. Just because there may yet remain some few isolated instances of fully reverent and abuse-free Novus Ordos, well that's just like there being some fornicators out there who through sheer luck, somehow manage to avoid all of those various ills. The fact remains that sooner or later, the odds must eventually catch up with all of them, and then watch out! Here come the abuses!

A fair follow-on question to that would be, "How is it that the Novus Ordo engenders abuses; what is the mechanism?" The big difference between the Catholic Mass and the Novus Ordo (in their official forms) is that the Catholic Mass is thickly-laden with the deep mysteries of God, of transubstantiation, of His friends the Saints and their continual intercessions for us, of the awesome and frightful mysteries of eternal Heaven and Hell, whereas the Novus Ordo has none of that. For transubstantiation, it substitutes a caricature of the prayers of consecration with its doubtful and defective form, and for the remainder the caricature of an option of either a flawed copy of the Roman Prayer, an even more flawed version of a disused Eastern Prayer (which, by the way, was never used even in the East at such a point in the Mass), or other "Prayers" of unabashedly human (and amateurish at that) origin. The merits and intercessions of the Saints are entirely removed, as are all references to Grace, the Soul, Purgatory, Hell, and indeed everything which the natural man cannot grasp. Prayers for the Saints to apply the merit of their personal sacrifices to our deep spiritual need are replaced with prayers that we all have a nice day. The whole ceremony is reduced to a collection of amateur greeting-card verse, the most extreme expression of total banality itself. It is poetry by committee, uninspired and uninspiring.

Such a banal and theologically empty service constitutes a religious vacuum, and as is well known, nature abhors a vacuum. **Something** has to be done to liven it up, and since a return to the Catholic Mass is almost always ruled out, that spells abuses. Some, such as Altar Girls, invalid breads, or hard rock music, are done purely for shock value. Others, such as Lay Eucharistic Ministers or audience participation in recitation of the banal "prayers" are done with the idea that such things will involve the people more and make them participants, as if that could hold their interest. Still others, such as many of the most extreme examples where clowns and donkeys and hand puppets are used, are done purely for entertainment purposes. They have even given Communion to a dog. That the dog doubtless received on the tongue is scant consolation.

Didn't some expert in Aramaic claim that translating "for you and for many" as "for you and for all" in the consecration form of the Mass was harmless because the original language Jesus used had no distinct words for "many" and "all?"

Yes, his name was Joachim Jeremias who wrote a book entitled *The Eucharistic Words of Jesus* in which he claimed that neither Hebrew nor Aramaic had distinct words for "many" and "all" and so therefore Jesus could have meant "all." Clearly, his book is simply an "artifact," a falsified thing brought into existence solely for the purpose of providing those who are pushing for the new liturgy some pseudo-scholastic "source" whom they can quote in defense of their indefensible position. By saying what he said in that book he lost all respect within the scholastic community, since the Aramaic and Hebrew languages do in fact have distinct words for "many" and "all" as do all known languages. One thing to bear in mind is that the *Catechism of the Council of Trent* states in its section on the **Form of the Eucharist**, "The additional words *for you and for many*, are taken, some from Matthew, some from Luke, but were joined together by the Catholic Church under the guidance of the Spirit of God. They serve to declare the fruit and advantage of His Passion. For if we look to its value, we must confess that the Redeemer shed His blood for the salvation of all; but if we look to the fruit which mankind have received from it, we shall easily find that it pertains not unto all, but to many of the human race. When therefore (our Lord) said: *For you*, He meant either those who were present, or those chosen from among the Jewish people, such as were, with the exception of Judas, the disciples with whom He was speaking. When He added, *And for many*, He wished to be understood to mean the remainder of the elect from among the Jews or Gentiles.

"With reason, therefore, were the words *for all* not used, as in this place the fruits of the Passion are alone spoken of, and to the elect only did His Passion bring the fruit of salvation. And this is the purport of the Apostle when he says: *Christ was offered once to exhaust the sins of many*; and also of the words of our Lord in John: *I pray for them; I pray not for the world, but for them whom thou hast given me, because they are thine."*—pages 227–228, TAN Books edition, *italics* in original. To me, that ends all room for controversy as to whether "for all" or "for many" is to be used in the consecration formula. Rome has spoken; the cause is finished.

Not content with that, some have recently tried to revive this argument by claiming that "for many" might be some sort of idiom in the Aramaic language for "for all," but that is a very weak claim for which there is no scholastic support whatsoever, and which furthermore flatly contradicts the Church's teaching, as seen above.

But didn't the liturgy grow and change throughout the history of the Church? If such an amount of change was lawful then, how is it not lawful now?

One must concede that there was a certain amount of flexibility and fluidity which existed in the liturgy of the opening centuries of the Church. Even the canon of the Mass could not have had its present form in the first century since it names saints who come several centuries later. The same was true with Scripture, much of which was still being written over the course of that crucial first century. But as details came and went and changed, those details which proved fruitful to the edification of the saints and which were in agreement with the living knowledge of the Fathers were retained while the other details were allowed to fall into obscurity.

By the time Pope Gregory the Great put the final finishing touch on the canon of the Mass, it had been pretty much honed to perfection, and after another thousand years, it was formally canonized, having proven itself over that time. As other details came to be honed to perfection, they too have ceased to change. The main difference between the changes possible in those early years and the changes attempted now as a result of Vatican II is exactly comparable to the difference between an infant experimenting with crawling and walking versus a grown person whose walking and running abilities have proven useful and while still retaining them, suddenly deciding to experiment with crawling and other infantile modes of transport.

For example, in the earliest days of the Church, there was no particular location for the Eucharist to be stored between Masses. It could be near the altar, or near the baptismal font, or hidden in some back place, or just wherever else was convenient. It wasn't that anyone was trying to be different or creative, but merely that no one had given the matter any thought. At some point as the High Middle Ages approached, somebody decided that the most appropriate and fitting place was on the altar. There was true merit in this discovery and it caught on. Soon all new churches being built had the tabernacle being placed on the altar. However, those older churches where the Eucharist was stored at just any place were allowed to keep their configuration.

So it remained until Vatican II. Now the tabernacles are being ripped out and being placed in locations which have already proven inferior to the place on the altar, in terms of the edification of the faithful. As one can see, the fact that some flexibility existed in the old days does not in any way justify attempts to seem (and it really is only seeming) as flexible as things had been in the oldest days of the Church.

It is not that change can never occur, since obviously it has over the centuries, but that it must be extremely slow, incremental, virtually inconsequential within any particular era past that of the Apostles, such that any push influenced by any passing fashions of worldly thought will be compensated for in future eras that equally push in different directions, and all of these random walk pushes will be kept inconsequential; a sudden major shift in any direction, all within a single generation, or even more, a single decade, is a sure sign of rupture, discontinuity.

Why allow the liturgical reform of Pius V while forbidding the liturgical reform of Paul VI?

The attempt to set the two liturgical reforms as being parallel is nothing but a deception. What Pope Pius V called a "reform" was merely the Mass as it had already been for centuries. The changes only applied to various local forms which had sprouted in the previous 200 years, which amounted to a small prayer here, or an additional rubric there, or even a saint who is only locally honored. For the reform of Pius V, no new prayers were written, only each prayer as said in most regions was universally applied, and even that much was only done where the different traditions were less than 200 years old. If anything, the reform of Pius V strengthened the uniformity of worship throughout the Church.

By contrast, the deform of Paul VI smashed "the Mass" into smithereens, with each one being a unique religion unto itself. It did the very opposite of what the Pius V reform did, and had the very opposite effect. Many prayers were mutilated or even rewritten for the Paul VI "Mass" by people who obviously had no understanding or appreciation of the liturgical history of the Church. Pope Gregory the Great would have no trouble recognizing the Mass of Pope Saint Pius V, for apart from a few minor rubrics and the commemoration of saints or feasts unknown in his time, it is the Mass precisely as he knew it. On the other hand, he would find the Novus Ordo Missae to be totally unfamiliar and unrecognizable as Catholic worship. Without a doubt he would certainly condemn it.

Finally, the Tridentine Mass was not written at that time, but merely confirmed then as the universal Mass of the Church. That Mass had already existed for ages. It is called Tridentine only because the Council of Trent formally confirmed the Mass as it had been said from the beginning clear up until then.

But no pope should be able to bind future popes, or else their authority is not truly equal. Each pope stands in the Shoes of the Fisherman and his authority is supreme. If Pope Pius V could bind his successors with *Quo Primum*, then his successors would have had less authority than he did, and would not be true Popes. Why couldn't a later Pope (such as Paul VI) replace or undo the liturgical decrees of a previous pope (such as Pius V)?

First of all, popes most surely **can** bind their successors in certain things. For example, when Pope Pius XII infallibly proclaimed the doctrine of the Assumption of the Blessed Virgin Mary in 1950, he thereby bound the consciences of all future popes who are therefore in no position even merely to teach, let alone bind the entire Church to, the contrary.

For another example, there is the Gelasian decree in which a fifth century pope attempted to name for all time which books constituted scripture and which did not. Was he attempting to bind all his successors to the same set of Biblical books? Of course he was! Could a later pope validly change that list by adding new books to scripture, deleting any long accepted New Testament writings, or rewriting any Bible books? Of course not!

The unification of scripture at that point so very clearly resembles the unification of the Liturgy under Pius V. Prior to that point there was still some amount of local variation between various dioceses and even parishes. In the case of Scripture, there were a number of congregations which still used as Sacred Scripture the Epistle of Barnabas, the Revelation to Peter, the Shepherd of Hermas, the Didache, and even the letters of Polycarp, Ignatius, and Pope Clement I. There were even some outright forgeries, such as the Revelation to Paul, the Acts of Andrew, or the Gospel of James, which were also beginning to receive some recognition in a few quarters. On the other hand, some Christian communities still had their doubts about the Revelation to John, the letters of Peter, John, and Jude, and the letter to the Hebrews.

The Gelasian decree settled once and for all on the exact list of New Testament Scripture as we have it today. Likewise, there were still a number of local variations in the Liturgy, resulting from some prayer or rubric being introduced here, but not there, or being omitted or changed there, but not here, and worst of all, some priests, for fear for their lives, were beginning on their own initiative to

Protestantize their Liturgy, so as to avoid any trouble, by deleting prayers and rubrics. The *Quo Primum* decree merely did for the Liturgy what the Gelasian decree had done for Scripture.

Quo Primum was meant to be as permanent as the Gelasian decree since it fixed for all time a part of the Revelation of God, which is higher than Faith and Morals (and those would have been high enough) because Revelation is the source of Faith and Morals. But now within the Vatican organization, there is a plot to silence or ignore *Quo Primum*. Those asking for the "extraordinary form" have learned not to mention *Quo Primum*. The leaders of the Vatican organization know that *Quo Primum* can never be revoked, so their current strategy is that whenever anyone mentions it, the conversation automatically terminates immediately. Any possibility of granting permission for the "extraordinary form" of which they were talking about promptly disappears.

But *Quo Primum* defines a matter which is, if anything higher than Faith and Morals, and for that reason can never be abrogated. That is the one reason that the Novus Ordo People of God must schismatically separate themselves from Traditional Catholics. We bring up *Quo Primum*, and they cannot answer that.

Finally, the idea that a Pope limits the authority of the later Popes by promulgating an irrevocable doctrine, which no later Pope has the authority to undo, is fallacious. The Church does not invent Truth; She discovers Truth. Even before Pope Pius XII confirmed the doctrine of the Assumption of Mary, no pope was ever at liberty to promulgate a denial of that doctrine. The reason is that the doctrine is based on a historical fact. God does not go back in history and cause Mary to be assumed into Heaven or not simply because some pope decides to confirm one doctrine or the opposite two thousand years later!

All twentieth century popes, from Pope Saint Pius X onward have supported the Liturgical Movement, which culminated in the promulgation of the Novus Ordo Missae! How do you explain that?

When Pope Pius X condemned those Modernists who labored even within "the bosom of the Church" in his encyclical, *Pascendi*, everyone may have thought that the enemy had been routed and driven away. The reality of course is merely that the enemy merely moved underground to a number of other places, including most notably, the Liturgical Movement. Pope Pius X seems to have gone to his death without ever discovering this particular hideout of the enemy. Had he but known the true intentions of its members, he would have promptly condemned it and shut it down, or else completely re-staffed it.

The ostensible purpose of the Liturgical Movement was to prepare new vernacular translations of the Mass to go alongside the Latin in the Missals for people to use, and to prepare the overall structure and format of Missals, including introductory notes on how the Mass is conducted and what themes each day's readings focus on, and even the artwork to be used. They were also supposed to make recommendations regarding what hymns would be acceptable (and for what occasions) for use in the Mass. As a sidelight, they were also expected to do some research into the ancient Liturgical sources so as to be best guided by them. That is why one can look at such prefatory matter in their Roman Missal and often see such statements to the effect that "in ancient times, such-and-such was done, or prayer was said, etc.," which may or may not be true. In any case those bishops granting the edition of the Missal an Imprimatur were typically not competent to judge the scholarship of those passages.

It is important to differentiate between that innocuous public image of itself the Liturgical Movement presented to many, including the popes, versus the nefarious actions of its personnel behind the scenes. In order to gain a reputation as "great scholars," a certain proportion of their research work was no doubt legitimate, exploring the actual origins of various Liturgical practices and details. However, a considerable portion of their "studies of the ancient sources" were scholastically dishonest, misquoting and misapplying ancient texts so as to make it seem as though they were discovering that the ancient Christians were practicing a sort of "Novus Ordo" worship.

Among themselves, they knew themselves to be one of the main, if not **the** main place, where dissenters were hiding within the bosom of the Church. When one explores the actions and words of some of the early leaders of the Liturgical Movement such as Virgil Michel and Fr. Gerald Ellard, one sees (at least in retrospect) a master plan to supplant the Mass with a ceremony very much like the Novus Ordo (and this described back in the 1940's and 1950's) and seek out many novelties, especially those calculated to reduce "offense" to the Protestants and other heretics and schismatics.

When Pope Pius XII wrote *Mediator Dei*, he was already having a number of doubts about the direction the Liturgical Movement was taking and made a point of specifically condemning a number of the things they were contemplating such as the turning of the altar into a table, forbidding the use of black vestments, or resorting to any other "forms of antiquity." Unfortunately, Pope Pius XII evidently lacked the scholastic competence to detect the false scholarship by which those of the Liturgical Movement were merely injecting into their "antiquarian" model the design they had for what the new "Mass" would one day be. He may never have known just how totally inauthentic this design really was, and how much it was really modeled on Thomas Cranmer's "Mass," and that of various other Protestants and Old Catholics.

However, his papal instincts and infallibility being fully intact, he did sense that going "back" to such an "antiquarian" model would be catastrophic to the devotional piety of the faithful of the Church. He wrote *Mediator Dei* primarily **to** the Liturgical Movement hoping thereby to redirect it to more constructive ends. Little did he realize just whom he was dealing with. He had been led to believe that he was dealing with honest scholars who had found a lot of interesting details about the worship of the early Church, most of which would be of mere academic interest. But who he was really dealing with was a coterie of deliberate liars who had already fabricated a new "Mass," which was in reality merely an early draft of what would later emerge as the Novus Ordo Missae and which would have been totally unacceptable to him and unrecognizable to the ancients. He was dealing with two-faced dissidents who smiled to his face and said "Oh yes, Your Holiness, we shall revise our plans accordingly," but who then returned to their camp and continued the damage they were already preparing, totally unaffected by the guidance the Pope had given them.

Feeling that he had gotten the Liturgical Movement back on track with his encyclical, he finally allowed a small revision they proposed to be made to the Liturgy, namely the revision of the Holy Week Liturgy. I strongly suspect that this approval was a grievous mistake, obtained under false pretenses at the advice of his secretary Montini (future Paul VI), and that it was on account of this subterfuge (along with many others which were at last coming to light) which led to his refusal to grant the Cardinal's hat to Montini, sending him to Milan instead.

When Pius XII died, there was no "true" Liturgical Movement which died or changed, rather the existing Liturgical Movement simply began to show its true colors as never before. Suddenly, they could do whatever they wanted. The seemingly innocuous 1963 Vatican II document, the Constitution on the Liturgy, became a carte blanche permission for these subversives to replace the Mass with a ceremony of their own design. I truly believe that had Popes Pius X, Benedict XV, Pius XI, and Pius XII truly known the nature of the Liturgical Movement, they would have shut it down, or at least replaced its crew with reliable persons of unquestionable orthodoxy.

Do you mean to claim, therefore, that the Novus Ordo Missae is intrinsically invalid?

No, only that the Novus Ordo Missae lacks intrinsic validity, which is not the same thing. It has happened on occasion that some traditional writers, having proven its lack of intrinsic validity, then continue as if its intrinsic invalidity has been proven. Such serious sloppiness on their part can only be counterproductive since it makes them seem deceptive, and that unnecessarily so: The fact that the new "Mass" lacks intrinsic validity (to say nothing of its uncatholicity) alone is sufficient reason for Catholics to avoid it; there is no need to "prove" it to be intrinsically invalid, even were such a thing possible.

Whenever one finds Papal references to the corruption of the Canon of the Mass, and especially the words of the consecration, one repeatedly finds that changing the words is always condemned as being sinful but that validity is only lost where the meaning is changed, not where the meaning is kept, or at least simply becomes ambiguous, provided that an orthodox interpretation is possible. Such an ambiguous consecrational formula does however become doubtful, which is not the same as being necessarily invalid. Those who regard the New Mass formula as disallowing an orthodox interpretation ought to consider the following extreme and probably nonexistent, but theoretically possible scenario:

A validly ordained priest, using valid matter and having valid intent is saying the New Mass in the vernacular. Out of his own eucharistic piety, he says the New Mass consecrational formula as written except he also subvocalizes some additional words (subvocal words in parenthesis) by saying: "Take this, all of you, and drink from it: This is the cup of my blood, the blood of the New and Everlasting Covenant, (the Mystery of Faith); it will be shed for you and for all (of the many) so that sins may be forgiven." Such a formula, although conspicuously illicit and sinful, would certainly seem to be valid, since the exact meaning of the original is retained despite the use of the new formula. Even the subvocalizations might not be necessary provided that he clearly means the above by the words he says.

It is therefore reasonable to allow the New Mass a kind of "extrinsic validity," that is, the priest <u>may</u> have the ability to supply to it the validity which does not intrinsically exist therein, through his own eucharistic piety.

You wouldn't allow the Novus Ordo alongside the Tridentine Mass, would you? It doesn't seem fair to want equal rights for the Tridentine Mass, but then not be willing to grant the same to the Novus Ordo Missae.

Such statements display a false assumption that both rites are equally good and Catholic, and therefore deserve equal treatment. The Novus Ordo Missae is without a doubt a non-Catholic and totally inferior service which Catholics have no more a right (let alone duty) to participate in than at any other non-Catholic religious service. The only reason any Traditional Catholic would ever advocate having both allowed equally side-by-side is that then everyone would see the difference and voluntarily choose the Tridentine Mass and allow the Novus Ordo Missae to die a natural death. Alternatively, they may be merely arguing for the best they might conceivably get while the present delusion holds sway.

A fair question to ask is "could a reliable pope allow the Novus Ordo Missae to continue on an Indult basis?" In its existing form he certainly cannot, owing to the frequent invalidity it has. Were the alternate "Canons" ruled out, and any other fixes made to keep out the heresies, it might at least be valid, but then new problems present themselves. If such a Mass is made to look reverently Catholic by using Latin, incense, bells, facing the altar (instead of the people in attendance), and having Gregorian chant, at that rate why not just go the rest of the way and use the Tridentine Rite? If such things are not done, then the ceremony continues to have a decidedly non-Catholic flavor to it which resembles an Episcopalian or Lutheran service, and the Church has no business creating such an irreverent atmosphere.

One could argue for using an Indult to have something like it on a temporary basis while gradually introducing those who have only known the Novus Ordo to the details of the Tridentine Mass. That is purely a judgment call for the next reliable pope to make.

Why make this big deal about Latin? I happen to like the New Mass in its vernacular, which I find much easier to understand than all that old Latin.

If you have actually read this book, you should be aware of the fact that the use of Latin versus any other language is an extremely small point within the list of issues the Traditional Catholic community has with the Novus Ordo Church of the People of God. It is only those who are not Traditional Catholics who caricature Traditional Catholics as being obsessed with the use of Latin, as if that were the only issue.

As to the case where someone actually likes the Novus Ordo Missae, it is not a question of which Mass one likes; it is a question of which Mass is Catholic! One well-known and basic requirement of a Catholic is to attend Mass each Sunday and Holy day of Obligation. The Novus Ordo Missae, like a Lutheran, Presbyterian, Episcopalian, or schismatic East Orthodox service, is not, and could never be, the lawful object of a Catholic's Sunday obligation. Nor can attendance at the Novus Ordo fulfill that Sunday obligation. All Catholics are morally obliged to attend the Tridentine Mass, or alternatively the Uniate Eastern Rite Masses (in the case of the Eastern Rite Catholics).

The difference is not merely one of language or liturgical styles. It is one of worshipping God in a manner which pleases Him, and in a manner which He himself directed, as opposed to "worshipping God" by using a ceremony invented by men whose hatred for God and for His holy Church is a documented fact. It is the difference between being united to the Barque of Peter and the Eternal

City versus being in formal schism (and heresy) within a "Church" which is changeable, wobbly, totally ephemeral, and destined to collapse. Novus Ordo "worship" does not please God nor obtain any graces or favor from Him. The Novus Ordo "worshippers" have become just like the children in the street who say, "We played the flute for you, but you didn't dance. We told sad stories for you, but you didn't cry,"—Matthew 11:16–17.

If the Vatican organization returns to the Tridentine Mass for its Western Rite portion, would that mark the end of the crisis?

If that were done, but Vatican II allowed to remain on the books unchanged, with its advocacy for false ecumenism and religious liberty, then that is probably the greatest challenge which the Church could face in the years to come. The problem is that the return to the Catholic Faith on the part of the Vatican organization must be total for the current crisis to end, but all too many Catholics may be content with merely a return to the Catholic Mass. The battle to rehabilitate the Vatican organization does not stop until Vatican II is revoked and the other issues discussed in my concluding chapter have been addressed.

Could gradual change have created a Novus Ordo?

Since the liturgy undergoes some slow and organic change over time, suppose that over some very long period of time things gradually evolved to the Novus Ordo. Would that make it valid? Could that even happen, and if not then why not? The fashions of worldly thought that made the Novus Ordo something the world desires at that time are very much continually in a state of flux. Like buffeting winds, ever changing direction, they push this way, then that. Like the stalk that bends gently in the wind rather than being broken off, the Church yields to (but sometimes also opposes) the wind of the world, but this "swaying" of the Church is measured in micrometers, whereas what happened with the creation of the Novus Ordo Missae is as vast a move to one particular side, in one particular direction as a complete uprooting and transplanting somewhere else miles away.

Now if, let us say, a "Novus Ordo" pressure could be sustained in its one direction for some several thousands of years or more, then perhaps something akin to the Novus Ordo might indeed appear (assuming things don't stop short of that before going beyond some "pale"). But that is simply impossible to the fallen world. Fashions ever change, even as hemlines ever go up and down like a pogo stick. It won't happen; it can't happen. Even now, if the Vatican II followers were to invent a new liturgy, it would differ at least as much from both the Novus

Ordo and the Catholic Mass as they two differ from each other. Let another fifty years pass and the result would be entirely alien to all seen thus far. But in the real Catholic liturgy, incremental changes in one age, in one direction are eventually compensated for by other incremental changes in other ages, in other directions, much like what mathematicians call a "random walk." The ages are in effect "remembered" in the incremental changes, such as new saints, but no age dominates over the rest (other than the Apostolic age).

QUESTIONS REGARDING VATICAN II

Would not the Holy Spirit prevent an Ecumenical Council from doing the clear and serious harm Vatican II has evidently caused?

While God has never mandated evil, the fact is that He has allowed it, and every evil which has taken place has been permitted by God to happen. Through abuse of free will, men have always had the power to do evil and create misery for their fellow creatures. One would think that an Ecumenical Council of the Church, where one would normally expect infallibility to hold, should never be the cause of such trouble since that has never happened before in the history of the Church. However, it cannot be truthfully said that such a thing has never happened before. It's just that one must go back a great deal further before one finds the previous occurrence of it.

For example, one might go to the Biblical book of Numbers, Chapter 11 verses 3 through 34. The parallels between that episode and the present crisis in the Church are quite striking, almost like reading a prophecy of current events. At God's direction, a council of all of the elders of Israel is convened, after which the elders wander off in all directions, never to be heard from again with the exception of two faithful elders who never left the camp of Israelites. Israel then goes through a crisis of having no manna to eat, but they have instead quail meat for breakfast, lunch, and dinner until they get so sick of it that it comes out their noses.

In this passage, the reason for all of this also comes out. The Israelites had gotten tired of the miraculous manna which God had sent them to feed on in the wilderness. So they began to desire a return to Egypt and slavery. How very like today where many Catholics were getting bored with their Catholic faith and wanted something new, so in judgment God let them have the spiritual "quail meat" of the Novus Ordo religion and false sacraments.

I have actually seen where one writer claims that "the amazingly quick collapse… that has been manifested since the Second Vatican Council and the rapid changes which followed in its wake" actually supplies "proof of the weakness underlying much

Traditional Catholic observance and practice," as found in the Church before Vatican II. That is like blaming God instead of the Hebrews who got bored with the Manna. Perhaps God should have made His Manna come in six different flavors! Actually, a far more accurate reason for the quick breakdown is that heretical tendencies on the part of many bishops and cardinals as well as John XXIII were present and had already caused some rather considerable damage to the Church, even though the Holy Spirit was still protecting Her official pronouncements from error.

How is it that it took until this book to find out what had happened at Vatican II to separate the Catholic Church from the Vatican organization; and on the other hand, what makes you so smart as to be able to find those tiny clauses about the Church "subsisting in" the Vatican organization and actually figure out the true import of those statements?

Actually, the phrases about the Church subsisting in the Vatican organization have long been widely known, and much ink has been spilled over their interpretation. Abp. Lefebvre is known to have devoted entire speeches and homilies to discussing those statements. The difficulty which he and all other Traditional Catholics have had in applying that discovery correctly has been their handling that statement as if it were Dogmatic rather than disciplinary.

Part of the reason for this mistake is the title which the main document with that error has in most vernacular languages instead of its true Latin title. In the vernacular, it is called a "Dogmatic Constitution on the Church." Such a title is a complete misnomer since there is nothing about it which is actually dogmatic. Its true title is "*Lumen Gentium*," which means "Light of the Gentiles." The other part of the problem is the tendency on the part of those who are implementing changes in the Spirit of Vatican II to treat this document (along with all other Vatican II documents) as being some new sort of super-dogma.

Taken as dogma the statement, as stated in the Vatican II documents, "The Catholic Church subsists in the Catholic Church," is not only heresy, but absolute nonsense, gibberish. So taken, the traditional community has reasonably (if mistakenly) concluded that the Vatican leaders and their many local functionaries must have already lost their Catholic infallibility (and authority), and even their minds, at some previous point. Their inability to find or agree upon this unidentifiable previous point has been one of the difficulties which now sets loyal Traditional Catholics at odds with each other.

It is the lack of any such previous point, coupled with the overall fact that Vatican II was only and strictly a "Pastoral" Council which helped me to realize that up until that point the Holy Spirit was still protecting the Vatican hierarchy from all of the heresies and pet theories held by the bishops, cardinals, and pope at that moment. Although many of them harbored heresies and even in some cases were consciously working to destroy the Church, the Holy Spirit still prevented them from promulgating false or invalid or uncatholic sacraments or teachings. At most, some recent teachings were ambiguous. What few changes which had been made in the celebration of the Mass, although they set very bad and dangerous precedents, did not in any way in and of themselves threaten its validity nor tamper with Christ's words at the consecration.

As a further example of how this statement can be understood either in a dogmatic sense (in which case it is heresy) or a disciplinary sense (in which case it legally detaches the Vatican organization from the Catholic Church), let us take a look at some teachings of John Paul II as criticized but amply summarized by Fr. Cekada in his booklet *Traditionalists, Infallibility, and the Pope*:

1. Christ's Mystical Body is not exclusively identified with the Catholic Church. (*Osservatore Romano*, 8 July 1980)

2. The one, holy, Catholic, and apostolic Church is present, in all its essential elements, in non-Catholic sects. (*Letter to the Bishops on "Communion," 1992*)

3. The Catholic Church is in communion with non-Catholic sects. (*Ibid.*)

4. The Catholic Church shares a common apostolic faith with non-Catholic sects. (*Osservatore Romano*, 20 May 1980)

5. Non-Catholic sects have an apostolic mission. (*Osservatore Romano*, 10 June 1980)

6. The Holy Ghost uses non-Catholic sects as means of salvation. (*Catechesi Tradendae*, 16 October 1979).

Taken dogmatically and as literally stated here, these teachings of John Paul II are heresy, precisely the same heresy at heart which drove the council fathers to write that the Church only "subsists in" the Church.

Now, let us see how those statements read if interpreted in a disciplinary sense:

1. Christ's Mystical Body is not to be exclusively identified with the Vatican organization (anymore).

2. The one, holy, Catholic, and apostolic Church is present, in all its essential elements, in non-Vatican-approved Catholic religious orders

and groups (such as the SSPX, SSPV, CMRI, Trento Priests, and so very many others all around the world).

3. The Catholic portions of the Vatican organization (such as FSSP, Institute of Christ the King, all other "extraordinary form" priests, and perhaps still some few in the Eastern Rites) are in communion with non-Vatican-approved Catholic religious orders (whether they admit to it or not).

4. The Catholic portions of the Vatican organization share a common apostolic faith with non-Vatican-approved Catholic religious orders.

5. Non-Vatican-approved Catholic religious orders have an apostolic mission.

6. The Holy Ghost uses non-Vatican-approved Catholic religious orders as means of salvation.

As you see here, when interpreted in a disciplinary sense, the "heresies" in fact become an extraordinarily accurate description of the actual state of affairs. In this sense, John Paul II was more right than he knew!

Therefore, since they cannot have defined a heresy, their statement cannot have been a doctrinal or moral statement, but only a disciplinary one. It is impossible for them to change the nature of the Church or of any of the Church offices which they were holding up until that point, but it has always been possible for them to change their own personal relationship to their offices and thereby to the Church as well. **Vatican II did not change the Catholic Church even one tiny bit (other than to make it considerably smaller in numbers) because that is an eternal institution which can never be changed, but it did change the Vatican organization quite drastically!** First, it detached the Vatican organization from the Catholic Church, thus rendering the Vatican organization fallible and changeable in ways that the Church is not, and second, it opened the door for and introduced to that organization a false new religion, and third, it granted a universal charter for Catholic orders to exist and sustain themselves without any connection to the Vatican organization, and fourth, in granting this charter to faithful Catholic priests and bishops it recognizes them as having the jurisdiction of the Church, complete with the four marks and all other characteristics of the Church.

This sequence of events is essential in reconciling the fall of the Vatican organization with the promise of Christ to be with His Church always. Some pope did not just wake up some day and say to himself "I'm sick of always having to teach the truth all the time; I think I'll go teach some heresy today," and inexplicably find himself able to do that without any resistance from the Holy Spirit, the Cardinals, the Bishops, the Roman Curia, or anyone else. More importantly, the Church cannot defect nor disappear merely because a pope decides to make it so. There absolutely **had** to be a formal, material, legal, and public loss of Catholic authority on their part **first**. He and they had to resign, in at least some partial manner, from their sees in order to be free to propagate their own heresies and other pet theories, and they managed to do it in such a fashion that it would not be immediately obvious how they did it. And simultaneous to that they would also have to provide some means and manner for the Church to continue without them. Providence demanded that much.

I do grant that the council fathers really did intend to promulgate the heresy that other religions share our apostolic faith and mission, and for that reason insisted on saying "subsist in" even though other more conservative council fathers intervened in favor of saying "is" at those crucial points. The intention was heresy, but the result was their own at least partial departure from their sees. The Holy Spirit, in guaranteeing correctness on faith and morals (infallibility) does not guarantee intent, but result. The words of the Vatican II documents must not be evaluated according to what the Council fathers intended, but what the words themselves literally state, which is one of the most basic tenets of Canon Law. Once they had legally resigned (at least in part), the Holy Spirit could leave them to their strong delusion, "that they should believe the lie."

The real beauty of this explanation is how everything falls right into place. So many questions get solved all at once: How did the Vatican organization become able to promulgate error and invalid sacraments? How come the Holy Spirit no longer protects it from error? On what legal or canonical basis do "independent" Traditional Catholic priests operate their parishes? What guarantee is there that the various groups described here operating outside the Vatican organization will never fall into error? How could the recent popes and an ecumenical council be allowed to fall into error? Where exactly is the Church today? Who are the true Catholics? Can each and all of the traditional groups be truly Catholic, and how?

Why can't that statement in *Lumen Gentium* about "elements of sanctification" subsisting outside the Vatican organization apply to Protestant ministers or schismatic East Orthodox patriarchates as the Council Fathers obviously intended?

It is important to focus on what the grammar of that expression literally implies. An "element of sanctification," whatever that is, cannot be merely a "sanctified element." The grammatical difference between those two expressions directly parallels the difference between an "object of light," and a "lighted object." In each, the first implies "source" where the second does not. While the phrase "object of light" may be somewhat strange, it clearly refers to a source of light, e. g. a lamp, a flame, or a light bulb. A "lighted object" is merely any object positioned near a functioning light source; it has light shining on it, but it is not the source of the light that it can merely reflect, diffuse, or transmit.

Certainly, it is possible for a soul to be sanctified while outside the Visible boundaries of the Church, providing that the soul in question is invincibly ignorant of the truth. Such a soul would then properly be called "sanctified." However, that soul lacks the power to provide any sanctification to anyone else. His good example or behavior, as he abides by Natural Law and has true and perfect contrition for his sins, by which he unites himself to the soul of the Church, even while outside the body of the Church, only serves others as a guide towards false religion. No matter how sanctified such a soul is he simply lacks the power to be a source, or "element," of sanctification. He can no more sanctify others than could a dog serve as the Captain of a ship.

The quote from *Lumen Gentium* continues, "Since these are gifts belonging to the Church of Christ, they are forces impelling towards Catholic unity." The "these" which are gifts belonging to the Church are the "elements of Sanctification and of Truth." It is interesting to note that neither the Protestants nor the East Orthodox are "forces impelling towards Catholic unity." The case of the East Orthodox especially merits study.

Here is a vast group of Christians, with seven valid Sacraments, a tremendous devotion to Mary, and an adherence to all but a very few tiny particles of the Magisterium of the Church. Yet despite certain recent Roman overtures of unparalleled (and, I might add, unjustifiable) generosity, to accept them into their communion pretty much "as they are," most of the schismatic East Orthodox will have no part of it. As it happens, their reasons for refusing to be "forces impelling towards Catholic unity" are directly traceable to one of those tiny particles of the Magisterium they reject. They have no belief that the successor of Peter, the Bishop of Rome, is to have universal jurisdiction. They are willing to have him be the "Western Patriarch," perhaps even have a "first among equals" status (whatever that means), but they believe in a Church composed of a confederation of several groups none of whom have any real jurisdiction over another. They don't believe in submission to the Supreme Pontiff.

By contrast, consider the factions of the First Great Western Schism in the fourteenth century. That, too, was a schism similar to the schism between East and West, but unlike that schism, this one was between groups which each held to the

entirety of the Magisterium. They all believed that there was to be **one** man, a bishop of Rome, a successor of Peter, a Supreme Pontiff who has universal jurisdiction, and to whom all Catholics must submit. The only question was regarding the identity of this individual. It was their mutual adherence to all of the teachings of the Church which made their reuniting possible. Ironically, it was one of the antipopes who actually convened the Council of Constance which solved these issues, terminated the schism, and elected a new single successor to lead the Church as all factions desired.

The various factions of the contemporary Traditional Catholics are in exactly the same status as the various factions of the Fourteenth century Church. They **are** "elements of Sanctification and of Truth," and "forces impelling towards Catholic unity." Just as in the case of the Fourteenth century Church, it is impossible for any Traditional Catholic group to say to another, "We believe this Doctrinal or Moral Truth which you deny; you and we must therefore part company." Traditional Catholics all seek a time when there would once again be a Bishop of Rome, a Successor of Peter, who is clearly recognizable as such, to teach and govern the entire Church as is his right and duty. They show their loyalty to the papacy by adhering to all the teachings of those whose hold on that office is beyond doubt (the reliable popes).

"Elements of sanctification," therefore, can only be bishops and priests in union with and submitted to the Supreme Pontiff. For such to exist, or "subsist," outside the visible confines of the Vatican organization necessarily implies that the boundaries of the Vatican organization no longer coincide with the boundaries of the Roman Catholic Church, and that the two (long historically united) are formally severed from each other as ontologically distinct entities.

Leaders and representatives of the Vatican organization frequently refer to their organization simply as the Catholic Church, and yet you give more weight to those few references in *Lumen Gentium* and other Vatican II documents stating that the Catholic Church merely "subsists in" their organization than to all of those other references they make to the contrary. How do you justify that?

A simple illustration should suffice. A married man could remove his wedding band and put it in his pocket and walk into a bar and claim to all of the lonely women that he attempts to picks up there that he is single, free, and available. However, if the records at the courthouse say he is married to so-and-so, who is alive and well at such-and-such a place, then that should similarly take precedence over all of the times he claims a single status to all of his girlfriends and lovers. The documents of Vatican II are the Vatican organization's most formal and weighty attempt to define its new existence as an organization in some small portion of which some small portion of the Church merely "subsists." Furthermore, the entire membership of the Vatican organization leaders and functionaries participated in and gave their consent to the production and promulgation of those documents (even though some few did so grudgingly owing to their desire to adhere to the Church). Therefore, any and all attempts on the part of individual leaders and representatives of that organization to claim identity with the Church are in precisely the same category as that married man's public protestations of a state of singleness.

Did not Cardinal Ratzinger refute your claim about the infamous "subsists in" clause in *Lumen Gentium*?

Let us start by looking at the actual comments of Cd. Ratzinger which he wrote in opposition to Liberation Theology. While one can and should praise his effort to oppose that heresy, the approach used shows that he either did not understand the Vatican II text or else was committed to concealing its true meaning:

> "…In order to justify [his position], L[eonard] Boff appeals to the constitution Lumen Gentium n. 8 of the Second Vatican Council. From the council's famous statement, 'Haec ecclesia (sc. unica Christi ecclesia) Catholica subsistit in ecclesia Catholica' (This Church— namely the sole Church of Christ—subsists in the Catholic Church), he derives a thesis which is exactly contrary to the authentic meaning of the council text, for he affirms: 'In fact it (sc. the sole Church of Christ) may also be present in other Christian churches' (p. 75). But the council had chosen the word subsistit—subsists—exactly in order to make it clear that the one sole 'subsistence' of the true Church exists, whereas outside her visible structure only 'elementae ecclesiae'—elements of the Church exist: these being elements of the same Church tend and conduct toward the Catholic Church (Lumen Gentium, n. 8)."

And this proves nothing. I did not get my discovery regarding what the Vatican II text says from reading Leonard Boff or any of the other many commentators, of which there are many, who have discerned within that same text the dis-identification between the Catholic Church of all history and their Vatican organization, but merely from the same place he and they all got it, namely from a close and careful reading of the conciliar text itself. No matter how much one might try to explain it away, there is no way a thing can "subsist in" itself, let alone possess differing boundaries from itself. While there most certainly is some room for different interpretations of the text, no legitimate interpretation would allow "subsist in" to be the same thing as "is," nor indeed anything but "is not."

Leonard Boff of course employed this same grammatical observation for very different ends, namely, so as to declare an overlap between his leftist ideology and the Church. Given what several popes have had to say about communism and socialism, there could never be an overlap between any leftist ideology and the Mystical Body of Christ. The two are intrinsically mutually exclusive. In one sad sense however, Mr. Boff does have a point, namely that his leftist ideology most certainly does exist (but not "subsist") in a portion of today's Vatican organization. But it does not exist in any of those portions of today's Vatican organization ("Motu Proprio" and perhaps some Easter Rite) in which portions of the real Church subsist.

Even Cd. Ratzinger's comments seen above show a certain ambiguity and two-sidedness to it as he first seems to claim that "subsists in" is some special fancy sort of way of saying "is" ("that the one sole 'subsistence' of the true Church exists"), but then also admits that it also means "is not" ("outside her visible structure…these being elements of the same Church").

Did not the Vatican II Schema *Lumen Gentium* also promulgate the heresy of Collegiality?

The schema does contain the claim that a "College of Bishops" is a permanent group which "exists all the time," even though a footnote (added after much wrangling about it at the Council) states that they do "not always act in full act," whatever that means. While that may sound harmless enough to the casual reader, a subtle distinction may help to clarify matters.

The Church has always used the word "Body" to describe what all the Catholic Bishops around the world constitute. The word "College" only describes them while an Ecumenical Council is in session. While so convened, the pope's authority and infallibility, to a very large measure, are shared with the College of Bishops, to the point that his role with respect to them ceases to be that of Monarch and becomes merely that of President, and in some cases even less. The Councils of Nicaea and Constance are examples where, in the first the Pope did not even bother to attend, and in the second three papal claimants surrendered that claim and another Pope was elected.

Again, as in the case of those "subsist in" statements, this claim is heretical if taken as a dogmatic truth. Furthermore it flies in the face of the plain facts of history, namely that only a few short periods of time has the Church been in a Council (even if one counts the various lesser councils and synods) and all the rest of the time there was no Council in session and the bishops were merely a Body, not a College. And once again, there is a purely disciplinary interpretation to these statements of the Vatican II schema.

The disciplinary interpretation is this: The statement "The College of Bishops…exists all the time," is not an attempt to state a moral, doctrinal, or historical fact, but a mandate. From that point onward (until further notice) the Council is an ongoing entity. In other words, it is as if Vatican II did not end with its official close in 1965, but is still in session to this day! Is that possible? Don't bishops all have to be gathered in one place in order to be in council? That used to be the case, but now with such contrivances as the telegraph, the telephone, and now e-mail and computer networks, all bishops can readily communicate with each other as if they were still face-to-face in one room.

Could that be the reason why so many things, such as the removal of the tabernacle to a place of dishonor, can be attributed to Vatican II even though the 16 official schemas say nothing of the kind? The once-Catholic Bishops still a College, as mandated in this schema; Vatican II still (secretly) in session? The so-called "Post-Conciliar Documents" which are obviously the bitter fruit of that ongoing Council have been misnamed; there is nothing "Post" about them at all! This is also the beginning of what it is about the Second Vatican Council which deprives the Vatican leadership of the monarchical role which Christ gave to Peter and his successors. The schema on Collegiality would further enlarge on that point, but these statements in *Lumen Gentium* alone are enough to turn "Pope" into merely "President," and even that only when he bothers to preside.

Even during the Council, Abp. Lefebvre and many other Council Fathers picked up on the fact of that disciplinary impact, and what it would mean. Writing about the Third intervention which took place in October of 1963, concerning Collegiality, Lefebvre states, "It was clear that this was the aim envisaged—to **set up** a permanent collegiality which would force the Pope to act only when surrounded by a senate sharing in his power in an habitual and permanent way. This was, in fact, to diminish the exercise of the power of the Pope. The Church's doctrine, on the other hand, states that for the College to be qualified to act as a college with the Pope, it must be invited by the Pope himself to meet and act with him. This has, in fact, only occurred in the Councils, which have been exceptional events."—*I Accuse the Council,* page 13 (See Bibliography).

"Pastoral" doesn't really mean anything at all; the Second Vatican Council was really just like all the others.

On the contrary, "Pastoral" means a great deal. By its very nature it precludes all Moral or Dogmatic considerations, and therefore promises to pronounce no anathemas. For a Council, even an Ecumenical Council of the Church, to be "Pastoral" simply means that it is to be concerned strictly and solely with disciplinary, procedural, and administrative matters only. One must concede that it has always been theoretically possible for the Church to convene even an Ecumenical Council while limiting it to such purposes. It has been long understood that infallibility would not apply to such a Council, even if it were a General and Ecumenical (worldwide) Council of the Church, since infallibility only applies to Faith or Morals, not discipline. Such a Council would still, of course, be authoritative and binding on the faithful until its disciplinary measures should be revoked by a later pope or Council.

The fact that the Council refused to follow through as advertised by attempting to pronounce "decrees" and "dogmatic" constitutions is one of the great oversights and tragedies of the modern Vatican leadership. There is a special class of statements which historically have been formally promulgated as *de fide* teachings of the Church. Before any statement could ever be admitted to that exclusive and special class, it must first be subjected to rigorous tests of doctrinal correctness and historical accuracy.

Even after all of that, the statement only gets through by permission of the Pope, who is at liberty to refuse to promulgate it even after it has passed all those other tests. Since Vatican II was convened as merely a Pastoral Council, none of these strict tests should have been necessary. Disciplines which prove disadvantageous to the Church can always be revoked or amended by new disciplinary rulings. One ancient Council had ruled that there shall be no new religious orders founded. Within ten years, a later pope abrogated that ruling and gave permission for the founding of a religious order. Since none of the usual tests should have been needed, none of them were provided. For anyone to come along afterwards and claim for the documents of Vatican II a doctrinal or moral infallibility, instead of mere disciplinary impact, is to allow these documents to have "cheated" in that they would gain access to that exclusive and special class of *de fide* statements of the Church without ever having to have passed the rigorous tests of doctrinal and historical accuracy which all other such statements have had to pass.

Instead of coming through the Shepherd at the front door of these rigorous tests, they leaped over the wall by means of various shenanigans and Church politics. Repeatedly at the Council, Fathers who took exception to the ambiguous or heterodox or even heretical wordings of proposed documents were silenced with the admonition, "But we are not holding a dogmatic Council, we are not making philosophical definitions. This is a **pastoral** Council aimed at the world as a whole. Consequently, it is pointless to frame here definitions which would not be understood." In other words, "Don't bother with trying to understand the fine print, Your Excellency/Eminence/Holiness, just sign here, if you please." How can documents formulated under such circumstances possibly ever have the weight of infallible doctrinal or moral authority? (They can't, of course; therefore, they don't.)

There is one last claim to their being doctrinal and moral instead of disciplinary which must be dealt with. Michael Pavel wrote an article in which he attempted to prove that Vatican II was not merely a pastoral council but dogmatic like all the others. His argument is best described as a reaction to Michael Davies' argument, or rather a slightly caricatured version of Michael Davies' argument. The "Michael Davies" of Michael Pavel's article argued that the documents of the Second Vatican Council can more or less be ignored because they are strictly of a disciplinary or procedural or administrative nature only. Michael Pavel then turns that around and shows numerous places where John XXIII and Paul VI clearly intended these "decrees" and other documentary miscellany from Vatican II should be binding on the faithful (of the Vatican organization) and **must** be obeyed, and so therefore he concludes that they cannot be merely disciplinary.

The problem with that line of reasoning is that disciplinary rulings are also always expected to be binding on the faithful as well as dogmatic rulings. The only difference is that disciplinary rulings can be revoked by the Church whereas dogmatic rulings are confirmed forever and irrevocable. Actually, Michael Davies' true argument is much more sophisticated than as presented here (or in Michael Pavel's arguments). For one thing, Michael Davies explains that disciplinary rulings of course also must be obeyed, the only exception being where a disciplinary ruling flatly contradicts faith or morals, (and apparently also) if it is detrimental to one's faith.

As one last nail in the coffin of the notion that Vatican II could ever be taken as part of the Extraordinary and Infallible Magisterium of the Church, I present the words of Paul VI himself as spoken by him in an address on January 12, 1966: "Some ask what authority—what theological qualification—the Council has attached to its teachings, knowing that it has avoided solemn dogmatic definitions backed by the Church's infallible teaching authority. The answer is familiar to those who remember the conciliar declaration of 6 March 1964, repeated on 16 November 1964. In view of the pastoral character of the Council, it has avoided pronouncing in an extraordinary way, dogmas carrying the note of infallibility. Nevertheless, its teachings carry the weight of the supreme ordinary teaching authority."

But isn't every council followed by a period of doubt and confusion on the part of the faithful? Won't people eventually get used to having Vatican II around and soon let things return to normal? Maybe the decline in religious interest is temporary.

This claim has sometimes been made in response to the figures cited by Traditional Catholics that Vatican II and its associated chaos has caused in decreased mass attendance, baptisms, marriages, religious vocations, and increased marriage annulments. I have never seen a better example of putting two different things, not merely apples and oranges, nor even apples and baseballs, but apples and hand grenades, into similar sized and shaped little boxes painted the same color, and therefore referred to in the same way.

True, a certain amount of chaos has followed each Council, including Vatican II, but there the similarities end. With reference to each Council from Nicea to Vatican I, the chaos, confusion, loss of faith (as measured by every possible criteria), and disunity (doctrinal and rubrical as well as organizational) was **always** amongst those who **rejected** the Council, **never** amongst those who **accepted** it. With Vatican II, one finds that precisely reversed. **All** of the chaos, confusion, loss of faith, and disunity is amongst those who **accept** Vatican II and who show that by trying to implement, each one of them, their own interpretations of its directives.

But didn't Archbishop Marcel Lefebvre finally end up signing even those last two Vatican II schemas? I read somewhere that he did.

At the end of Vatican II, all bishops, cardinals, and other prelates were required to sign a document by which they testified to their presence and participation in the Council. Marcel Lefebvre himself was among those who signed this document. Fortunately, we have his own explanation of those events as he understood them at that time: "This idea of interpreting the signatures as signifying an approbation of the conciliar documents was born in the badly intentioned brain of Father de Blignieres."

"The approbation or refusal of the documents was obviously accomplished for each document in particular. The vote was in secret, accomplished on individual cards, and made with a special pencil, which permitted the electronic calculation of votes. The cards were then collected by the secretaries from the hand of each voter.

"The large sheets which were passed from hand to hand among the Fathers of the Council and upon which everyone placed his signature, had no meaning of a vote for or against, but signified simply our presence at the meeting to vote for four documents.

"One would really have had to have thought that the Fathers who voted against these texts changed with the wind by trying to make believe that they would have approved of that which they refused but a half-hour beforehand."— *The Angelus,* January 1991, page 5.

QUESTIONS ABOUT THE CRISIS

Obscure technical distinctions between the Vatican organization and the Catholic Church notwithstanding, I just find it too hard to believe that God would allow that, which so many honest, sincere people take to be His Church, to promulgate invalid sacraments.

Believe it. Have you ever heard of an interdict? It's been a few centuries since we have had one, so you may find the concept somewhat unfamiliar. On occasion, the Church has been obliged to punish a parish, diocese, or even an entire nation with an interdict. An interdict is a refusal of all sacraments to the people within the parish, diocese, or nation. It is a kind of excommunication, not of an individual, but a large group of people for some serious crime or other serious reason.

The Catholic Faith was purchased for us at a most tremendous cost, first that of our Savior Himself who was crucified, and also in the huge cost in human suffering on the part of multitudinous saints and martyrs who endured torture, death, and dismemberment in order that not one particle of our faith be allowed to pass into oblivion. Most modern Catholics on the other hand became so apathetic about their faith that when it was taken from them, they couldn't care less. Indifference to the Truth is probably one of the most offensive things to God, even worse than being violently opposed to it. "I could wish that you were cold or hot. So then, because you are lukewarm, and neither cold nor hot, I am like to vomit you out of My mouth."

Those who have willingly gone along with the New Religion are therefore to be placed under an interdict and thereby deprived of the sacraments. This is carried out in practice by giving the faithless ones false, invalid sacraments, the validity of which they never really cared about anyway. Those having the faith of the martyrs have stuck with the Catholic Church and sacraments, even at the cost of losing their friendships, their families, their reputation or good name in the community, and anything else which the Vatican organization has ever felt empowered to take from them when they were forced to leave it. Since they cared enough to seek out the true and valid Catholic sacraments of the Church, they are not under the interdict. Even now, anyone can get out from under that interdict simply by resolving to go exclusively to the authentic valid and Traditional Catholic Rites.

But aren't things getting better now? I know that many crazy abuses and so forth have happened, but isn't the pendulum starting to swing back to normalcy?

This rumor keeps coming around again and again, and so far, it has always been false. People keep thinking that their indefectible Church cannot go any further off course. Somehow it has got to start getting better! This false rumor gets a new breath of life every time some positive action takes place at the Vatican, such as the promulgation of *Humanae Vitae* (in 1968), Paul VI's admission regarding the Smoke of Satan being in the Church or regarding the auto-destruction of the Church (in 1972), the election of the supposedly conservative John Paul II (in 1978), some of the condemnations of Hans Küng and Edward Shillebeeckx and his apology for liturgical abuses (in 1980), the 1984 and 1988 indults, the promulgation of the new Catechism (1992*), Veritatis Splendor* (in 1993), *Ordinatio Sacerdotalis* (in 1994), *Evangelium Vitae* (in 1995), etc. but each time it is a beginning which has come to nothing.

Since the Vatican organization is no longer identical to the Roman Catholic Church, there is no reason or necessity for it to return to normalcy. The persons who spread that rumor only say it so as to put people back to sleep, spiritually, so they won't notice that their ship is still sailing for disaster, and for that matter, they are on the wrong ship in the first place! No such "swinging of the pendulum" is needed within the Traditional Catholic community, nor indeed even possible to that which is the true Church! The Traditional Catholic ship, unmistakably united to the Barque of Peter, always sails exactly right on course.

The only way the Vatican organization can ever be put back on course in any real or lasting way is for Vatican II to be revoked. But all this talk of the Church being like a pendulum, or like a door swinging each way as if it needed to be boarded up with cedar planks (Song of Solomon 8:9) is sheer nonsense. The world may indeed be that way, but the Church is rooted firmly on the Rock of St. Peter. Anything which swings back and forth, or which obviously needs to swing back, as the Vatican organization does, is quite obviously not rooted on the Rock of St. Peter.

Hasn't the Church always had trouble? Why should we regard today's troubles as some sort of more serious crisis than before?

It is true that the Church has always had troubles and difficult times of various sorts. There has always been a certain criminal element which was trying to ruin or destroy the Church. Vatican II was a quantum leap in making that criminal element much more serious than ever before. A good way to illustrate it is with a town which is suffering somewhat from the presence of gang activity. So long as the gang is merely on the fringe of society, selling drugs or prostitution on street corners when no one's looking, the problem can be readily helped by granting more authority to the leadership of the town, such as by hiring more police officers or allowing the judges to give longer sentences.

Imagine how much more serious the crime problems in that town would be if a key member of the criminal gang were to be elected as Mayor, and as Mayor were to go on to appoint other members of his criminal gang as trial judges, Chief of Police, etc. This is exactly what happened to the Church (or more precisely, the Vatican organization) at Vatican II. It used to be that you could follow the leader because he spoke for the true interests of the Faith, but that is no longer so.

Times have changed. You can't turn back the clock. The old ways don't work anymore. Why hang on to those obsolete structures and methods which may have served in the Dark ages well enough but are ineffective in these modern times?

Wrong, wrong, wrong! The old ways work now every bit as well as they ever did. The nature of Man, as created by God and injured by Original Sin, is exactly the same as it has always been in any known time of history (excepting only Adam and Eve before the fall). The bastions of the Traditional Catholic Faith of which I have written about here are all places where the old ways are used, and used successfully. Many of the traditional Mass centers and parishes in the traditional orders I have written about have schools attached to them. These schools are taught by real Catholic nuns, guided by real Catholic priests, and run as all Catholic schools were run only a generation or two ago.

I strongly encourage anyone to check out any several of these schools for themselves. I have found that the children are courteous, respectful, studious, well-behaved, having a well-bred sense of right and wrong, always willing to share, and always playing fair in their games. These are not kids with problems and hang-ups, but happy, playful children who live in a well-ordered Universe and who know it. I have yet to meet any responsible parent, who upon seeing any of these Traditional Catholic schools, would not want to send their own children there. What these children have is the true Grace which comes through the true Sacraments given in full union with the Church Christ established.

The moral crisis experienced in the 1960s and 1970s was every bit as much happening in most parts of Brazil as it was in the United States. In most parts of Brazil, Communism was running rampant (helped along by ex-Catholic clergy involved in "Liberation Theology"), Mass attendance, baptisms, marriages, and religious professions were on a steep decline while marriage annulments, conversions to other religions, violence, pregnancies outside of wedlock, and abortions skyrocketed. The citified places may have been somewhat more affected by these trends than the countrified places, but all were affected.

All, except one: The diocese of Campos. In Campos, they kept all the old ways: the old Mass, the old Sacraments, the old catechisms, the old rules. Not surprisingly (to this writer anyway), they had the same old success as well. Visit any school in any other part of Brazil and what do you see? Graffiti on the walls, vandalism, kids cutting classes, young girls pregnant (just like in America…). But visit any school in the diocese of Campos (especially while Bp. de Castro Mayer was still in charge, up until November of 1981, however a considerable amount of his influence yet survives even now) and what do you see? Clean, nice facilities, where doors don't have to be locked, truant officers don't have to be hired, and pregnancies are almost unheard of among the young girls.

The old ways still work, but (obviously) only when applied. When people started to say that the old ways weren't working any more, the fact was that they were already slipping out of common practice. Part of the problem is that a large proportion of the new generation never really understood the old ways but thought they did. Where the older generations thought of discipline as correction and instruction, this ignorant generation thought of discipline as getting revenge, or getting even, or taking it out on the children, or even child abuse. Naturally, when they applied discipline in these twisted ways, they failed to have the success of those who used discipline as loving correction.

Might not the problem simply be the bureaucratic powers in the Vatican, and the Pope's "handlers," effectively imprisoning the Pope, rendering him incapable of taking the necessary steps to heal the Church?

Fr. Malachi Martin once ventured this theory in an interview in 1990. According to him then, the bureaucrats had completely taken over in the Vatican, to the point that the Pope cannot exchange a greeting with some national leader, or anything else whatsoever, without having his speech approved and signed by six different bureaucrats, that he can't get rid of these bureaucrats as it is not up to him to decide who any of them are, that they comprise a "superforce," a power that cannot be dislodged by any means whatsoever. He even claims that John Paul II wanted to come to Ecône and personally consecrate the bishops that Abp. Lefebvre had chosen, but was prevented, an interesting thought, to be sure, but irreconcilable with the hot and emotionally charged language of *Ecclesia Dei* that evidences nothing but open contempt for Lefebvre and his heroic act.

But such a view of this runs up against both Catholic theology and practical facts. If that were really the whole problem, then all the Devil had to do was get the bureaucrats on his side and then it's all over for Christianity; the infallibility of the Pope vanishes behind an ocean of perpetual bureaucratic rivalries and infighting. Again, for such a madcap scheme to work there has to be an ontological distinction between the Mystical Body of Christ and the bureaucrat-laden society, such that the latter, not being the former, is in no wise protected by the Divine promises God made to His Church. On the practical level, there is in fact a great deal which the Vatican leader (whether really "Pope" or not) would be able to do about it, if only he wanted to do it. The fact is that the Pope (or his loose contemporary equivalent Vatican leader today) is surrounded by the people he chooses, pure and simple; if he does not want them, then he can simply get rid of them unilaterally, and all without consulting anybody else. And if they say, "but you need us; there are all these things we are doing that need to be done," no that is not so. If saboteurs in a factory were causing all of the widgets it makes to be defective, unsaleable, there is no point in having the factory continue at all, wasting raw materials by producing defective widgets, until all the saboteurs can be found and rooted out, enabling all further widgets to be made correctly.

Any real Pope's authority extends to every Catholic alive, and not merely to his immediate subordinates. The CEO of some corporation would have to go down through some entire management chain to get a wicked employee removed, as only that employee's immediate supervisor has the authority to do it, but the Catholic Church is not that way. If the CEO of a company were to tell little Billy, the 5-year-old son of someone working for the company "I order you to stop stealing for your parents, he would be usurping parental authority, despite the soundness of the advice. But (assuming we have a Catholic family) if the Pope were to come and tell little Billy "I am the Pope and I am ordering you not to steal," that is no usurpation of parental or any other kind of authority; he acts entirely within the rights granted to him by virtue of his universal, direct, and episcopal jurisdiction. Those who do not carry out specifically Catholic policies relevant to their offices exactly as directed should be fired on the spot, and "with cause," which is to say, no severance pay, no continuing benefits, no forwarding recommendations for other prospective employers, and not even a chance to clean out their own desk. Any real Pope, who is a real Man, would do that, and do it right away rather than allow such a situation to fester for years or even decades.

Why is there almost nothing said in this book about the various visions and apparitions which talk about this, or the new apparitions?

There are a number of subjects which I have deliberately avoided in this book, apparitions being one of them. Fátima, for example, and particularly the mysterious and unknown, yet perennially famous "Third Secret" may very well indeed be all about the "current crisis in the Church," but the "current crisis in the Church" is not about Fátima. So much has been written about Fátima by people who know so much more about it than I do that there is nothing I could add to it. Since private revelations, no matter how accurate some of them have proven to be, are not binding on the Faithful, I cannot in good conscience base my book on any of them. For the sake of argument, suppose that by some bad fortune, the "Third Secret" were to be one day revealed to be something way out in left field, and Fátima thereby discredited, what would become of every "Catholic" book which is based on it? Not that I seriously believe that could ever happen, but can one afford to be so careless when trying to understand the current crisis in the Church? In point of fact, I am actually quite impressed with some of the better-known private revelations, including Fátima, and hope that the reader may indeed take some time to become more familiar with them.

The new apparitions, such as Međugorje, are an entirely different ball of worms. Only one of these (Betania) has ever been approved, and that only by the questionable Post-Catholic Vatican leadership. Some of them (such as Bayside) have been so far off base as to be disapproved even by the current Vatican leaders (and rightly so). Since all of these say really off the wall things and/or deny obvious truths regarding morals, doctrine, or even the current status of the Church, they really prove to be of little use except as examples of human or even demonic inventions. Furthermore, visionaries tend to be sources of division as different Catholics are unable to agree on which visionaries to trust. Unity will be achieved in the Church only by fixating on the Universal and Historic Magisterium of the Church **and nothing else** because that, unlike Madame So-and-so's latest message from the Blessed Virgin Mary, is something all Catholics are constrained to agree with and adhere to.

Why is there almost nothing is said in this book about the plots of Masons, Jews, and Communists who have infiltrated the Church, nor about the "Three days of Darkness," the "Antichrist," or the "Man of Sin" or other End-time prophecies about what has brought about this current crisis?

The specific nature of the plots of various persons who seem to be members of these or other groups hostile to the Church are not terribly important to me. The Church has always had enemies trying to overthrow it by various means; none of these people have come up with anything new. The real question in my mind is "Why should God now allow the enemies of the Church to have the apparent success they have had at, and since, Vatican II?"

It must be admitted that conspiracies of a sort **do** in fact exist. There are many powerful figures in the media, in politics, in finance, in industry, and even seemingly lacking all of these things as they quietly function in the seminaries and other local institutions of the Vatican organization, who work together to try to destroy the Church. However, I find that meditating too long on such things only promotes a paranoid state of mind in which no one and nothing can be trusted, thinking becomes impossible, and one finally becomes immobilized and unable to do anything for the Kingdom of God for fear of falling into the hands of the Enemy.

There have always been conspiracies and there shall always be conspiracies, until the End of the World. Were we Catholics ever to find a truly effective means for rooting all of them out, I suspect that such an effort would only place God Himself in the rather bizarre and awkward position of having to intervene in order to protect His enemies from His friends. I don't for a moment imagine that would ever happen. Just as the Church is the Body of Christ, conspiracies might be properly spoken of as the "Body of Satan." Each "Body" shall persist in this world for as long as the spirit which moves it is permitted influence here, Jesus forever, and Satan until the End of this World.

Our part is to be that Body of Christ, and to live in a manner which is appropriate for that Body. If we do that and help all others that we can to do the same, we have discharged our entire duties in this earth. We leave it to God to set the times and seasons for the limits on the conspiracies to be expanded or contracted as suits His plan.

For some reason, excessive prophetic speculation seems to have the same deleterious effect on those who engage in it as speculation about conspiratorial plots. In either case, one ends up getting that feeling that one is stumbling about through a dark room littered with dangerous objects to trip over or fall into. No amount of meditating on these things ever seems to turn up the lights in the least.

A somewhat more interesting question is "Why should so many Traditional Catholics be so concerned with these plots or prophecies?" Remember that although the Vatican organization legally separated itself from the Catholic Church in 1964, virtually no one at the time, or for quite some years after that time realized it. In ignorance of that fact, Catholics simply made an unspoken assumption that the Vatican organization was still identical to the Roman Catholic Church. Indeed, with such assumptions rooted in their minds, they have found themselves at a loss to explain how the "Catholic Church" could become so uncatholic, and yet on the other hand what justification could ever be found for obtaining their sacraments outside it.

Even all of the books I recommend in my bibliography were written as if it were the Catholic Church which was doing all of these uncatholic things, when in fact it was merely the Vatican organization and not the Roman Catholic Church at all. The prospect of some pope just waking up some day and saying to himself, "I am sick and tired of always having to teach the truth all the time; I think I'll go teach some heresy today," and being able to get away with it, with no resistance from the cardinals, the bishops, the Holy Spirit, or anyone else, is truly frightening. Considering what horrible and sordid persons some popes in the past have been, one marvels that none of them have ever done such a thing before. Now that we finally seem to have it happen, one has to wonder why "popes" who have such apparent good will should be the first to be able to do this. One instinctively knows that there must be more to this than meets the eye.

In the absence of the knowledge of the separation between the Vatican organization and the Catholic Church, one grasps at straws as they try to explain to themselves what has happened. Either the conspiracies somehow just got dramatically smarter than ever before (which hypothesis leads one to be concerned with the plots of the Masons, the Jews, Communists, or the Illuminati etc.) or else we have entered some special time in prophecy leading up to the End of the World, such as the "Three days of Darkness," the "Final Apostasy" or else the arrival of "The Antichrist" (which hypotheses leads one to be concerned with End-time prophecies), or some combination of the three. The latter two at least leave God in control while the first has it that Satan has just figured out how to outsmart God, at least for the time being.

Now that we can know that the Vatican organization is no longer identical to the Catholic Church, and precisely when and how that visibly took place, there really is no need for those other two hypotheses. Their importance can only be expected to decline in the minds of most traditionalist Catholics, excepting only such few who are of a particularly nervous or excitable disposition.

Are the Uniate Eastern Rites still a good safe haven for Traditional Catholics over the long term?

The disastrous Liturgical ruination "mandated" during the 1960's and 1970's had no application to the Eastern Rites. They were as of yet left alone and allowed to continue as they had all along, and as such were in the beginning by far the main place the Church still subsisted within the Vatican organization during those years. Unfortunately, the Devil, having done what he can to the Latin Rite, then turned his attention to the Eastern Rites. Abuses which were unheard of even as late as the mid-1980's in the Eastern Rites have since sprouted and multiplied, and the Eastern Rites are now progressively being vernacularized and corrupted. As of the turn of the millennium, their level of damage was quite comparable to the level of damage the Latin Rite had suffered by the year 1966. The Eastern Rites were still valid (and perhaps many still are, but mere validity is not the only thing required); their basic texts and the vernacular translations thereof borrow from the schismatic East Orthodox churches they were meant to rival, and they are increasingly doing without the iconostasis, and pews are becoming a frequent find. While pews are perfectly appropriate in the Latin Rite, they have no place within the Eastern Rites, except for a very few seating places for the elderly, infirm, and nursing mothers. The more recently published Eastern Rite (St. John Chrysostom Byzantine) liturgical books even contain directions to the people in attendance to "sit," thus rendering official the abuse of having pews. Further damage is certainly on the way, and the Eastern Rites are those who might be most affected by the Balamand agreement, should they be foolish and ignorant enough to take it into account. The Eastern Rites are rapidly losing their status as a safe haven, although some Eastern Rite parishes as yet remain (however barely) within acceptable bounds.

For a time, the SSPX had their "Transalpine Redemptorist" project which exists to preserve faithful Uniate Eastern Rite Catholic Churches. Since then, the Society of Saint Josaphat has fulfilled a similar function. For information about them, I suggest their Website, https://www.saintjosaphat.org/.

What about the rumor that Cardinal Liénart was a Mason?

Even the significance of this strange rumor is probably lost on most Catholics. However, for the record, the rumor is simply not true. Some years ago, the *Angelus* offered a considerable sum of money as a reward to anyone who could provide evidence (for example membership rolls of a Masonic lodge, photographs of the Cardinal with other Masons inside the lodge, or wearing distinctive Masonic garments, etc.) that the Cardinal who consecrated Marcel Lefebvre had been a Mason. The reward was never claimed, for the reason that no such evidence exists. That the Cardinal was probably quite sympathetic, or even friendly towards Masons and Masonry, or even their confederate or stooge, is quite probable, owing to the liberal, anti-Catholic actions of his later life. But as to the claim that he was a formal member, that is definitely not true.

Of a little more interest is the real reason some people spread this rumor. The rational goes like this: If the person who consecrated Lefebvre is a Mason, then perhaps out of Masonic villainy he deliberately withheld a valid intent to consecrate Lefebvre as a bishop. This would cast doubt on the validity of the ordinations of all priests of the SSPX that he ordained. One can readily see that the only people to gain from spreading this "conspiracy theory" are those of the Novus Ordo establishment who seek to undermine Lefebvre's work and divide the Church.

Really, this rumor is so ridiculous that I hesitated to even put it in this part of this book. For the sake of argument, let us suppose the extreme worst case, that Cardinal Liénart was a Mason. But Bp. Charles-Maurice de Talleyrand-Périgord in France had also been a Mason and many French bishops came from him, but the Church never questioned the validity of their orders. The Masons don't really believe in the Catholic doctrine about valid Orders, but anxious to install their agents in vital places within the Church they would do nothing but "play it absolutely straight" as to making bishops as properly as possible. The Church accepted those, and accepted the ordination and consecration of Marcel Lefebvre by Achille Liénart have always been accepted as valid by the Church, all long before Vatican II, and so cannot be validly challenged now.

The main reason I include mention of this rumor is that this is the perfect example of why conspiracy theories are not discussed in this book. It is too easy for those who really **do** belong to the conspiracies to manipulate certain Traditional Catholics by spreading certain conspiracy theories of their own.

Why don't any of the modern apparitions speak of the Traditional Catholic community?

For starters, how can we be so certain that none have? Indeed, several known prophecies have spoken of the Church as being reduced to a small but faithful "remnant," certainly in the End Times, but not ruling out the possibility for other times leading up to it. Indeed, if one meditates on the Church as reduced to a faithful remnant, certainly in the End Times, and quite probably at other previous times such as now, one must wonder how much of the conventional canonical structure the Church might have, with the Pope in hiding or else held incommunicado by Antichrist (with "much to suffer"?), Catholics fleeing in all directions as to render conventional diocesan sees irrelevant, and so forth. If this is not the End (and indeed I personally doubt that it is) then at least it is quite a substantial and detailed "dry run" for when the real event occurs.

Apart from the apparitions which have been approved by the pre-Roman Schism Catholic Church (before 1964), I strongly doubt that any of the well-known apparitions have any source in the supernatural realm, although it is possible that some may be positively Satanic in origin. Our Lady, assumed into Heaven and dwelling in the very sight of God, knows where the Church is, but "Our Lady," as presented in the mouths of these new seers, knows nothing more than the seers themselves (which isn't much).

An example to illustrate this point would be the numerous and bulky "messages" by Don Stefano Gobbi. Turning to the only mention (and that without even a name) of any of the great heroes who are helping the Church through this difficult period, one would surely expect to find Our Lady praising this hero and recommending his canonization. But let us see what we find in message 385, dated June 29, 1988. "The heart of the Pope is bleeding today because of one Bishop of the holy Church of God who, through an arbitrary episcopal ordination carried out against his will, is opening up a painful schism in the Catholic Church."

Clearly, this is a reference to Archbishop Marcel Lefebvre, but notice something missing? History recounts that there were **two** bishops performing an episcopal consecration, Marcel Lefebvre and Antonio de Castro Mayer. Yet the message only says **one**! Clearly, Gobbi's "Our Lady" is limited to the same information sources that Gobbi himself had access to. That source was the news media, both Vatican and secular, TV, radio, and print, which all unanimously stressed Lefebvre to the total exclusion of de Castro Mayer. It was all "Lefebvre, Lefebvre, Lefebvre!" without so much as a word about who that *other* bishop was, or even the fact that more than one faithful bishop supported and participated in that action. That right there alone is proof positive that Fr. Gobbi's messages absolutely **do not** have their source in the supernatural or preternatural! It is interesting to note that some short time after putting the first edition of my book on-line, the Vatican organization, in a rare moment of sanity, declared Fr. Gobbi's messages to be not the least bit supernatural in their origin.

Furthermore, John Paul II was hardly bleeding his heart that day. When Philosopher and personal friend of Paul VI Jean Guitton paid a visit to Lefebvre several days prior to the consecrations, he asked "Monsignor, John Paul II is actually in Vienna in Austria. Let us suppose that he decides to take a helicopter and he arrives at Ecône during the ceremony. Would you then consecrate the four bishops?" Lefebvre answered him saying "Well, of course not. I would throw myself in his arms." He then reminded Jean Guitton that he was not a schismatic and **not** rebelling against Peter. John Paul II never took that helicopter. Instead, he spent the day in some now long-forgotten bit of administrivia which at the time he felt was much more important.

THE TRADITIONAL GROUPS

What are the three basic groups?

Traditional Catholics seem to come in three basic types: "Motu Proprio" (those who use the "extraordinary form," formerly known as the "Indult"), Sedevacantist, and Society of Saint Pius X, to list them alphabetically. The Motu Proprio Catholics are those who have cut some deal with the Modernists in Vatican

City to retain their Catholic worship with their blessing. The Sedevacantist Catholics are those who have analyzed the present situation theologically and ascertained that the Vatican leadership has no claims on our Catholic obedience, owing to a visible loss of office on their part. The Society of Saint Pius X Catholics are those who maintain a delicate balance between the first two positions. A few other Catholics may also adopt some combination of any two or all three of these positions, or even choose to stay out of these questions altogether, just holding on to the traditions of our Faith.

In defending so many different kinds of Traditional Catholic orders and not taking sides in their disputes, aren't you acting just like the Protestants who agree with each other in "the essentials" but feel free to disagree with each other about "the non-essentials;" doesn't that make you therefore a pan-traditionalist?

When Protestants try to get Ecumenical, since they are not really willing to unite around the true faith of the Catholic Church, they will always have things about which they disagree. Their idea of solving this problem is to settle on a few key doctrines which they see as "the essentials" on which all members must agree, and then allowing them to retain their various opinions about everything else which they label "the non-essentials." One problem with that approach is that often they are not able to agree as to which doctrines or morals are essential and which are not.

For example, they might all agree that Trinitarian theology, Heaven and Hell, God's creation of Man, the nature of Sin, and the inerrancy of the Bible are all essential, but one will come along and say that Baptism in water is essential while another will say it's not, or one will say that Signs and Wonders (e. g. Tongues, Prophecies, Healings, Miracles, etc.) either must or must not occur while others will say that those things don't matter.

Another problem with that approach is that one quickly finds that the fewer "essentials" one requires religious bodies to have in order to be a member of their ecumenical club, the more religious bodies one can include in their club and therefore the larger they become. It is only a matter of time before their standard becomes so broad and so general and so vague, their "essentials" so very, very few that practically everyone is by definition in their club and together they have virtually nothing left to say to those few remaining outside it.

That is not what I have done here. The disagreements which Catholic orders have are quite important. In many cases they are much more important or "essential" than many other minor teachings of the Church (in whatever sense that any teaching of the Church could ever be regarded as more or less important than another). The liberty which the various orders described here have to disagree with each other is solely and strictly based on the fact that the Church has never provided a formal, doctrinal, and definitive answer to these specific questions. There are many precedents of course, but some of these precedents point towards one solution, others towards another, and other precedents toward yet another. There are also Catholic doctrines which, when examined against the implications of the various solutions proposed, can enable us to see which ideas can be accepted by the Church and which cannot, though relatively little investigation towards this application has been made thus far; it is the hope of this author that considerable unifying progress can be made in this manner. But when the next reliable Catholic pope answers these questions, all Catholics will submit to his rulings and these disagreements will no longer be permitted.

It is interesting to note that all of their differences boil down to questions regarding how a Catholic is to interpret and respond to the present crisis. All of the older questions have long since been settled by the Church and one finds among the various bishops, priests, orders, and lay faithful of all of these groups a valid Catholic position in total uniformity regarding everything the Church has defined as morals or doctrine, which really is quite a bit. It is that uniformity regarding the established teachings of the Church which sets apart the Catholic Church in all of its present groups from all others.

I am not a "pan-traditionalist" since there are many "traditional" forms of worship which are not Catholic at all, such as the Voodoo worship traditions of Benin or even the noble traditions of the Jews. Even some groups which try to pass themselves off as "Catholic," such as Old Catholics, the Patriotic Church of China, the Eastern Orthodox, the TFP group after it separated from Plinio Corrêa De Oliveira, and so forth are still rejected here (although I still harbor hopes for the rehabilitation of the TFP) because of their refusal to adhere to established teachings of the Church. I am not a "pan-traditionalist" but a "pan-Catholic." I believe in the Catholic Church in its entirety, not merely this or that portion.

I find it difficult to see any "Oneness" amongst the various groups you write about since they seem to bicker with each other so very much and even say horrible things about each other.

Remember, the "Oneness" of the Church does not necessarily imply that all Catholics are just "buddy-buddy" pals with all other Catholics. One also has to be aware of the principle of false animosity. If you wanted to make person A and person B enemies of each other, you simply go to person A and tell them that person B has been saying all sorts of terrible things about him. You then go to person B and say the same things regarding person A. Until persons A and B should figure out what has happened, they will be at odds with each other, and indeed may even add to that enmity themselves by actually speaking out against each other. The entire feud is based on nothing and means nothing.

It is also a fact that exterior factors can put Catholics at odds with each other. For example, if two nations should go to war against each other, and if both have Catholic soldiers enlisted in their armies, you could, and typically would, have Catholics shooting at each other at the command of mere secular rulers. Indeed, the most virulent divisions currently within the Church are similarly traceable to such exterior causes. The division between Catholics who are within the Vatican organization and Catholics who are outside it is directly caused by that Vatican organization because they tell those Catholics within it to "have nothing to do with those who are outside our organization, and in exchange we might allow you a special privilege of professing and practicing your Catholic faith under our roof and with our blessing." Catholics who are outside that organization have quite reasonably and sensibly responded in kind, and thus is created that division.

Similarly, the division between the SSPX and other groups operating outside the Vatican organization is based on the hope that they might continue to be defended by the laws of the Vatican organization and therefore in a better bargaining position with the current Vatican leadership, providing that they distance themselves from all sedevacantist groups. In like manner, the division between the SSPV and other sedevacantists, is based on the hope that they might again one day be recognized by the SSPX, providing that they continue to distance themselves from the Thục-line bishops which Lefebvre had criticized shortly before the division between the SSPX and the SSPV happened.

Another factor, and the only one which represents a certain degree of true schism within the Church, is the different attempts to understand the nature of the current Church crisis and ascertain just what is to be done about it. While all Catholics must be (and in fact still are in these confusing times) united in belief regarding each and every teaching of the Church as promulgated and defined by the reliable popes, the current crisis has brought up many new questions, and there does not presently happen to exist any living single universally recognized Catholic authority which is both willing and able to settle these questions authoritatively. Persuasive sounding arguments have been made for all sides of such questions as "do we have a Pope, and if so in what sense," "can there be marriage annulments, and if so on what basis, and need the Vatican organization be involved," and "at which point do we draw the line and say that such-and-such an edition of the Roman Missal is the last official one, rendering all previous editions obsolete and therefore seeing all later editions merely as products of the non-Catholic Vatican organization."

There is also a tendency to react against the false ecumenism of Vatican II. It is important for the Church avoid all pretense of unity with false religions, as Pope Pius XI taught in *Mortalium Animos*. However, in the zeal to avoid the false ecumenism of Vatican II, Traditional Catholics have sometimes neglected that true ecumenism which is internal to the Church and essential to the principle of the Church being One (along with being Holy, Catholic, and Apostolic). It is not a trivial thing to draw the lines where they truly belong. With the Vatican organization being so lenient with false religions as it is, traditionalists often respond by preferring to risk erring on the side of being too strict rather than too lenient. At least they have the satisfaction of knowing that no one within the circle they draw is not a Catholic. And all would agree that there almost certainly exist some real Catholics outside the narrow circle they each have drawn, even though they might be uncertain as to precisely who. Of their own parish or congregation or society, they can be certain; of anyone else's they can only guess.

Not being aware of the full and precise state of affairs brought about by Vatican II, and also, the practical implications of existing Catholic ecclesiological doctrines, many such Catholic orders tend to feel as if they are functioning in an irregular fashion. While each may be satisfied with the reasons for their own seemingly irregular functioning, they are often uncertain as to the reasons for the similar seemingly irregular functioning of the others.

Drawing from that is also a fear for the future of the other groups, on behalf of their followers. That fear seems to have a certain logic to it. After all, if so great an organization as the visible Church (for which they mistake the Vatican organization) should be allowed to fall so rapidly into error, what is to stop any of these smaller groups from succumbing from the particular dangers each of these groups face? What is to prevent those approved by the Modernists from being swept up into the modernism which is so pervasive throughout the remainder of the Vatican organization? What is to prevent the SSPX from coming to believe the heretical position that popes can be routinely disobeyed? How can we know that the sedevacantists will not become so accustomed to functioning without a living pope that they cannot recognize a true, valid, and reliable pope even when one should appear? Each group is aware of the precautions they have taken against their own particular danger, yet they see all too readily the weaknesses in the precautions the other groups are taking. They all lack knowledge of the fact that their stability is guaranteed not by the precautions they have taken, but by the promises of God to His Church, which all Traditional Catholics from each of these groups together constitute.

Still one more factor to consider is that each faction seeks to unite all Catholics under their opinion. This exactly parallels the situation in the First Great Western Schism when all who favored the Pope in Rome from the Pope himself on down sought to unite all Catholics under their Pope, and at the same time all who favored the Pope in Avignon from that Pope on down sought to unite all Catholics under their Pope, and again the same for those who favored the Pope in Pisa. Notice that the Council of Constance did not favor one over another but instead elected a new Pope acceptable to all. Nobody had to see their rival win, and every Catholic got what all truly desired. Thus, I am sure it shall be with the present situation.

Finally, one must remember that extremely few ever possess the gift to see into the heart of another. All traditional priests and bishops are quite conscious of their own efforts at remaining orthodox and catholic in their teaching and practice, but none of them can see within the hearts of the others. When another starts to reach conclusions about the unsettled questions which differ from what one has concluded, it is very easy to suppose that the other fellow has taken an uncatholic turn in his thought processes.

It is easy to see that none of these "divisions" entail any real long term threat to the unity of the Roman Catholic Church since, when the next reliable Catholic pope rules on the unsettled questions, all Catholics will accept the Pope's ruling, and in their union with the Pope they shall be united to each other as well. One can properly pray that those who were right will forgive those who were wrong and avoid gloating over them, and that those who were wrong will swallow their pride and accept what is right. Also, as more and more Catholics come to understand the true natures of the Catholic Church and the Vatican organization, the simulated authority of the Vatican organization will be seen for what it is and the division between those inside and those outside the Vatican organization will fade along with all other divisions. The Church is eternally One, even when She seemed to be Two or Three, and even now when She seems to be Several.

How can you justify the sedevacantist's rejection of the Pope?

First of all, sedevacantists do not reject the Pope. What they reject is any claims to the papacy on the part of certain men whose teaching and public example differ quite markedly from the that of St. Peter and all the reliable popes. If only Paul VI, John Paul II, Francis I, and indeed all Vatican leaders from John XXIII onward were to have done the works of true and reliable popes! If any of them had withdrawn all of the Novus Ordo sacramental forms and return all the ancient sacraments, have the rebellious clergy deposed and the loyal clergy conditionally ordained or consecrated as applicable, revoke the entirely of Vatican

II and all that follows from it or even merely claims to follow from that, and make and enforce peace with and among all Traditional Catholic bishops presently functioning outside the Vatican organization, then every Catholic sedevacantist would have acclaimed and loyally stood behind him as the Vicar of Christ's most obedient sons!

That may sound like a lot to ask, but really, having a monarchical rule over the Church means that all that a true and reliable pope needs to do personally is mandate these things with the full weight of his Petrine authority. If the Novus Ordo bishops and cardinals want to get together and attempt to outvote him, all they demonstrate by so doing is their already present lack of submission to the Supreme Pontiff. If they care to find some way that Vatican II gives them the right to outvote the Pope, then they demonstrate the difference between the structure of the Vatican organization (as mandated at Vatican II) and that of the Catholic Church (as mandated by Christ).

So long as the current Vatican leadership continues refusing to do those works which a true and reliable pope must do, sedevacantists look at him and they look at the Chair of St. Peter, and it is as if the two have never met. They may not have understood the anatomy of how any authority had been lost, but they were quite correct in discerning that there had to have been some sort of a loss of authority. So long as the Vatican leader upholds Vatican II, he deprives himself of universal jurisdiction and undermines, weakens, or even eliminates the authority he otherwise would certainly have. And without that absolute Catholic authority, why should he be divinely granted the infallibility exclusively reserved to those successors of Peter whose authority is truly supreme and universal?

For example, John Paul II taught on many occasions that divorce, contraceptives, abortion, "mercy-killing," infanticide, homosexual "marriages," and priestesses are all evils which must be condemned. Great! Wonderful! Let us all stand up and give the man three cheers and a round of applause! For some reason, John Paul II himself maintained a guarded orthodoxy regarding these controversial issues. Good for him on that; even the sedevacantists give the man credit where credit is due. Certainly, all true Catholics must agree to what he taught about these things. Yet this same John Paul II himself, by upholding and carrying out the Vatican II directives on "Religious Liberty," has gone to each and every nation which still based their laws on Catholic Morals, and told them to rewrite their constitution so as to allow other religions to come in on an equality with the Catholic faith. Many of these other religions teach that divorce, contraception, abortion, "mercy-killing," infanticide, homosexual "marriages," and priestesses are acceptable! He was also all warm and fuzzy ecumenical with other religions that also preach these horrible things to be just fine, going on to affirm (as does Vatican II itself) that they also save. By so doing he deprived his own perfectly good advice in these areas of the authority it rightly deserved.

Sedevacantists are not disloyal to the Pope; they merely fail to see in the current Vatican leadership the lawful object of that loyalty which is rightly owed only to the Vicar of Christ. They have treated this period starting with the death of Pope Pius XII as an interregnum, which is quite an appropriate course of action, even in the event the Church should someday conclude that any of the doubtful popes were true (but very weak) successors of Peter.

On the practical level, sedevacantists are a great deal more concerned with the fact that a great many bishops, archbishops, and cardinals have entirely vacated their sees, or indeed any claim to being Catholics at all, and the rest fare only marginally better. Their real fight is not only against the apostate Vatican leadership itself but also such local "diocesan" bishops who harass them on account of the true sheep who are departing from those false shepherds. Nevertheless, the principle remains true that a commander is, to some degree, most principally responsible for the behavior of those who answer to him.

I hear that sedevacantists don't pray for the Pope.

That is entirely false. First of all, lacking a living Pope at the moment, they certainly do pray and desire that God would one day soon send a true and reliable Catholic Pope who would undo all of the mess decreed at Vatican II and restore all things in Christ. If one wants to say that they don't pray for the current apostate Vatican leadership, even that is not true. They pray that the man should repent of the heresies he harbors and become a true Catholic, for the good of his own soul and for the good of the Church.

At the Mass, in the Canon of the Mass, there is a point at which the priest may be uniting his prayer of that Mass with the prayers of the Pope. The sedevacantist priest saying Mass simply proceeds the way all priests do when the Church is between popes, namely they omit any name of the Pope with whom they unite their prayer since there isn't one at the time.

What sedevacantists refuse to do is unite their Mass prayers **with** him since they and he are praying for contrary things. Sedevacantists (as with all Catholics) pray for the restoration of the Church, and of Christendom, and for the Kingdom of God, and for souls to come to the Catholic Church and be taught, edified, and saved. They pray that the world may come to a uniform knowledge of the Truth but the Vatican leadership prays that the world may become peaceful by having everyone set aside any search for Truth, and for other outright leftist and Communist goals. Furthermore, these apostate Vatican leaders have personally prayed with the leaders and representatives of false religions, so that whoever prays with him might be, indirectly, praying with them.

Doesn't your claim regarding the fall of the Vatican organization mandated at Vatican II negate all the other sedevacantist theories?

It is still conceivable that some future reliable pope may rule or conclude that authority was lost at some previous point, during either the reign of John XXIII or the first part of Paul VI, due to heresy on their part. The problem with opining that authority was lost at any previous point is that there are so many such previous points at which authority might have been lost that Traditional Catholic sedevacantists have difficulty agreeing as to which of these points was where "the Pope" lost his office.

There is quite a list of possible places:

a. the election of John XXIII in 1958,
b. his announcement of intending to hold a Council in 1959,
c. his suppression of the Third Secret of Fátima in 1960,
d. his Pact of Metz concluded with the Communists in August 1960,
e. his change made to the Canon of the Mass in November 1962,
f. certain statements he made at the opening of Vatican II in 1962 about the Faith being one thing and its expression another,
g. the manner in which he allowed the heretical modernist faction the upper hand at the Council in 1962,
h. the promulgation of *Pacem in Terris* which contained certain statements dangerously open, even favorable, to heretical interpretation in 1963,
i. the election of Paul VI in 1963,
j. the decision of Paul VI to continue the Council which John XXIII declared to be finished and ordered to be stopped in 1963,
k. the promulgation by Paul VI of certain encyclicals which contained dangerous statements definitely tending towards heresy in 1963 and 1964,

…The list could go on.

Unfortunately, the loss of authority at any of these points is such an invisible and debatable thing that, in the absence of any formal proclamation from a future reliable pope, it would have to be regarded as mere theological speculation. Furthermore, if the failure was complete and not merely of the man at any of those moments, then what acknowledgment is there in any of these failure moments any clear provision for the continuation of the real Catholic Church, and especially outside the Vatican organization? Where and how did the Vatican organization visibly and officially cease to be the Church? Clearly, somewhere amongst these events there was a very real defection from the Church and then the Faith on the part of a vast majority of those who formerly comprised the Catholic hierarchy. What was needed was a way to remove, or at least indicate the possible removal of, all of them from their offices in the external and public forum. The Pope could not excommunicate all the bishops since he was as far gone as they; the bishops and cardinals could not get together and conclude that the Pope had excommunicated himself and thus hold a conclave since they were collectively as far gone as he. This author would venture the theory that Vatican II was God's way of accomplishing that which no one was willing or able to do.

How can you justify the SSPX's disobedience to the Pope?

Their disobedience is quite strictly limited to those disciplines which contradict Faith and Morals. Insofar as the teachings and disciplines of the Vatican leadership are in accord with Catholic Faith and Morals, or at least reconcilable with it, they are quite happy to work with him and in fact among his most ardent supporters. It is unfortunate that this loyalty is scarcely reciprocated at all, and even that much only by Cardinal Ratzinger who **had** gone to bat for them at least twice before being elected to become Benedict XVI, and of course the friendly overtures conducted by Cd. Hoyos. Disciplines that forbid the conveyance of a valid apostolic succession (by attempting to condemn the 1988 and 1991 episcopal consecrations) run counter to the Catholic doctrine that the Apostolic succession is meant to persist until the end of time. Disciplines that require attendance and participation in non-Catholic worship (such as the Novus Ordo) run counter to the Moral principle that Catholics must only attend Catholic worship.

The main real reason this group comes under so much fire is on account of their size. They are, after all, larger than all other Traditional (Western Rite) Catholic groups put together. It is because of this considerable size that most Catholics, when they think of traditionalists, think of the SSPX. Sometimes, their position is seen as an unstable compromise, straddling the fence between that of the sedevacantists and that of the "Motu Proprio" crowd. But actually, their role is quite different from the roles of those other groups. Their unique role requires that they remain just close enough that diplomacy is possible, just far enough that diplomacy is necessary, and sizable enough that diplomacy is urgently sought after.

So often it seems that the SSPX is always just barely escaping a charge of being schismatic or excommunicated from the Vatican organization on what strikes me as little more than obscure technicalities of Canon Law.

That is true; only a couple small technicalities of Canon Law separate them from outright "excommunication" from the Vatican organization. However, it is precisely for such strange and exceptional situations as theirs that those technicalities were written into Canon Law in the first place. The technicalities exist in order to protect any group of Catholics from any injustice at the hands of potentially corrupt Vatican officials. If anything, the new Code of Canon Law actually strengthens those technicalities. Under the new Code, unlike the old Code, Lefebvre doesn't have to prove that he **was** right in consecrating those four bishops, only that he **thought** he was right. How ironic that the same man who strengthened those technicalities has wished that he could have eliminated them, at least with respect to the SSPX!

How can you justify the SSPX's refusal to study the theological implications of the current crisis?

If the SSPX were to study the theological implications of the current crisis, they would become sedevacantists, which would not allow them to serve in their diplomatic role in the potential rehabilitation of the Vatican organization. It is their refusal to fall either into the camp of the sedevacantists or the "Motu Proprio" crowd which makes possible their future "zigzagging," as Abp. Lefebvre himself had done.

Did not Padre Pio predict that Abp. Lefebvre would "tear apart the community of the faithful" with his disobedience?

An unknown Novus Ordinarian writer, plainly acting in bad faith, once put forth the claim that Padre Pio and Abp. Lefebvre had met under the following unfortunate circumstances:

> Among the many, many people who came to see Padre Pio was Archbishop Lefebvre who, later clinging stubbornly to Catholic Tradition, as he called it, questioned the authority of Vatican II and was removed from office by Pope Paul VI.

The archbishop had a meeting with Padre Pio in the presence of Professor Bruno Rabajotti. This witness reported that at a particular moment Padre Pio looked at Lefebvre very sternly and said: *'Never cause discord among your brothers and always practice the rule of obedience; above all when it seems to you that the errors of those in authority are all the more serious. There is no other road than that of obedience, especially FOR THOSE OF US WHO HAVE MADE THIS VOW.'*

Padre Pio could give this advice because he had had to obey some rather questionable orders himself. His attitude was to put this in God's hands because He would find a way for truth to triumph. It seems Archbishop Lefebvre did not see things in quite the same way even if he did respond to Padre Pio with:

'I will remember that, Father.' Padre Pio looked at him intensely and, seeing what would soon happen, said: *'No! You will forget it! You will tear apart the community of faithful, oppose the will of your superiors and even go against the orders of the pope himself and this will happen quite soon. You will forget the promise you made here today, and the whole Church will be hurt by you. Don't set yourself up as a judge. Don't take powers that do not belong to you and do not consider yourself as the voice of God's People, as God already speaks to them. Do not sow discord and dissension. However, I know this is what you will do!'*

Fortunately, we have the Archbishop's own response to this false accusation:

For several years now this slander, a fabrication from start to finish, has been circulating in Italy. I have already refuted it, but lies die hard; there is not one word of truth in the page of that magazine you photocopied for me.

The meeting which took place after Easter in 1967 lasted two minutes. I was accompanied by Fr. Barbara and a Holy Ghost Brother, Brother Felin. I met Padre Pio in a corridor, on his way to the confessional, being helped by two Capuchins.

I told him in a few words the purpose of my visit: for him to bless the Congregation of the Holy Ghost which was due to hold an extraordinary General my Chapter meeting, like all religious societies, under the heading of *aggiornamento* (up-dating), meeting which I was afraid would lead to trouble...

Then Padre Pio cried out. *'Me, bless an archbishop, no, no, it is you who should be blessing me!'* And he bowed, to receive the blessing. I blessed him, he kissed my ring and continued on his way to the confessional...

> That was the whole of the meeting, no more, no less. To invent such an account as you sent me the copy of calls for a satanic imagination and mendacity. The author is a son of the Father of Lies.
> Thank you for giving me the chance to tell once more the plain truth.

While the above-quoted lie is attributed to a "Pascal Catanco" or "Cataneo," the name is plainly a pseudonym and no such individual has ever been identified with any of Padre Pio's known friends, associates, or acquaintances. No one has proved willing to step forth and say, "I am the one who first wrote that," since five minutes interrogation of such an individual would readily show that they never knew Padre Pio, but were merely cashing in on his good name. How fortunate we are that the Novus Ordo liar was foolish enough to have published this lie while Abp. Lefebvre was still alive and in a position to answer it.

How can you explain Abp. Lefebvre's continual zigzagging between calling the Vatican leadership alternately the Successor of Peter and the Antichrist?

As an Archbishop, charged with the mission of rehabilitating the Vatican organization, he needed (and his present successors in the SSPX leadership need) to be responsive to events in the Vatican. Unlike the sedevacantists, for whom the Vatican leadership almost never does anything right, and the "Motu Proprio" crowd, for whom the Vatican leadership almost never does anything wrong, Abp. Lefebvre and the SSPX respond to recent events in the Vatican. It is like a parent punishing a child who is naughty and rewarding a child who is good. A parent who always spanks or always spoils teaches the child nothing, but a parent who rewards good behavior and punishes bad serves as a good parent and teacher. It is this quality which gives the SSPX a unique parental function with respect to the Vatican organization that neither the "Motu Proprio" crowd nor the sedevacantists can fulfill.

Lefebvre would punish Paul VI, and then the John Paul's by referring to them as Antichrist and pulling his priests and religious just a little bit further away from them whenever they made a move which betrayed the Church and the Faith. When Paul VI or the John Paul's would make a move in the right direction, or which in some way affirmed the Roman Catholic Church or Faith, Lefebvre would reward them by referring to each of them in turn as the Successor of Peter and by bringing his priests and religious just that little bit closer again. Unfortunately, since his death the succeeding leadership of the SSPX has implemented this strategy rather haphazardly at best, as they don't appear to have fully understood their founding Archbishop.

In the early days of the founding of Ecône, he was perfectly happy to give Paul VI his blessing, since that blessing was being returned. As trouble mounted up with the apostate French Novus Ordo functionaries, he distanced himself from them but continued to show his loyalty to Paul VI. As Paul VI got into the act, Lefebvre began to take a hard line against him, ultimately culminating in his comments during the Mass at Lille. When, a short time later, Paul VI finally granted Lefebvre the desired audience, there was a brief peace between them which ended shortly thereafter when Paul VI forgot that momentary reconciliation and renewed his unjustified attacks on the SSPX. The last couple years of Paul VI were almost certainly the one period of time that Lefebvre had come the closest to deciding to become a sedevacantist. Certainly, many of his seminary students at that time unintentionally picked up on the feelings he had right then loud and clear.

The leadership of John Paul I being so short, Lefebvre does not appear to have formed much of an opinion about him one way or the other. When John Paul II took office, his prompt willingness to receive Lefebvre in an audience, and his deep sympathy for Traditional Catholics brought Lefebvre much closer to the Vatican leadership, and into being willing to work with Cardinal Seper, despite the latter's catastrophic statement that "They are making a banner of the Mass of St. Pius V!" Lefebvre might have been more put off by the promulgation of the 1983 Code of Canon Law had he not been then quite busy removing the sedevacantist priests from the ranks of the SSPX. Cardinal Ratzinger's appreciation of Lefebvre for removing the sedevacantist priests and the 1984 Indult may have brought Lefebvre just that small bit closer to the Vatican organization, but as continued negotiations achieved nothing, Lefebvre stopped his move to the arms of the Vatican.

When John Paul II performed his Assisi fiasco in late 1986, it may have been at that time that Lefebvre privately resolved to consecrate bishops to succeed him. When he announced his intention to consecrate bishops and they responded with unheard of gestures such as sending Cardinal Gagnon to visit the SSPX, and negotiations which at last seemed to be going somewhere, Lefebvre brought his priests and religious as close to the Vatican organization as they had ever been since the early days of the founding of Ecône, but as negotiations broke down and he realized what was really going on, he hardened his stance and renewed his determination to consecrate the bishops. In the context of these events and his mission to rehabilitate the Vatican organization, his zigzagging actually makes a great deal of sense.

How can you justify the "Motu Proprio" crowd's unity with the current apostate Vatican leader and his local functionaries?

They have been given a very special, but rather odd, favor, in that they have been permitted to keep their Catholic Faith and worship despite their membership in the Vatican organization. Towards them (and no one else), one could reasonably argue, he seems to behave *just enough* like a pope that they might not be sinning by praying in union with him in that limited context. Really, they pray not so much in union with him personally, as they do with the office they see him as ostensibly representing, although personal loyalty to the man, owing to personal and other reasons, may also have been factor at times.

If there are any soldiers fighting on the front lines and personally facing the dangers of the war to rehabilitate the Vatican organization, it is the "Motu Proprio" priests and lay faithful. It is they who have written all the letters, signed all the petitions, appealed all the cases to the Vatican, and they who have made it possible for any portion of the Catholic Church to continue to subsist within the Vatican organization, at least thus far.

How can you justify the "Motu Proprio" crowd's (and even that of Abp. Lefebvre on some occasions) attempt to make the Tridentine and Novus Ordo Rites coexist side by side?

Granted, a request for the true and a false religion to coexist side by side is quite untenable, from a doctrinal standpoint. However, from a practical and diplomatic standpoint I cannot think of a more effective way to terminate the existence of the Novus Ordo. Think of it this way: A car dealer who has had a long and mostly successful career buying and selling Fords or Chevrolets is suddenly told that from now on he must only deal in toy cars. Grudgingly, he goes along with this, even though his customer base shrinks rapidly. At the same time, a black market for real cars develops. Perhaps he is granted permission to sell a real car or two once in a while, if some customer puts in an order, pays a very high price for it, and is willing to wait for delivery. Only on that basis can toy cars continue to sell. Now suppose that the new rule he must follow is that toy cars and real cars must be equally available side by side on the same showroom floor, and for the same price. Guess which ones will sell!

If, for example, the decree went forth that all parishes in the Vatican organization must have an equal number of Tridentine and Novus Ordo Masses, and at equally convenient times etc., for a period of three years, and after that the parishes may increase or decrease the number of either kind of Masses depending on how well attended they are, one can easily see that the Novus Ordo would die a natural death inside of ten years.

How can you justify the "Motu Proprio" crowd's membership in a non-Catholic church?

Understand that the Vatican organization is not a church, strictly speaking, but a human organization, most directly comparable to a secular and non-Catholic nation, with a state religion of Novus Ordoism and a tolerance for other religions. Membership in that organization is no different than citizenship in a secular State. The fact that Yemen is an Islamic State does not rule out the possibility of some of its citizens not being Muslims, albeit treated as second class citizens and separated from the rest by a kind of "dhimmitude." Likewise, membership in the Vatican organization does not make every "Motu Proprio" Catholic a practitioner of Religious Liberty, Ecumenism, or the Novus Ordo Missae.

How can the traditional clergy possess jurisdiction without being assigned conventional territorial dioceses?

It is a matter of Divine Law that the jurisdiction of the Pope is over all the Church, no exceptions. It is also a matter of the Divine Constitution of the Church that the bishops, taken together collectively, also possess jurisdiction over all the rest of the Church, no exceptions (though they don't have jurisdiction over each other, except where a Pope has set up, via Ecclesiastical Law, some inter-episcopal hierarchical arrangement, i.e. an Archbishop over other bishops, and normally, bishops also don't have jurisdiction over each other's flocks). Obviously, there are meant to be many bishops (always plural, seeming to imply at least two as a rock-bottom minimum at all times continuously), each with their own respective "particular flocks" over whom their authority is direct, personal, and episcopal, much as the Pope's jurisdiction is over every Catholic. The particular manner of divvying up all the individual Catholics to this or that particular flock of this or that particular bishop, however, has always been a matter Ecclesiastical Law. The use of geographically territorial "dioceses" is a very old custom, and overall an extremely effective one which has served the Church quite well for nearly all of Her history, and will certainly serve it again sometime. By it, the place of residence of the individual Catholic determines who his bishop is, direct, plain, and simple.

In our present circumstance, with no Pope who is willing to assign real Catholics to real and historic dioceses, the fact remains that the bishops, taken together collectively, still possess jurisdiction over all the rest of the Church, no exceptions, no change, and regardless of what form their actual jurisdiction may take, or what their numbers are. As it is, we live in highly mobile times in which territorial dioceses become somewhat unworkable, even not considering our present ills. Son or daughter goes away to college, which is in another diocese (therefore under another bishop), then they meet someone and get married and may settle in the place of one, or the other, or someplace completely else. After some years, the main breadwinner takes another job in another city, a promotion to be sure, but it means taking up residence in yet another city, under yet another bishop, and so forth. And even that does not begin to hold a candle to how it necessarily will be once the final Antichrist of Biblical prophecy arrives, a time in which Catholics are persecuted and forced almost continually to flee for their lives, ever moving hither and yon, and having to take recourse to whatever Catholic clergy as they can encounter wherever they go. In such times, far more so even than in ours, territorial dioceses will be utterly moot, since the priests and bishops will also be continually on the run. Who would be so rash as to claim that they lose their episcopal authority and jurisdiction on account of such circumstances?

Granted, the circumstances in our day are not that dire; there can be some stability as to where to go and find Fr. So-and-So, or his bishop. But I see in our practical inability to resort to conventional dioceses (despite a good and heroic good faith attempt on the part of the early Thục bishops) a Providential rehearsal for that most dire End Time, when, again for all practical purposes, the territorial dioceses (other than that of Rome, which could be taken as becoming a merging of all territorial dioceses and therefore global in scope, if one wanted to account for things canonically) become impossible or at least wildly impractical. Will we respect our episcopal and priestly leadership better at that time than we do now?

Yes, one can say that we choose our bishop, but once made that choice should not be changed, or only most rarely and for the gravest of reasons. But even under the conventional diocesan system, you still choose your bishop merely by choosing your place of residence. And even now, most Catholics might have only one priest who is reasonably accessible to them, and is turned to as their regular confessor, and for all other sacraments besides the Mass as needed. That priest's bishop is the individual Catholic's bishop, direct, plain, and simple.

So what does it take to be a Traditional Catholic bishop with real and canonical jurisdiction in these present circumstances? For starters, the bishop must be Catholic. That one detail alone excludes all East Orthodox, Old Catholic, and others actively belonging to yet other schismatic lineages, and Novus Ordo as well. We don't have to attempt to discern anyone's heart, but merely their open profession of Faith. The profession of the Catholic Faith requires that one uses the Traditional Mass, the Traditional Sacraments, the Traditional teachings, and the Traditional disciplines. Those who do this thereby profess the Catholic Faith and are therefore to be counted as Catholics. Those who use the Novus

Ordo equivalents to these things thereby profess their membership in a different and non-Catholic church and are not to be counted as Catholics. Whether such an individual of apparent sincerity and good will may well be justified in the sight of God belongs to exactly the same category as whether a sincere and good-willed Protestant or Jew or anything else might well be justified in the sight of God. For those claiming membership in some alternate Uniate Rite of the Church, that would require more familiarity with that Rite than can be provided here, as to its authentic traditions. A good indicator would be that the cleric in question continues to use the "Latinized" versions of the Sacraments and to refuse recognition of or by their schismatic counterparts but instead actively proselytize out of the schismatic churches and into the Church. A bad indicator for a given cleric would be his making his Mass look as much as possible like that of his schismatic counterpart, intercommunion with said schismatics, mutual recognition, and a refusal to proselytize.

Next, the bishop has to be really a bishop, sacramentally speaking. Those who are not validly consecrated, but who nevertheless proceed with simulating sacraments they are not capable of, cannot be counted as Catholic bishops, even were their profession of Faith Catholic. This rules out nearly all Novus Ordo bishops (a few very old timers and some transfers from other Rites or regularized from schismatic but valid lineages), and an unknown quantity of those active in the other Rites, the changes of which came decades later, and may not in all, or even very many cases, have imperiled validity. This does however admit of one exception, and that is one who is appointed to an episcopal office but not consecrated (or at least, not consecrated right away), who therefore remains sacramentally a priest, but juridically a bishop. For example, Fr. Davide Pagliarani is presently the Superior General of the SSPX. Sacramentally he remains a priest, but juridically he is a bishop, whose authority over the SSPX and all of their attached consecrated Religious and lay Faithful is truly episcopal and canonical. Being a priest, he performs no episcopal sacraments himself, but has resort to his auxiliaries, the sacramental bishops of the SSPX (previously, one of those auxiliaries, Bp. Fellay, was also the Superior General, and therefore truly in every sense sacramental and juridical, "the bishop" of the SSPX).

Finally, the bishop needs to be licitly a bishop. All but the most extreme and miscreant home-aloners generally acknowledge that the traditional successions have to be considered "licit" by virtue of the extreme crisis in the Church, the need for a valid and lawful episcopal succession, the needs of the individual Catholic souls who will have recourse to these bishops and their succession, and the priests they ordain, for their own sacraments and practical parish life. If there were nothing else, at least Epikeia would nevertheless of itself justify these steps taken. But this is not enough. Historically there has never been such a thing as someone being "licitly" a bishop but lacking any and all canonical mission, apart from those suspended, and it is accepted that those who, after years of faithful service, have fallen ill and descend into their final convalescence, may retire, effectively ending their own share of the canonical mission. But for bishops to run seminaries, ordain priests, and perform all the usual episcopal functions for a given community ("flock") of Catholics, all without a canonical mission, that is altogether unheard of. Even auxiliary bishops possess the canonical mission insofar as they are appointed to serve here or there as auxiliaries, and the same goes for all bishops not suspended, excommunicated, or retired.

On the scale of the whole and entire universal Church, it is one thing that an individual cleric might, in such an emergency circumstance such as ours, perform the duties of a priest or bishop merely on the strength of supplied jurisdiction due to common error, that is, the person who approaches the cleric for something he has not the ordinary canonical authority to do, but assuming (mistakenly) that he does, and if the cleric functions appropriately as one bearing the rightful authority would, the action takes its juridical effect. But if every truly Catholic cleric around the world has only this supplied jurisdiction as a basis for their "ministry" then there is no Church. And by now, all of those with more conventional claims to jurisdiction have died off, or at least retired from all active service. By simple process of elimination, either at least some of the Traditional Catholic bishops have to have the canonical mission, or else the Church is truly finished, and from this point can only be imitated, not continued.

So, let's look at what it takes to possess the canonical mission. Key to that is the state of being "sent." Now, "sent" has two basic parts to is, namely the "by" part and the "to" part. The "to" part is somewhat easier so I start with that. To be truly apostolic, a bishop must be sent to some specific assignment within the Church, historically most commonly a diocese, but could also be Abbot or Superior General of a Religious Order or Congregation, also as curial officials, deans of Pontifical Seminaries, Vicar Generals, those set over Military (or other) Ordinariates, mission territories, Societies of Pontifical Rite, and so forth. Are there such assignments available today? Yes, the various Traditional Catholic societies require heads, for any society lacking a head must eventually dissolve and cease to be, and in fact every such Catholic society is headed up by a bishop, or at least someone with a bishop's authority. There must be an identifiable congregation, of identifiable persons, over whom a bishop is set, and (especially in such times as ours) must also be accepted as bishop by those comprising said society. With a bishop set over a given community of Catholics and accepted by them as their bishop, said society functions as an "ecclesial district tantamount to a diocese" and that bishop's responsibilities with regard to said society are precisely those of the responsibilities of a diocesan bishop to his diocese.

Now we look at the "sent by" part. A bishop has to be sent by the Church, which means the authority of the Church. All authority in the Church is rooted in the Pope, as successor of Peter. This does not mean that every bishop must be personally vetted by a living Pope at the point of his appointment to an office; indeed, a great many bishops down through history have been appointed by others, usually delegated by the Pope, and sometimes further down the hierarchical chain, but necessarily in all cases by at least legitimate bishop(s). It is easy to defend the claim that whatever can be delegated can also devolve, especially should sufficient necessity require it. Lacking a functioning Pope, functioning Patriarchs, Metropolitans, or even Archbishops, the prerogative must therefore descend to the remaining legitimate bishops. And appointment by legitimate bishops (in this circumstance) to an office set over real people who accept it constitutes the one so appointed to be himself a legitimate bishop, and thus the canonical and valid succession can continue as long as necessary.

The only difficulty is the express policy of Pope Pius XII as given in 1951 making all episcopal appointments to have to go through himself personally. One probably could argue that the situation is desperate enough to justify an appeal to epikeia and the devolutions as described, but there is another possibility, a truly elegant one if applied:

In resolving the conflict of the multiple papal claimants, the Council of Constance could have brought about the election of the Pope (Martin V) on its own authority by exactly this manner of devolution, but as brought out by Billot, "among scholars it is now held almost with certainty that the election of Martin V was not effected by the Council of Constance's own authority, but by faculties expressly granted by the legitimate Pope, Gregory XII, before he renounced the papacy, such that Cardinal Franzelin justly and correctly says, that is, of course, 'why, in humble praise of Christ the King, the Spouse and Head of the Church, we marvel at providence because He put in order that vast confusion occasioned and sustained by greed and ignorance, *saving all laws*, demonstrating very clearly that the indefectibility of the rock upon which He built His Church so that the gates of hell may not prevail against her depends not on human effort but on divine fidelity to His promises and on His omnipotence in governing.'" If one takes the Roman claimant Gregory XII as having been the one "real" Pope among the three claimants (a proposition by no means agreed upon at the time, and even now existing only as a consensus that has never been formally confirmed), then the faculties he granted to the Council "saved all laws" by avoiding any need for any devolution in the first place.

To our situation today, if one takes my theory as described in Chapter 3 to be the actual ontological origin of our present crisis, with the legal power to institute our present situation, then in that same document the right of any real bishop to make a real bishop was also explicitly created (abrogating Pius XII's 1951 decree), when *Lumen Gentium* stated, later on, that "Episcopal consecration confers the office of teaching and of governing together with the office of sanctifying," which is to way not only mere sacramental power, but magisterial and ruling power as well, which can only exist with the canonical mission. And this is given without any mention of any condition of a papal vetting, but only that their authority "can be exercised only in hierarchical communion with the head and the members of the college." A "college" is also how this same document describes the Catholic episcopacy as being, not only during that council, but continuing onwards thereafter. So the reference here is to all the Catholic bishops together collectively. And once again no real need to resort to devolution, but rather that "all laws are saved" again, a proposition also by no means agreed upon by all at this time, but at least worthy of quite reasonably and probably becoming a clear consensus in ages to come.

Why don't you count the followers of Fr. Leonard Feeney as Traditional Catholics?

When Fr. Feeney broke off from the Church by refusing to obey (or even go on an all-expenses-paid journey to Rome, complete with an audience with the Pope (Pius XII) so as to plead his case), the establishment he stepped out of was still identical to the Roman Catholic Church. When he left the Vatican organization, he left the Church. Like all others who left the Church, he subsequently veered into error, namely the claim that a soul cannot be "saved" under any conditions whatsoever unless literally baptized with literal H_2O water. The Church had already explicitly taught otherwise, namely that a soul can be not only justified, but saved, by a Baptism of Blood or Desire.

Indeed, Feeneyites are the perfect test case, since they were in the beginning reacting against another heresy, namely that any vaguely well-intentioned soul automatically thus obtains a valid Baptism of Desire. Their motives differed little from the motives of Traditional Catholics. The reason they have wandered off into error and Traditional Catholics have not is NOT that Traditional Catholics are "better people," but simply because the organization they left was still identical to the Catholic Church, so that in leaving it they left the Church and thereby departed from God's protection from error. Once Vatican II was on the books, the Vatican organization ceased to be identical to the Roman Catholic Church, so departing from it (as necessary for most Traditional Catholics) no longer meant departing from the Church, thus the characteristics of the Church also continue to be with the Traditional Catholics.

I've heard that [such-and-such Traditional Catholic organization] is like a cult. What do you say about that?

One has to put such charges in perspective. By the current secular viewpoint, one would have to regard the Catholic Church has having always been a "cult," to say nothing of some explicit names of certain parts, such as the "Cult of Mary" and so forth. Let us examine a couple parallels one could draw:

David Breese, who wrote "Know the Marks of Cults," writes "One of the marks of a cult is that it elevates the person and works of a human leader to a messianic level. The predictable characteristic of a member of a cult is that he will soon be quoting his leader, whether Father Divine, Prophet Jones, Mary Baker Eddy, Judge Rutherford, Herbert Armstrong, or Buddha as a final authority. A messianic human leader has used the powers of his intelligence or personality and with them imposed his ideas and directives on the ignorant."

Could that not, with equal validity be applied to Jesus Christ Himself, since He, as a man walking about on the earth some 2,000 years ago or so, nevertheless claimed to be God, accepted worship and adoration from His followers, and whose every word is final and utterly authoritative? Or again, what about another common mark of cults which is to conclude that everyone who is not a member of it loses out? The Church does, after all, teach that "Outside the Church there is no salvation." Is that not cultish?

(If you happen to be Protestant, or Moslem, or Jew and are enjoying that comparison of Christ to the likes of Rev. Moon, etc., or the Church to a cult, bear this in mind: Protestants also say the same things of Christ, that He is God and His words are authoritative, and the same thing of Christianity in general as Catholics say of the Catholic Church. For Moslems, it would be Muhammad, who, although not claiming to be God (or Allah, in their theology), nevertheless makes himself greater than Jesus as the only "Seal of the Prophets" or for Jews it would be their roughly similar views of Moses and the Torah.)

One must face and come to grips with the fact that the Church Jesus Christ founded has from the beginning possessed what the insipid, lukewarm, banal, and worldly sorts would all have to categorize as "cultic qualities." For the Church (the Traditional Catholic community taken as a whole) to seem any different, now **that** would be out of order! Therefore, when for example, certain detractors and opponents declare "Oh! The SSPX is a cult!" all they are saying is "Oh! The SSPX is just like Jesus Christ's Church has always been!"

All of which brings us to the real issue here: "How is Jesus Christ different from Rev. Moon or Jim Jones or any other bogus cult leader?" If Jesus were truly like these hoaxers and frauds, then all of Christianity would be (and have been from the beginning) a total hoax and a fraud. Is there a real difference? The claims made are quite similar. What differs? The fundamental difference is results. Jesus delivers the goods and makes good on His promises; the others do not. Jesus works real and certifiable miracles; the others do not. And I don't mean the silly and mass-produced hysteria-induced "miracles" of the Pentecostals and like sorts such as Novus Ordo "Charismatics" and others making unsubstantial claims, but real and substantial miracles which bear examination and stand up to all tests, including that of time.

Jesus, alone of all of those claiming any messianic or Divine qualities, has come back from the dead in His resurrection, and then again as His Mystical Body the Church He comes back from the death mandated at Vatican II. That latter is a broad-based, slow-motion miracle which has been taking place before our very eyes, and absolutely no less miraculous than His original physical resurrection way back then. Destroy, corrupt, or eliminate any cult leader's organization, and it is destroyed, corrupted, or eliminated, never to recover. But do the same to Christ's Church and even though many may be misled and many defect and many die, that Church continues, undestroyed, uncorrupted, and uneliminated.

Furthermore, while real miracles (of the dramatic sort which flagrantly violate physical or natural laws and are not publicity hoaxes, but real) have always been quite scarce, but what few of them which still take place all take place among Traditional Catholics. When someone dying of a terminal disease receives the Last Rites and recovers to go on and live many more years than medically expected or possible, it turns out that he received his Last Rites from a Traditional Catholic priest who administers the true Sacrament of Extreme Unction, never from the Novus Ordo "Anointing of the sick." Another interesting fact is that the only exorcists today who are truly effective in ousting Satan from a Satanically possessed soul are all Traditional Catholic clergymen, or at least those who insist upon using only the traditional Rite of exorcism. All of them! Think about that. Even Fr. Gabriele Amorth, who served among the Novus Ordo modernists, would use only the traditional exorcism Rite since, as he stated publicly, the Novus Ordo exorcism Rite simply does not work.

The only reason one hears more about "miracles" in the non-Traditional Catholic community is that there is where the hoaxers and publicity seekers are. Traditional Catholics are not what they are because of "signs and wonders," but because they have carefully studied the teachings of the Church and the historic facts relevant to that Church and made the decision and commitment to do the right thing. As Scripture states (Mark 16:17) the signs and wonders "follow the believers," not the other way around.

QUESTIONS ABOUT RELIGIOUS LIBERTY

Precisely what mental gymnastics, logic stretching, and special pleading can you possibly come up with to reconcile Religious Liberty with Catholic doctrine?

Let me first quickly review the statement in the schema on Religious Liberty at which point Abp. Lefebvre balked and ceased signing any more documents in that council (*Gaudium et Spes* was the only other document he hadn't signed so far and there were no more to come). That is the statement which said, "This Vatican Council declares that the human person has a right to religious freedom. This freedom means that all men are to be immune from coercion on the part of individuals or of social groups and of any human power, in such wise that in matters religious no one is to be forced to act in a manner contrary to his own beliefs. Nor is anyone to be restrained from acting in accordance with his own beliefs, whether privately **or publicly**, whether alone or in association with others, within due limits." This passage itself is highly suggestive of at least several heretical notions, for example if no individual or group of any sort has a right to coerce another person in religious matters, does that not seem to imply that parents have absolutely no right to demand that their minor and dependent children go to Church or behave themselves while there or avoid stealing candy or obey any other of the ten commandments?

The particular notion in this passage which I intend to take on is the notion that everyone of any belief has the absolute right to express their beliefs in the public forum, even for the purpose of proselytization to their false religion. As brought out quite clearly in the booklet *Archbishop Lefebvre and Religious Liberty* (See Bibliography), this is the exact notion which Abp. Lefebvre found just too heretical to sign his name to. Numerous popes and councils and all of Church tradition have all universally affirmed that "error has no rights" and heretics must not be allowed to confuse the public by being allowed to spread their heresies through the public forum. It is this notion I will attempt to reconcile. I do this, not believing that this "reconciliation" would prove valid, but as a demonstration of what would be necessary to make the best effort towards a good faith attempt to reconcile them. And bear in mind that EACH and EVERY such anomaly of the Novus Ordo beliefs, not only blathered in Vatican II itself, but ever since – a vast swath of material, would require such far bulkier still "reconciliations" as this one in order to assert the claim that the Novus Ordo religion can be reconciled with the true and Catholic Faith.

How can this be reconciled? On the one hand, Church tradition teaches that something is a manifest evil, and on the other hand this document seems to teach that the identical self-same thing is a manifest good. First of all, I am not talking about Religious Tolerance which deals with the nature of some compromise accepted by the Church in this or that society simply because it's the best it can get there, or which is politically feasible, but what is spoken of in that document as an absolute right on the part of the heretic, independent of any social or political considerations. A little analysis would help clarify the issue. There is in this situation a conflict between two goods, namely the good for the heretic who is after all a soul to be saved, and the good of the public, or society at large. By referring to these two goods as first and second principles in this response, I am not referring to their relative importance but the chronological order in which I introduce them into the discussion.

For the first is the principle that the conscience of an individual ought not be violated, even when malformed through upbringing in a false religion. What the Church must seek to do is to reform the conscience to bring it in line with reality as made and described to us by God. Typically, that is a gradual process, and until it is complete there are bound to be some false scruples which ought to be respected. One does not in good conscience force an individual who was raised in a false religion which forbids the eating of meat, to eat meat until such time (if any) as that person's conscience comes to be completely reformed on that particular subject. Nor does one in good conscience scandalize such a person by eating meat in their presence. (Romans 14:1–3, 14–17, 21–23) To violate a conscience is to destroy it, to sear it, as it were, with a branding iron (1 Timothy 4:2), thus rendering it insensitive or nonexistent.

The second principle is the necessity of protecting society at large from false, heretical, and destructive notions. There are many cases where the Church can respect this first principle without violating the second, such as in the case of the Jews who do not go out making proselytes and haven't throughout most of Church history. As long as they kept to themselves, married amongst themselves, didn't bother anyone else, and made absolutely no attempt to convert anyone who did not actively seek them out, the Church should have had no problem with allowing them to live within a Catholic society. On the other hand, a Protestant fundamentalist may believe that he is morally obliged to go out and make converts and publicly deny Catholic doctrines such as the Real Presence or purgatory or the sinlessness of

Mary. At this point, the second principle kicks in. A heretic can do a large amount of harm to society (i. e. a great many other souls) by teaching his errors publicly, and must therefore be stopped, both in the interest of social order as well as the eternal welfare of those who could lose their souls from listening to him.

Classically, the Church has always subordinated the first principle to the second since the second concerned not only the great many more souls in society, but also the future of that society and of the Church in that society, whereas the first only concerned a single soul of an individual who, for all anyone knows, may well be very possibly acting out of bad faith rather than truly obeying his conscience anyway. The ideal thing would have been to silence the heretic, except in public debates between the heretic and a qualified doctor of the faith who would also happen to have debating skills at least on par with the debating skills of the heretic, so that everyone in the community can see for himself that the heretic is completely off base and laugh him out of town. Possibly the heretic himself might even be brought to the true Faith by the sheer force of logic and the Truth. Brute practical necessity has usually forced the Church to try first to explain the truth to the heretic, and if that fails, to order the heretic to remain silent under the threat of dire consequences, or if that fails, to have him permanently removed from society, utilizing if necessary, the Law of the secular authorities. Regrettably, some secular authorities had the heretic publicly put to death by slow torture and thus made them into false martyrs.

With all of this in mind, I now tackle the Vatican II document. The document itself does not in any way make it clear that the principle of respect for an individual's malformed conscience is and must still be the lesser of the two principles, and concern for Catholic society at large the greater principle. One could (and indeed must, if this notion is to be reconciled with Catholic doctrine at all) take it as an assumed that the lesser principle is still subject to the greater, and that this document is merely concerned with seeing to it that the greater principle does not simply negate the lesser, but that, without surrendering the greater principle, one nevertheless does their level best to accommodate the needs of the lesser principle as well. In the case of the Protestant fundamentalist heretic who feels obliged in conscience to preach his errors publicly, I maintain that society must be protected, but if the Church is truly all that concerned about respecting the malformed conscience of the heretic, I see no reason why they could not arrange for providing some contrived situation in which the heretic can be fooled into thinking he has discharged his duty to propagate error while in fact he has not. Allow him to preach his errors to an audience who cannot hear him or who are otherwise known to be incapable of believing his errors but are hired to act as though they are very interested to hear what he has to say.

I realize that such an example sounds very far-fetched, and furthermore I know of no instance where it has ever taken place, but the theoretical possibility of such a scenario quite literally represents the only way such an objectionable notion as that in the Vatican II document could ever be served in a truly Catholic way. Either you must accept that admittedly ludicrous and absurd attempt at a reconciliation, or else you are constrained to agree that the third phase (1965) of Vatican II has propagated religious error and cannot be regarded as a product of the Roman Catholic Church. There are no other options. "Error has no rights," even though sincerely erroneous people do have rights, provided that others can be protected from their errors.

If you don't believe in Religious Liberty, then does that mean that you don't think people should have the freedom to worship God any way they please? What if a nation should decide to mandate a religion other than Catholicism?

First of all, keep in mind that many nations already **do** mandate other religions. The communist nations all mandated (and what few remaining still mandate) Atheism as their state religion. The Muslim nations all mandate Islam. Protestant nations have at various times mandated Protestantism. Not surprisingly, the Church is strongest in such nations where She is not allowed since they treat Her precisely as She was treated in ancient Rome during the first few centuries. Unfortunately, Her existence in those countries is undocumented and we may never know the extent of Her influence in those places for a very long time. Under the test of persecution, just like the test of scholarship, most false beliefs soon melt away and disappear, but the true religion only thrives all the healthier.

People may call us "bigoted" or "prejudiced" because we believe that a Catholic nation should not permit any other religion to be recognized within its public and governmentally approved forum. We Catholics cannot afford to be afraid of being called names. In point of fact, it is only the proselytization by false religions which is to be limited, **not** the private practice of them. Many other religions, such as Judaism, Jainism, or Hinduism do not proselytize at all, but only gain members through the cradle, or by attempting to explain themselves to those who, on their own initiative, actively seek them out. It is therefore quite possible for such religions to exist peacefully within a Catholic state, but certain others such as Buddhism, Islam, and Fundamentalist Protestantism, do proselytize, and a Catholic state ought to limit that aspect of those religions, at least in the public forum. In particular, official public and civic recognition cannot be given to these or any other false religions by the Catholic state, in precisely the same sense that no school of Obstetrics which teaches the "stork theory" of child delivery should ever be accredited.

This is not a matter of trying to legislate people into the Church. The laws of any nation, by their very nature, can only regulate the exterior acts of a person, never their interior life. It is only one who has an interior life as a Catholic who is saved by being a Catholic. It is unrealistic for anyone to think they could mandate the interior life of another. Indeed, any attempt to force grown people to be "Catholics" without their interior consent would only produce false Catholics of the sort who gave us Vatican II. But we **can** regulate what people do exteriorly, in the public forum, and in particular limit their attempts to lead others astray.

Someone may want to say, "What are you Catholics afraid of, that people might find something out?" The response is, "All that we are afraid of is that these cranks and crackpots, having been unable (and often unwilling even to try) to convince the experts of their insanities, may then turn their efforts to those who are not qualified to see the flaws in their arguments, and thereby succeed in leading many astray and causing strife and social disorder." What would happen to the economy of a nation in which everyone was at liberty to choose for themselves what constituted legal tender? If one could say currency and another say bottle caps and another say tiddlywinks and so forth, and all such opinions had to be treated as being of equal validity? That would be sheer chaos!

A Catholic nation should protect its citizens from divisive, seductive, and stupid ideas and belief systems in the same manner as responsible parents should protect their children from them as well, or that laws against false advertising protect us from ineffective or dangerous "patent medicines." When everyone in a society has the same idea as to which way is up and which way is down, then each person knows where he or she stands in it. People who feel free to worship God any way they please could quite readily "please" to worship God by committing human sacrifice or ritual molestations, or any of a great many other abominations. In the name of "religious freedom," in the United States of America, cults have been allowed to brainwash and kidnap minor children from their parents, certain families have been permitted to get their children "high" on hallucinogens, and small children have been allowed to die rather than receive a blood transfusion or other medical treatment. None of those horrors could ever take place in a Catholic nation.

This is not a question of making everyone be the same like "cookie-cutter Christians" or members of a religious cult; it is only that everyone should live in the same Universe. People will differ in personality, interests, priorities, hobbies, and tastes as God intended, even as people within a single economy (as opposed to the "economic diversity" mentioned above where everyone decides for himself what constitutes money) will have different amounts of money, different ways of earning it, and different ways of spending it, but all are in agreement as to what **is** and what **is not** money.

It doesn't seem fair to me that the Catholic religion should be given special treatment by the civil governments.

On the contrary, it is God who is totally fair by desiring that all civil governments give recognition only to the Catholic religion. He will judge all persons by the same standard, namely that standard taught by the Catholic Church. He plays no favorites. Even if God should excuse a particular soul from a specific duty on account of a legitimate ignorance or other reason on the part of that soul, the soul in question is not being judged by a different non-Catholic standard from the rest. Rather, that too is just another part of the Catholic standard. Catholic Moral theology teaches us that we are responsible for attempting to seek the Truth and to live in accordance with what we have learned, but also that we cannot be held responsible for any Truth which we have genuinely sought, but which Providence has failed to provide.

The Protestant will not be judged by Protestant standards; the Buddhist will not be judged by Buddhist standards, but all shall be judged by the Catholic standard. What could be fairer than to see to it that all souls are equally informed of the one standard by which they shall all be judged?

It seems quite arrogant of you to insist on the Traditional Catholic standard as the only right standard.

Such dogmatism does not originate with me. Whenever Moses said, "Thus says the LORD,…" there was no room for anyone to say rightly, "I think the LORD might be saying something else; tell you what, Moses, you go on thinking that the LORD told you that and I will go on believing that the LORD told me this and we can just respect each other, agree to disagree, and try to get along by talking about other things, how about that?" When Jesus said "I am the Way, the Truth, and the Life," such a statement escapes total arrogance only by virtue of the fact that it is true because Jesus is God. The reliable popes have continued this arrogant-seeming dogmatism by pronouncing anathemas on all who disagree with their teaching magisterium. I in turn merely reiterate the teachings of the reliable popes, especially as they apply to the present situation of the Church.

Such seeming arrogance of the Church in Her persistent insistence on being the final and authoritative arbiter as to what is true and what is false, what is right and what is wrong, and what is good and what is bad, is more than justified. It is no different than the "arrogance" of a person who insists that everyone is obliged to believe that $2 + 2$ equals 4, in a room full of fools who think that $2 + 2$ can equal anything anyone likes. I can no more deny any Catholic teaching than make $2 + 2$ equal to 5.

QUESTIONS ABOUT QUESTIONING THE POPE

Aren't you being schismatic by recommending that Catholics attend parishes which are not under the diocesan bishop, and by casting doubt on the papacy of the Vatican leadership, or the obligation to obey him?

Let us start by reviewing the definition of what it is to be schismatic. There are two parts of the definition of the term. The first pertains to one's relationship to the Supreme Pontiff and the second pertains to one's relationship to fellow Catholic believers. With regards to the first, refusal of subjection to the Supreme Pontiff is schismatic as is giving such subjection, proper to that for a pope, to an impostor. Since there is a true and proper doubt that the Church even has a living pope at this time, only those who opine that the current Vatican leader is a pope have to deal with any obligation to submit to him as an authority and even that only insofar as it is not contrary to Faith or Morals. Sedevacantists refuse submission to the Vatican leader for precisely the same reason they refuse submission to Reverend Sun Myung Moon. If a man isn't a pope, as the sedevacantists have good reason to believe that he is not, then he cannot rightly receive the submission lawfully due to the Successor of Peter. In such times where there is confusion as to who, if anyone, is the Pope (as the Church was so confused during the First Great Western Schism of the 1300's), it is the second part of the definition of schism which more directly applies.

This second part of the definition of schism is "refusal of communion with other members of the Church." By the word "communion" in this context, it is fellowship or association which is being referred to, not the Eucharist. It is the person who says to his fellow Catholic "I cannot eat with you; I cannot pray with you; I cannot discuss spiritual matters with you," who is being schismatic, not the one who is so snubbed. Contrary to the often-repeated big lies of those who are critical of the Traditional Catholic community, it is not the traditionalists who are being the "separatists," but the Novus Ordo People of God.

Traditional Catholics do indeed separate themselves from the world and sinful, worldly associations, but from fellow Catholics, and even sincere "Catholics-at-heart," persons who strive in their flawed way to avoid sin, they never separate themselves. It is the Novus Ordo People of God members who always schismatically separate themselves from Traditional Catholics, often while refusing to separate themselves from the worldly associations. Traditional Catholics are all supremely and serenely confident that they have the Truth and know they have nothing to fear from the stupid and ignorant arguments of the Novus Ordo. It is not and never has been the Traditional Catholics who have retreated from religious discussions, dialogue, or debate with the Novus Ordo People of God, but **always** the Novus Ordo People of God who cannot face Traditional Catholics in religious debate.

It is always the Novus Ordo People of God who are obliged take the position of "Don't confuse me with the facts; Bishop Beezlebub has already made up my mind!" The only thing even remotely schismatic within the Traditional Catholic community is those few (and rapidly growing fewer) voices who continue to argue for only one's own faction at the expense of all other groups, as if their group alone comprises the Church. As the Traditional Catholic community continues to grow and gather momentum, the voices of those schismatic few are increasingly getting drowned out by the Song of Praise rising from those Traditional Catholics who have come to realize, at least on an unconscious level, that the Traditional Catholic community IS the Roman Catholic Church. At any rate, that is a "schism" which is entirely internal to the Church, just like the fourteenth century schism, and not at all comparable to the schism between East and West which puts the East Orthodox completely outside the Church.

Aren't Traditional Catholics acting just like the Old Catholics by setting up their own hierarchy and separating from Rome?

"Once upon a time there was a group of Catholics who wanted to reform the Church, claiming it should return to the purity of the 'ancient faith' as practiced by the early Christians. Their platform included planks to prepare the way for the reunion (not conversion) of all Christian confessions, a reform of the position of the clergy, a reform of the Church with constitutional participation of the laity, the forming of parish communities, the abolition of celibacy (due to the shortage of priests), the abandoning of confession and the use of the vernacular in the 'service of the altar.' Protestants were included in their theological faculty. Their congress was attended by three Anglican Bishops and members of the Russian clergy. They also refused to accept the infallibility of the Pope in defining dogmas. Now my question for you is this: does the above description describe Old Catholics, New [Novus Ordo] Catholics, or both? It certainly doesn't describe *Catholic* Catholics."

That fine bit of wisdom from the pen of Frank Denke (*Angelus,* July 1990, pages 8–14) shows well where the **true** division lays. On one side of the great divide are the schismatic East Orthodox, the Eastern heretics, the Moslems, the Protestants, the Old Catholics, and the Novus Ordo, and on the other side of that same great divide is the Church founded by Jesus Christ, taught by all the reliable popes, and today existing as the Traditional Catholic community. One feature which only the Traditional Catholic community has which no dissenting group can ever have is the ability to point to an actual and well documented change in "the Church," and which is a well-known and widely witnessed event in the memory of hundreds of millions now living, coupled with a stand taken by its own members to hold fast to the traditions they and their parents were brought up in which all knew existed prior to that well-known change. By contrast, all these dissenters were forced to posit a fall from grace on the part of the entire Church so long ago that no one remembers it, followed by a long period during which there was no Church, or only a "church" that did not worship as they believed, during which Christ had no Mystical Body.

You "traditionalists" seem to me to be acting just like the Modernists who want to change the structure of the Church and feel free to disagree with the Pope.

Anyone who thinks that the Traditional Catholics of whom I have written about in this book are being in any way even the least bit disobedient to the Pope, or trying to undermine the authority of his office, has entirely missed the point and must go back to square one. The whole point and purpose of the Traditional Catholic community (as such) is that its people are Catholics who refuse to disobey the Pope. It is our solemn and irrevocable attachment to the Barque of Peter which compels us to attend the Traditional Catholic Latin Masses, even though most such Masses go unrecognized by the local Novus Ordo functionary. If that happens to entail anything which could be construed as being "disobedient" to the highly questionable Vatican leadership, so what?

When such an individual speaks against the traditions of the Church, allowing their "bishops" or "priests" to impose communion in the hand and altar girls, or even going so far as to push the bogus new "liturgy" and other "reforms" in the positively Satanic "Spirit of Vatican II," such an individual is clearly not functioning in the role of Peter confirming his brethren. If the post-Conciliar leadership of the Vatican organization were to be taken as being Peter at all, then it would be Peter openly declaring ("teaching") that he does not know Jesus, and if not, then the Chair of Peter is empty and the sedevacantists are right.

In the absence of any clearly and uniformly Petrine voice, the entire group of Vatican functionaries have become collectively like Aaron when he set up the golden calf for the Israelites to worship (Exodus 32). Traditional Catholics are simply those modern Israelites who refuse to worship the golden calf of the Novus Ordo Missae.

The notion that any Traditional Catholics would wish to preserve their present leaderless state and lack of a more conventional canonical structure is patently absurd, merely a lie spoken by the Novus Ordo enemies of the Church. Traditional Catholics know all too well the exact function of each ecclesiastical office in the Church as Christ intended it to be, and merely await the time those offices can come to be filled again. It is the villains of the Novus Ordo Church of the People of God who have redefined the offices of their "church." It is they who reduce their "Pope" to mere "President of the College of Bishops," they, who practically force their "Bishops" to run their dioceses in whatever manner has been voted on and approved by their "Bishop's Congresses," and they, who force their "Parish Priests" to run their parishes by the whims of their "Parish Councils," their "Liturgical Committees," and their "Finance Committee."

Isn't it schismatic to set up a parallel hierarchy?

There is actually no real need to worry about setting up any parallel hierarchy. The traditional clergy simply ARE the original hierarchy, or what's left of it. It is Vatican II, and *Lumen Gentium* itself which created a new "hierarchy" which is parallel to that established by Christ. They did this when they described that which is one, holy, catholic, and apostolic as merely "subsisting in," not "as" nor "throughout" nor "exclusively in" but merely partially in what in that same sentence is grammatically established to be an ontologically separate and distinct society or group or organization, a "parallel hierarchy."

However, even a parallel hierarchy need not be considered a schismatic thing. For example, the Eastern Rite Catholics already have their independent parallel hierarchy. The Eastern Rites have their own dioceses, many of which overlap Latin Rite dioceses. An example would be someone in the Pre-Vatican II days, and who was living in the Archdiocese of New York (according to the Latin Rite), but also living (at the same time and place) in the Byzantine Archdiocese of Pittsburgh. The Novus Ordo already created its own parallel "hierarchy" of sorts (not properly THE hierarchy in any theological sense), a pseudo-hierarchy which refuses to recognize many Catholic priests and bishops in its geographical midst.

What truly would make a parallel hierarchy schismatic would be to become a "hierarchy" which is by nature and definition not answerable to Peter, as for example the English "hierarchy" under Henry VIII who in effect said to the Pope and the Roman hierarchy he ruled, "We will be the hierarchy [sic] of England; you can be the hierarchy of Europe!" They just wanted to get out from under the authority of someone who was clearly and indisputably the Voice of Peter.

But didn't John XXIII promulgate heresy in his encyclical *Pacem In Terris, Paragraph 14* when he wrote that "Everyone has the right to honor God according to the just rule conscience and to profess his religion in private and public life?" How can you say that the official pronouncements of the Vatican were protected up until 1964 when *Lumen Gentium* was promulgated?

One frequently finds in the writings of John XXIII and the early writings of Paul VI statements such as this. These statements wherever found are always ambiguous, being able to be taken in more than one sense. There is a heretical sense it can be taken, but also an orthodox sense as well. Taking this statement in particular, a reference is made to a person's right to practice his religion publicly in accordance to his conscience. The ambiguity rests on the use of the word "religion" which can be taken as a reference to the true religion (Catholicism) which everyone does indeed have the right from God to practice publicly as taught by many reliable popes. Unfortunately, the same statement could also be read in a heretical sense if the word "religion" is taken to mean just whatever religion a person happens to believe in, whether true or false.

Such ambiguity in the writings of John XXIII places him in the same category as Honorius I who, in trying to establish a sort of peace between the Church of his day and a heretical faction, also promulgated an ambiguous formula acceptable to both the Church and the heretical faction. It was this action which caused later reliable popes to condemn that ambiguous formula. Pope Honorius I is a classic and proper example of what I have been calling an unreliable pope: Even though the Church accepts him as a valid successor of Saint Peter, one cannot authoritatively quote his ambiguous declaration against the more precise and definitive declarations of the reliable popes.

If the Holy Spirit protects the pope from teaching error, why are ambiguous statements allowed?

Ambiguity is allowed only because it is unavoidable. In a certain deep sense, **all** teachings of any kind are necessarily ambiguous. A prime example of this would be the Apostle's creed. One does not gain from a reading of it a clear answer to such questions regarding the Real Presence of Christ in the Eucharist, the Sacramental Priesthood, the Primacy of Peter, or even the routine authority of a bishop. The creed is ambiguous regarding these doctrines simply because they are not mentioned in it at all. Nevertheless, that ambiguity is precisely why so many of the more "high church" Protestants can recite that creed without a qualm despite their rejection of the above mentioned doctrines.

At the opposite extreme would be the case where a doctrinal or moral question is being asked throughout the Church, with factions arguing for different answers, and where the Pope speaks about the question in a manner which pretends to address the question while failing to do so. The Church in the days of Pope Honorius I had a certain doctrinal question which was brought to his attention. The culpably ambiguous response he gave to that question allowed people on both sides of the question to believe that he might be in agreement with them. Between these two extremes there is such a smooth continuum that nowhere can one arbitrarily draw a line and say, "Beyond this degree of culpability a pope's Charism of infallibility forbids him to be ambiguous."

A very common and typical example between these extremes is where a question is put to the Pope and the Pope says **nothing**. This might reasonably happen because the Pope may wish to take some time to pray about the matter, meditate on it, consult with his theologians or Curial experts, or even if he should happen to believe that the current moment is just not an opportune time. A good illustration would be, suppose a pope were to be infallible in matters of mathematics. If a page full of math problems were presented to him, infallibility does not mean that he will answer all of them correctly, for he is quite at liberty to leave it blank, answering none of them at all. It only means that whatever problems he solves (if any) **will** be solved correctly. Out of a hundred problems, he might answer two, and those two answers would be correct while the other problems are all left blank.

In that manner, ambiguity lives in all of those questions which are "left blank" by the Pope, regardless of where his reasons for leaving them blank may range from the perfectly innocent "because the question hasn't yet been raised," to the gravely sinful "because I am too weak and wimpy to take a stand."

What right do you "traditionals" have to judge the Pope? Didn't Vatican I say that "the judgment of the Apostolic See, whose authority is unsurpassed, is not subject to review by anyone; nor is anyone allowed to pass judgment on its decision. Therefore, those who say that it is permitted to appeal to an ecumenical council from the decisions of the Roman Pontiff, as to an authority superior to the Roman Pontiff, are far from the straight path of truth?" If it did, how can you violate that?

First of all, I maintain that no traditional priest, bishop, lay writer or speaker, or religious is known to have ever "judged" any popes. It is the reliable popes themselves who have condemned any and all who would attempt to do the very things which were done "in the Spirit of Vatican II" (See Appendix A: The Pope

Condemns Vatican II). "Traditionals" have consistently shown their unreserved loyalty to the Pope by their total refusal to participate in sacrilegious activities already condemned by the reliable popes in no uncertain terms, such as participation in Novus Ordo or other non-Catholic liturgies and worship. Being between reliable popes, we Catholics lack the juridical and formal authority to judge the doubtful popes in any way which could ever bind the conscience of another.

Even so, it is quite proper that the post-1964 actions of the Vatican leadership are subject to criticism, not only owing to the obvious nature of their frequent departure from the footsteps of Saint Peter, but also to the simple, clear, conciliarly established fact that they have partially resigned from their sees and have schismatically separated themselves from their Catholic predecessors. The fact that a man should serve one day as a pope does not give him unconditional and lifelong immunity from all criticism which would hold even if he resigns. The Vatican leadership must walk in the footsteps of Saint Peter in order to wield the authority of Saint Peter, and insofar as any of them have, we "traditionals" have never been critical of them in any way, but insofar as they have clearly deviated from the footsteps of Saint Peter, they no longer wield Saint Peter's authority, nor enjoy his immunity from criticism.

On the other hand, many who are highly placed in the Vatican organization have appealed to Vatican II in an attempt to overturn the judgments and teachings of numerous reliable popes. If you want to quote that Vatican I statement at anyone, quote it at those who have used Vatican II to repeal what Pope Pius XII said about the liturgy in *Mediator Dei*, or what Pope Pius XI said about ecumenical efforts in *Mortalium Animos*, and so on through Popes Pius X, Leo XIII, Pius IX, Gregory XVI, Pius V, and so many others.

MISCELLANEOUS QUESTIONS

Don't Catholic traditionalists tend to be disgruntled, cheerless, Neo-Nazis, Fascists, and other such unsavory types?

The Novus Ordo liars spread these rumors only because they know they cannot answer the theological claims of Traditional Catholics. It is simply the kind of name-calling they resort to since they lack any substantive claims against the Church. Really, such rumors should be taken no more seriously than the rumors spread by the Church's enemies back in the early centuries to the effect that Christians eat their children.

Such malicious rumormongering is not confined to the Novus Ordo. The secular press is fast becoming similarly aware that Traditional Catholicism represents all the truth and "hard sayings" of Christ they and their paying readership would much rather forget. When Mel Gibson stood up to be counted as a Traditional Catholic at the time of the release of his movie, *The Passion of the Christ*, many made the false claim that this film was somehow "anti-Semitic." Also circulated at the same time was the absurd notion that Vatican II had somehow "absolved" Jews of crucifying Christ, as if any Jews were to blame in the first place, other than those personally involved all those thousands of years ago.

It is true that Traditional Catholicism, being somewhat "outcast" by society at large (unjustly), may sometimes attract others, such as Neo-Nazis, bigots, or anti-Semites who are also "outcast" by society at large (justly), but when these types find themselves unable to spread their bigotry or anti-Semitism to Traditional Catholics, in time they either convert or leave. Sometimes, one comes across such miserable creatures as "ex-traditionalists." Scratch the paint off any one of them and underneath what you invariably find is a bigot who voluntarily left Catholic tradition because he was unable to inveigle Traditional Catholics into his bigotry.

Anyone who takes the time to actually go and meet Traditional Catholics will find that they are quite normal, happy, cheerful people who go to church, receive sacraments, go to picnics and barbecues, work, have babies, and lead a perfectly normal life just like all truly devout Catholics used to do before Vatican II. The only thing missing from their lives is the sinful ways of the godless secular world. For example, their young people are typically still untouched virgins on their wedding day. It is only those of the Novus Ordo religion who still mistake the Vatican organization for the Catholic Church who have cause to be disgruntled and angry. Though none of them would admit it, such a confusion logically leads them to conclude that God has deserted them and annulled His promises. When they see those happy, blessed traditionalists experiencing the joy and peace of truly Catholic communion, they in their envy project their own anger and disgruntlement on them.

Traditional Catholics come across to me as a bunch of kooks, crazies, cranks, and crackpots, pushing all the most absurd conspiracy theories, apparitions, visions, seers, or end-of-the-world gloom and doom, and even all the most pathetic attempts at a "theology." Why should I wish to be associated with that?

If one searches the Internet, one can find dozens, perhaps in time even hundreds of postings by some of the most bizarre and opinionated persons, passing themselves off as Traditional Catholics, and perhaps sharing our desire for the Latin Mass and true doctrine and Sacraments etc. To put this into perspective, such persons actually make up less than one hundredth of one percent of all

Traditional Catholics. It also has not helped that there are certain persons who are hostile to authentic Catholicism and who like to showcase these examples as being representative of Catholicism, in an attempt to repel their followers and readership from it.

Probably the best way to overcome this false perception is to attend any of the traditional Latin Masses going on these days and taking the time to get to know the priest and parishioners by and large. What one finds is perfectly normal people who simply, quietly go to church, raise families, live rightly, and show truly Christian love of neighbor.

But haven't bad things happened in the Traditional Catholic community? You yourself have mentioned the failings of several traditional clerics. Why should I risk having to live with that?

The sad fact is that there is no organization, group, club, church (either the True Church or any other), government, or nation which is somehow guaranteed to avoid having bad people who rise to positions of authority within it. While one would have liked the true Church to be somehow immune to that sort of problem, She never has been.

What one gets from traditional priests and bishops, is reliable Catholic Faith and Morals and sacraments. None of the traditional clerics, even the most corrupt, personally speaking, are known to have even once promulgated false doctrine. The same protection, which kept Pope Alexander VI from promulgating error despite his personal corruption, keeps our familiar traditional clergy honest as well, or else very swiftly discredited by all. When such bad people come to be in charge, the situation only becomes just like it was in Jesus' day when the Pharisees still sat in the seat of Moses (sort of a predecessor to the Chair of Peter), and of whom Jesus said "do as they say, but don't do as they do" (Matthew 23:2–3).

There is no guarantee I can offer that no Traditional Catholic priest or bishop will ever become corrupt, but in that same sense neither can such a guarantee be made with reference to any Novus Ordo presider, Jewish Rabbi, Islamic Imam, or Protestant minister.

Long ago, well before Vatican II, I was abused and mistreated by a Catholic priest. Why should I go back to that?

There is a hidden, and mistaken, assumption implicit within such an outlook. That assumption is the notion that Vatican II somehow changed such cruel and abusive priests (I must admit they did exist, and still do) into wonderful fine people who wouldn't hurt a fly. The notion that heresy can turn a bad person into a good person is, as one would have to realize upon even a moment's reflection, patently absurd. Furthermore, there is no fact or statistic to support the claim that Vatican II ever changed any of these horrible priests for the better.

Those horrid priests who abused and mistreated children at least still tried (sometimes even by means of their abuse and mistreatment) to teach those children the true Faith. When Vatican II came along and changed everything, these same horrid priests simply continued to abuse and mistreat children in all the same ways they already did, only now they are trying (sometimes even by means of their abuse and mistreatment) to teach those children a false and non-Catholic religion.

If anything, the total number of priests who engage in such behaviors has only skyrocketed since Vatican II, even while the total number of priests (including doubtfully ordained Novus Ordo presiders) has declined. Granted, some proportion of that increase might be traced to their being more victims willing to admit to their victimization than in days gone by, but even in the old days it was easy to detect the bad priests by their frequent transfer from parish to parish, always one step ahead of scandal.

The fact is that classically, very few such bad priests could ever get through seminary. It is not that hard for the faculty of a seminary to detect that a certain seminarian is likely to cause problems and bring scandal on the Church and that seminary. And after all there were any number of bright young men willing to take his place in the seminary classrooms. Today's modernist liberal Novus Ordo seminaries are so starved for seminarians that almost no one gets turned away. They have to take what they can get. Predictably, they only get all the more such creeps.

My heart is especially heavy for those who have been abused or mistreated by such priests. Such persons cannot help but have a certain distaste for anything Catholic, owing to their bad experience in the Church, and my heart goes out to them. Those bad priests will have much to answer for.

Aren't there many other independent bishops and priests out there, and don't some of them claim to be Traditional Catholics? What are we to make of them?

It is true that there are many validly consecrated bishops and many more validly ordained priests who trace their orders to schismatic sources such as the East Orthodox, the Old Catholics, or various other isolated bishops who, for whatever reason, departed from unity with Rome in the Pre-Vatican II days. It is also true that a few of them claim to be Traditional Catholics. Some do so implicitly by performing traditional Latin Masses, and others do so explicitly, by announcing their union with the Magisterium of the Church and/or their repudiation of whatever heresies their predecessors in their "apostolic succession" might have indulged in which had occasioned their original separation.

Francis Schuckardt was such a person, tracing his orders to the Old Catholics. Not all such need have his particular failings, and it is fair to ask what role his orders from a schismatic source had in the form his personal failings took. There are others of this category, for example the Society of Christ the King (not to be confused with the Institute of Christ the King, a Modernist approved group), who appear to be functioning reasonably enough as Traditional Catholic priests and bishops, despite their orders coming from Archbishop Carlos Duarte Costa of Sao Paolo, Brazil, who withdrew from Roman communion on June 6, 1945. The question is, can they be accepted?

First of all, in an emergency, anyone with a valid priestly ordination and willing to perform the Last Rites validly can administer that in any case where there is immediate danger of death. Second of all, it is a historical fact that some priests and bishops who trace their orders to schismatic sources have been welcomed back into the Church, and even on some occasions, allowed to continue serving in their priestly or episcopal role, once regularized by the Church. Given the juridical free-for-all mandated at Vatican II, it would hardly do for Traditional Catholics to stand on ceremony regarding the normal procedures for such a normalization, but the fact remains that those who obtain their orders from such a source will probably labor under a kind of "second-class citizen" status until such time as a future reliable pope officially regularizes them.

Part of this is justified from the standpoint of example. The particular priest or bishop may indeed do some good, in seeing to the needs of a desperate soul in their hour of need, yet their very ability to do this good thing is dependent upon the grievous sin of others, as if giving legitimacy to their sin of schism from Rome. Such a source of valid Holy Orders also forces those priests and bishops receiving them to choose between being soft on the heresies which made their ordination or consecration possible, and of being ungrateful to those to whom they owe their valid ordination or consecration. That places a certain quandary right close to the center of their spiritual lives.

Furthermore, it is all too easy for such priests and bishops to vary in a continuum from staunchly orthodox to flagrantly heretical or even invalidly consecrated or ordained. Unlike the clergy whose orders flow from the legitimate bishops of the Church, these types have no canonical mission from the Church and every juridical act they can carry out can only be by supplied jurisdiction due to common error. They have climbed over the wall rather than enter by the door, so one must question their good faith. It is safest to say that even the best, noblest, and most orthodox of such are already on the wrong side of the dividing line, although some few might reasonably be recognized on a case-by-case basis as Catholic ministers, and thereby permitted to do some real good.

These bishops from historically schismatic lines should not be confused with the hierarchical Catholic bishops I have written about whose departure from the Vatican structures came about after Vatican II authorized true Catholic bishops to operate entirely independent of the Vatican organization. Nor for that matter have all of those who departed from the Vatican organization after Vatican II done so for legitimate reasons. Some have departed so as to function as "married bishops," or to approve further outrages even worse than the Novus Ordo Missae. For example, the Immani Temple which was founded on July 2, 1989 by George Stallings, has done so for schismatic reasons, namely, to reject priestly celibacy, to allow divorce, remarriage, and birth control, to allow inter-communion with non-Catholics, and to abolish auricular confession.

You have written much here of unofficial ordinations and consecrations and unpleasant controversies; where is the love that Jesus Christ spoke of? Why couldn't this have read more like the New Testament?

The story of a hardworking goldsmith who devotes his entire life to making some beautiful work of art in gold would similarly differ from the story of those who everything short of have to resort to theft to prevent that golden masterpiece from being melted down and lost to all posterity. Christ Himself instituted the Church and all its sacramental treasures. The heroes of this account are not instituting a new Church but preserving the priceless public treasure entrusted to them by Christ while everyone else is so keen on destroying it.

Why is the individual Catholic at liberty to conceal his attendance at parishes within other factions of the Church?

Since the entire Traditional Catholic community is simply the One Catholic Church, the divisions between the factions, expedient as they may be for the purposes of God in restoring order, have no moral right to exist. Morally, the situation is exactly equivalent to how it would have been back in the pre-Vatican II "good old days" if two neighboring priests, both in union with Rome and their diocese, were to have had a feud going on between them where Fr. X refuses to give the sacraments to anyone he finds out to have obtained any sacraments from Fr. Y, and vice versa. Since that division has no moral right to exist (and presumably their bishop(s) should sooner or later get around to dealing with them about this), an individual who has received sacraments from Fr. Y is quite at liberty to conceal that fact from Fr. X, using silence or evasive answers. While Fr. X may be quite properly concerned with seeing to it that a parishioner is not receiving sacraments from a non-Catholic priest, so long as the parishioner has only received the sacraments from Catholic priests, Fr. X has no need or right to know **which** Catholic priests the parishioner has received sacraments from.

What about activity in such Catholic organizations as the Blue Army, the Legion of Mary, the Knights of Columbus, the Saint Vincent de Paul Society, or Sodalities, etc.?

These organizations were created by the Catholic Church, and as such belong to Her. Therefore, the Novus Ordo Church of the People of God has no right to them; in all justice they should become attached to Traditional Catholic parishes. Since these Catholic organizations are currently being held hostage by the Novus Ordo "clergy," they are not a safe place for real Catholics. However, those Traditional Catholics who are strong in their traditional Faith, wise to the deceptive argumentation used by the Novus Ordo, and who have a missionary spirit might perhaps be able to involve themselves in some of these organizations and even reach out for positions of authority within them.

The scriptural precedent for that is the example of the Apostle Paul who spent much time preaching in the (Jewish) synagogues (Acts 9:20, 13:5,14, 14:1, 18:19, and 19:8) because there were yet many persons to be found in the synagogues who wanted to find and serve the true God. We have the same situation today. There are still some "Catholics-at-hearts" trapped within the Novus Ordo establishment, and the greatest concentrations of these are often found with these sorts of Catholic organizations. What a prize it would be to bring back to the Church an entire Council of the Knights of Columbus or the Legion of Mary!

Another open door of missionary opportunity is teaching in the various religious classes for their children, such as Confirmation class. To do any good, one must firmly take the stand that they will teach the entire Faith out of the standard catechisms of the Church (something they can't object to without admitting their heretical position) instead of the current diocesan gobbledygook, or else they can just go and get someone else. In some cases, they will have to accept such an offer because they can neither inspire nor afford to pay any volunteers to teach these classes. By engaging in such missionary activity, one serves in that role of the angel by the empty tomb saying, "He is not here; He is risen!"

Why can't the Church "baptize" the more modern philosophies such as Existentialism, Phenomenalism, Teilhardism, and Marxism, etc. the way it once "baptized" the philosophies of Plato and Aristotle?

The ancient pagan philosophers did the best they could to formulate order to the universe, subject only to the common fallen condition of all Mankind. Their philosophies reflect logical thinking, rational ideas, sensible and practical results. By contrast, these and all other modern "philosophies" (actually "malosophies," to coin a word) were concocted in the presence, and in defiance of, a known and well-developed Catholic philosophy.

They are the products of Post-Christian European paganism, a paganism based on a rejection of known truths, rather than the ancient paganisms based on sincere though limited human effort. It is therefore quite impossible to bring forth anything positive from these deliberately erroneous systems of "thought," and the only proper thing to do about them is to anathematize them altogether. They can no more be "baptized" than can the Devil their creator.

Have you shown this book to any experts so as to get their approval or advice, or to try to bring them over to your opinion?

What experts? Seriously, what experts? There are none. Turn to the Vatican leaders and functionaries for advice on these complex issues? Very few of them can even give a simple straight answer as to what the Mass is. If I try to send this book to any diocesan "bishop," he will find himself out of his depth in trying to refute my claims. His only recourse will be to say, "I'm too busy" (which, given the way their lives are run for them, may not be that far from the truth) and sit on it indefinitely.

As a matter of fact, I have had several of the bishops and other persons I mention herein verify the viability of my theory, and I have had one of these bishops read the initial edition of the entire book so as to check it for any heresies, and made whatever corrections necessary. Even if, in the final analysis, my theory should prove to be wrong, at least it is not heresy. I have put my theory on the table for all to see, and after a reasonable time period of twenty years, it has gone uncontested. I continue to await a detailed, credible challenge.

Who is Edwin Faust and where does his quote come from?

Being the Church, the Traditional Catholic community is far larger than I could ever begin to encompass within this one small volume. For every person, group, school of thought, etc. there are many others I could not even get to, but are no less part of the Traditional Catholic community. As a symbol of that fact, the dedicatory page of this book deliberately features a quote from a Traditional Catholic author who went otherwise unmentioned throughout the book. The scope of all the many writers and voices who express the Catholic Faith was already far larger than I could document then, and though many of those covered (and not covered) in this book have passed away, their writings remain as a shining testimony to the Faith, scrubbed bright and clean in this darkest of all times, and have only been added to in the years from then until now. If even only a hundredth of this vast corpus should remain in future ages, even that will be quite vast, and I can only regret not having the resources to see to it that it all remains available for future ages.

But now in this revised edition, the point of my silence about the Edwin Faust quote can be made explicitly in this question, and curious readers can at last know about him. Edwin Faust is a Traditional Catholic journalist who has written articles for many publications over the years. I first met his writings on the pages of *The Angelus* in 1994; by the late 1990's he had transferred over to writing for *The Latin Mass*, and has also published articles in *The Remnant* and *Catholic Family News*, and of late now publishes on *The Fatima Crusader*. Despite his softlining stance, theologically speaking, he always manages to find truly worthwhile subject matter to write about and truly brilliant and savvy points to be made, and my esteem for him has never flagged. The quote used at the start of this volume comes from an article by him titled *Dutch Treat*, appearing in the August 1995 issue of *The Angelus*, page 36. In it, he is commenting on how the Amish, despite their flawed Protestant creed, at least really live their religion, as some sort of "sign of contradiction" against their pagan contemporaries. The immediate context of the quote is "The Amish also point to a certain paradoxical bump on the shape of reality, an unexpected contour that most tend to overlook or misunderstand. They underscore the fact that change is usually the result of lethargy; and stability, of intense effort."

What are your opinions regarding the issues that divide Traditional Catholics today? Are you "Motu Proprio," SSPX, sedevacantist, or what?

In a very real sense, I belong properly into all those categories, and in another sense, to none of them. I am an enthusiast for the Catholic Faith who rejoices in the gathering strength of all present groups within the Church. Each of these groups represent a sincere attempt to interpret the present-day crisis in the Church and respond in what each believes to be a Catholic response. I can and do respect the Catholic spirit of each as those in them seek to do the best they can, they all strive to live and believe and worship as Catholics and so must be so counted, and so I do.

But every position has its limitations. Beginning with the Motu Proprio position, I do appreciate the attempt to establish a beachhead of real Catholicism on the "shores" or "fringes" of a society which is in every other respect patently non-Catholic. They hope to create a seed of the Church which will grow and maybe someday take over the present-day Vatican organization. When I wrote the original draft of this work (back in 1996-1998), that seemed a very real possibility, as there were not only we traditionalists, but a vast majority of conservatives who could still remember and appreciate what once was, and would truly welcome and support most enthusiastically a return to all the old forms and teachings and ceremonies and morals. The simple election of a real Catholic to lead them could have quite possibly turned that whole rig around and returned it to its pre-Vatican II state.

The election of Ratzinger as Benedict XVI was almost certainly the last chance for that to happen. He had, over the years, become at least somewhat conservative, had (I think) a real appreciation for the authentic Catholic Mass and Sacraments, even if only for symbolic reasons rather than theological ones. But other than his famous Motu Proprio and the very occasional nod made in a Catholic direction, his tenure passed quite fruitlessly. Now, twenty years later, generations of even "conservatives" have grown up, knowing only the Novus Ordo. The Catholic Mass, if any of them should see it at all, only seems weird, foreign, and incomprehensible, and all they can seek to "conserve" now is the Novus Ordo of their youth. At the top, replacing Benedict XVI has been Bergoglio as Francis I, who has already populated a clear majority of the voting members of his pseudo "cardinalate" with like-minded clones, and he is showing no signs of slowing down, and it won't be long before clearly over two-thirds will be just like him, no one to be otherwise, no one to want otherwise (unless a small few who can and will always be outvoted). In short, the Vatican organization has about as much chance of being rehabilitated and restored to Catholicism as the United States has of resubmitting itself to the British Monarchy, not physically impossible, but we all know it's not going to happen. Really, who would even think of joining, say, the Lutheran church with the hope of thereby Catholicizing it? Much as I can respect the motives of those who intend that very thing, it has to be regarded as futile, and other than adding to the total number Mass locations it is of no real constructive value (plus, many of those locations and increasingly more as time passes do not even have a valid priest!)

And even their motives can be suspected in at least some cases. While some may well have the motive of inserting something of real Catholicism into the

Conciliar church, there are others, perverse as it sounds, whose aspiration in life seems to have been to grow up to be a Traditionalist worm wriggling on a Modernist's hook. Obviously, Peter and company are not the only fishermen around; the Devil has his fishermen as well. I guess it depends upon whom they are speaking to; to their congregations they are establishing a Catholic beachhead, but to their Novus Ordo functionary leaders they are the Traditionalist worms. Which side are they lying to?

Moving along to the SSPX/Resist-but-Recognize (I consider that a slightly more apt and favorable reference than "Recognize and Resist") and the sedevacantist, at least most on both sides of that divide seem to think that "the Church" can just descend into such complete decadence, not merely moral, but even doctrinal (which latter has never happened before), all without any resistance from the Heavenly quarter. I don't see how such a claim can escape being that either God has abandoned us ("I will be with you until sometime in the future when all of your Church leaders grow corrupt?") or else died, as many speculated upon back in the 1960's ("I will be with you until sometime in the future when I will drop dead?"). The only difference is that one group insists that such a fallen and heretical/apostate group is still the Church, meaning that the Church can defect by being made to abandon and adulterate Christ's religion in its dogmatic and moral content through a substantial corruption, while the other (in most cases) seems to insist that such a fallen and heretical/apostate group thereby loses its former status as the Church, and the real Church, at least as a visible and canonical society, can defect by being made to cease to be. I reject both propositions. And though I accept the Sede Vacante finding as true, but without something more of a formal pronouncement than was made in *Lumen Gentium*, or else by Abp. Thục in 1982 and 1983, it is not clear that such a conclusion can be considered as binding on the Church as a whole, despite its being unassailably true. In any case, I reach that conclusion without ever judging any of the men who have failed to function as a Catholic pope.

GLOSSARY

Absolute Sedevacantist
A sedevacantist who believes that all authority has been lost by the leader of the Vatican organization, along with his cardinals and bishops.

Aggiornamento
A word used by John XXIII to describe what he intended to do to the Church, namely to bring it up to date (as if the Church were not eternally new already).

Apostolic Visitor
A personal representative of the Pope who visits a seminary, religious order, or diocese in order to make an inspection of it and report back to the Pope. Such visitors often wield a considerable authority with the seminary, religious order, or diocese they visit, greater than that of the applicable Bishop or Superior General.

Cassiciacum Thesis
A claim that the man elected to lead the Vatican organization might be materially a pope, but not formally so, otherwise known as sedevacantism of the Materialiter/Formaliter variety.

Celebret
the formal and proper name of the permission given to a diocesan priest to say the Indult Mass. In a few early cases, Cardinal Mayer even granted celebrets to diocesan priests who had in effect gone over their bishop's head.

CMRI
An Acronym for the Congregatio Mariae Reginae Immaculatae, which is Latin for the Congregation of Mary Immaculate Queen. In this book, and in nearly all Traditional Catholic writings, it refers to the society founded in 1967 (made a priestly society 1971) by Francis Schuckardt and which was run by him until his expulsion in 1984, after which various other priests And bishops have operated from their facility in Spokane, Washington known as Mount Saint Michael's and their bishop's cathedral in Omaha, Nebraska.

Co-consecrator
A bishop who assists in the consecration of a new bishop. Normally, the Church uses three bishops to consecrate a new bishop, the one who is primarily responsible is called the consecrator, and his two assistant bishops are called co-consecrators. The purpose is to further guarantee that the new bishop is really being validly consecrated. Even if the consecrator has no valid intent or consecration, either one of the co-consecrators will still be able to supply validity. However, one bishop alone is enough to consecrate validly.

Commission
A group of Vatican prelates chosen to oversee a specific concern. In the case of the proposed commission to be placed over the SSPX, it would have been composed of seven members, of whom no more than two (and neither of those president or vice-president) were taken from tradition and the rest would have been faceless Novus Ordo Vatican bureaucrats. They would have had the job of settling disputes between, for example, a diocese where the SSPX is not wanted and the SSPX priest and faithful who want it.

Concelebration
The celebration of a Mass done by two or more priests together (by "priest" here, bishop is also included). Such an event implies spiritual union and agreement between the participating priests.

Conciliar Church
Among traditional circles, this term is sometimes applied to the entire Vatican organization, or more properly and commonly, to those portions of the Vatican organization which practice the New (Novus Ordo) Religion invented at the Second Vatican Council. This should normally be thought of as referring to the same thing as the Church of the "People of God."

Conclave
A meeting of bishops and/or cardinals (in recent centuries it has always been cardinals) in which a new pope (or Vatican leader) is to be chosen. This takes place after a pope dies or resigns.

Consecration
A ceremonial action by which the highest degree of the sacrament of Holy Orders, otherwise known as the fullness of the priesthood, is conferred on an individual. This action makes the person so consecrated a bishop. Also used

with reference to the host in the Mass which by being consecrated ceases to be bread but becomes the Body, Blood, Soul, and Divinity of our Lord.

Consecrator
One who consecrates, namely a bishop.

Conservative
In the Traditional Catholic Community, the word "conservative" is usually used in contrast to the word "traditional" to describe a person who, although trying to remain as Catholic in faith and morals as possible, feels bound to remain at peace with the Vatican organization, even at the expense of their Catholic faith. Such publications as *the Wanderer*, or *The Catholic World Report*, or *Fidelity*, or even *Soul*, best represent their position. Except where granted permission to worship using the "extraordinary form" for their Catholic (traditional) rites, they are outside the Church (however barely) but totally inside the Vatican organization. By contrast, "traditional" describes those who adhere to the fullness of their Catholic faith and therefore remain totally inside the Church, regardless of whether or not they may also remain within the Vatican organization.

Continuity
The state of being the same identical thing. Even though the molecules of a person come and go over the years as they eat, breath, and excrete, until little or no material they were born with remains with them, the person is still the same person. So it is with the Church. The Catholic Church (Traditional Catholic community) today is the same identical entity as that which Christ inaugurated in His day.

Eastern Rite Catholic
The Church in the East (Eastern Europe) which is subject to the Pope, but also subject to the discipline of their Patriarch, a successor to one of the other twelve apostles. Much of their worship may seem unfamiliar to those who are only familiar with the Latin (Western) Rite, but it is every bit as Catholic as the Western Rite.

Ecclesia Supplet
A Latin expression meaning the Church supplies, referring to the supplied jurisdiction traditional priests and bishops would indisputably have access to in cases where some are lacking a more direct form of jurisdiction.

Ecumenism
Among traditional circles, this term usually applies to false ecumenism, unless preceded by the word "true" to differentiate between the true ecumenism of every Ecumenical Council from Nicea to Vatican I, and of valid cooperation with non-Catholics in certain civil causes such as fighting abortion, versus the false ecumenism of treating other religions as being as good and true and salvific as Catholicism.

Ember Days
Special days of partial fast and abstinence which took place four times a year. These days come on the first Wednesday, Friday, and Saturday after the first Sunday of Lent, Whitsunday (now known as Pentecost Sunday), September 14, and December 13.

Epikeia
A Latin expression meaning equity, referring to a sense of proportion or common sense by which jurisdiction is not denied merely because the usual channels do not apply to a particular case, but reasonably should.

Formally
Refers to actual intent of the individual. For example, a formal sin is what a person commits if they know or believe that the action they intend to commit is sinful. With reference to a pope, the person elected by the College of Cardinals becomes formally a pope when he comes to understand just what being pope really means and accepts the office with its obligations, responsibilities, and prerogatives.

FSSP
An Acronym for the Fraternal Society of Saint Peter, a Society of Pontifical Right structured along the lines of the SSPX, but given approval by the Vatican and cooperating with them in their apostolic activities.

ICEL
Acronym for International Committee for English in the Liturgy. This group, headed by Frederick McManus authored all of the English vernacular editions of the Novus Ordo Missae and all other new false sacraments.

ICR
An Acronym for the Institutio Christi Regis, which is Latin for the Institute of Christ the King. In this book, and in nearly all traditional Catholic

writings, it refers to the Society of Apostolic life constituted in 1990 by Msgr. Gilles Wach and features a missionary emphasis.

Incardination

A process by which a priest, before he can be ordained, must be received by a diocese or religious order or congregation. If a priest is said to be incardinated in such-and-such a diocese or religious order or congregation, it means that the bishop intends to place the newly ordained priest in a parish within his episcopal jurisdiction. Without incardination, explicit or implied (tacit), he has no canonical mission and is not supposed to be ordained. Traditional Catholic priests are incardinated, either explicitly (SSPX) or implicitly (most others) into their particular society within the Church as determined by which bishop they answer to, except for the few remaining from before the fall resulting from Vatican II who were incardinated into some diocese or religious order or congregation.

Indult

A permission given by the Vatican hierarchy (if Catholic) or leaders (if non Catholic), or the thing so permitted. One often heard the phrase "Indult Mass" which refers to a traditional (Tridentine) Mass offered in union with the ex-Catholic Vatican organization and with its full permission, up until the 2007 Motu Proprio which replaced the "Indult Mass" with the "extraordinary form." Occasional reference may also be made to the Indult of Pope Pius V to say the Tridentine Mass and which permission was granted for all time in the Papal Bull *Quo Primum*.

Instauratio Catholica

The rights holders for the publishing arm of priests such as (then) Fr. Sanborn who published the magazines *Catholic Restoration* and *Sacerdotium*, but sometimes informally associated with Bp. Sanborn and the priests he works with; may or may not extend to Bp. Dolan and his priests.

Jansenists

People who adhere to the Utrecht Declaration which exaggerates the strictness of God in pardoning sin or permitting access to the sacraments, and which also teaches a sort of Calvinistic predestination which is the belief that individuals have no real free will but that God intentionally makes some people for Heaven and others for Hell. The term is often used as a mere epithet which is merely an abuse of language.

Jewish Shabbat
　　The worship practiced by Jews on their Sabbath which is Saturday (actually starting after sundown on Friday). This service includes blessings said over bread and wine which each begin by saying, "Blessed are you, Lord, God of all creation…" which the Novus Ordo Missae has in place of an offertory. (In a similar manner of concession to Protestants, the phrase "For the Kingdom, the Power, and the Glory are yours, now and forever" was copied from their King James Bible.)

Jurisdiction
　　The right, before God, to exercise authority over other individuals. The Pope, for example, has universal jurisdiction which means his authority is over all persons all around the world. Diocesan bishops have ecclesiastical jurisdiction over the persons within their diocese which is a geographical territory assigned to their care. The bishops who head religious orders (also called Abbots as are other heads of religious orders who are not bishops such as Mother Superior Nuns) have dominative jurisdiction over the members of their religious orders. Jurisdiction is ordinarily broken into the categories of being regular, ordinary, delegated, or supplied. In this context, the terms regular and ordinary are often interchangeable, but the terms delegated and supplied refer to lesser forms of jurisdiction. The jurisdiction of traditionalist bishops is Ordinary (regular, habitual) but not Local (territorial), and extends to the priests, consecrated religious, and attached lay faithful of their particular group or society within the Church.

Materialiter/Formaliter Sedevacantist
　　A sedevacantist who believes that the leadership of the Vatican organization holds the papacy in material sense, allowing him to nominate cardinals and prevent anyone else from being the Pope, but that he lacks a formal possession of the papacy which is needed in order to gain the charisms of absolute authority, universal jurisdiction, and infallibility.

Materially
　　Refers to the nature of the action as carried out. For example, a material sin is an action which violates a commandment of God or of the Church, regardless of whether the person is aware of its sinfulness or not. A person would sin materially but not formally if they violate a commandment which they are honestly unaware of. A person would sin formally but not materially if they commit an action which they believe to be sinful but in fact is not.

Those who practice the false new religion (which I describe in the next chapter) are materially ex-Catholics since they have gone over to a new religion which is not Catholicism, but formally, most of them never intended to leave the Catholic Church. With reference to a pope, the person elected by the College of Cardinals is materially a pope until he either accepts the office (becoming formally and materially, and therefore fully, a pope) or else refuses the office, or else as pope resigns, or dies.

Motu Proprio
Document originating from the Pope himself, as opposed to an encyclical which may often be prepared by the curia and merely approved or amended by the Pope. The 2007 Motu Proprio *Summorum Pontificum* replaced the former Indults with an "extraordinary form" which is simply the Catholic Mass, as grudgingly permitted by the Modernists. Where before there was an "Indult crowd" who had an Indult for the Catholic Mass, now there is a "Motu Proprio crowd" who use the "extraordinary form" under similar, but broader and more generous terms. Sometimes it is shortened into "Motu," especially in the phrase "Motu Mass."

MSM
An Acronym for Mount Saint Michael's, a facility from which the CMRI is run

Novus Ordo
Latin for "New Order," part of the name of their new pseudo Mass, "Novus Ordo Missae." The new Church really is a "New Order" with its new sacramental forms and is often simply referred to as the "Novus Ordo" in traditional circles. One would say of someone that "he is still in the Novus Ordo," if the person being spoken about still attends the New Church instead of the Traditional Roman Catholic Church. Contrast it with the original order ("Ordo") established by Christ all those centuries ago.

Oath Against Modernism
An oath required of every priest at the time of his ordination to oppose in every way the new heresy of Modernism. This rule was imposed by Pope Pius X and removed by Paul VI. Its full text is included in Appendix A.

Octave
An eight-day period beginning with a feast day during which the festival continued to be observed.

Old Catholic
> The schismatic church which follows Ignaz Dollinger and his false council in Munich. They reject the infallibility of the Pope and the indissolubility of marriage. Some of them (from the same lineage) name themselves "Old Roman Catholics" to distinguish themselves from regular "Old Catholics" to distance themselves from some of the errors of the "Old Catholics."

Ordinariate
> Approved parallel chain of command, such as the Eastern rites who have their own dioceses which occupy the geographical locations as the Western rite dioceses but apply to those belong to the alternate rite. Can also refer to aterritorial assignments such as a Military Ordinariate which has jurisdiction over military chaplains.

Ordination
> Usually refers to a ceremonial action by which the second highest degree of the sacrament of Holy Orders is conferred. This action makes the person so ordained a priest, empowering him to consecrate the host and wine in Mass, absolve from sin, and administer the Last Rites to a dying soul. In rare cases may refer to consecration to be a bishop, particularly if used in the phrase "ordained to the Episcopate."

Orthodox
> The word itself simply means "right teaching." It is also used to refer by name to certain schismatic churches in the East which reject the Pope, particularly when preceded by the words "East," "Eastern," "schismatic," or "schismatic East."

Papal Mandate
> the order given by the Pope to consecrate a bishop. Required as of April 9, 1951 for the consecration of any bishop. The current crisis in the Church has rendered this law inapplicable in order that the Church may continue to exist, though *Lumen Gentium* also legally implies its abrogation. Also sometimes called an "Apostolic Mandate."

Patriarch
> A successor to an apostle. The Pope, who is the successor of Saint Peter and therefore also called the Bishop of Rome, is called the Patriarch of Rome. Other Patriarchs reside in Constantinople, Alexandria, Antioch, and Jerusalem.

People of God

The name of the New Church with its New Religion, otherwise commonly referred to as the "Novus Ordo." Technically this term applies only to the non-Catholic portion of the Vatican organization. Since this name should be deeded to the New Church, those genuine Catholics whether inside or outside the Vatican organization ought not use this phrase to describe themselves, even though they are God's people. (Note: In the Vatican II documents, such terms as the People of God, the Catholic Church, the Christian Church, and even the Mystical Body of Christ are all used in various contexts to refer to each of the actual Roman Catholic Church as well as the newly-detached-from-it Vatican organization, much to the confusion of all. One just has to judge from the context of each appearance of each of these terms in the Vatican II documents which entity is referred to.)

Peritus

An expert, often a layman and many times not even a Catholic (nor even declaring themselves as such) who provided some sort of "expert" advice to the Fathers of Vatican II. Usually spoken of in the plural sense (periti) since quite a number of them were there at the Council.

Priory

A house for missionary priests. Often, priests living in a priory would travel hundreds of miles each Sunday to say Mass in scattered areas throughout a missionary territory, but the priory is their home living quarters.

Regularization

A process by which a parish, religious order, or even entire Church which has not been united to Rome is brought under the authority of Rome and granted jurisdiction from Rome. The talks of regularization between the SSPX and the Vatican have been deeply misunderstood; it was Abp. Lefebvre, speaking for Catholic and Eternal Rome, who was trying to bring the Vatican organization of Modernist Rome back under the authority of Catholic Rome (which would have thereby regularized it), not the other way around.

Reliable Popes

Those popes from Peter to Pius XII, excluding only those deemed unreliable (such as Honorius I) by later popes within that sequence.

Rite
    Refers to either a system of sacramental forms and disciplines, such as the "Eastern Rite" or "Latin Rite" or "Byzantine Rite," and usually used in this sense in this book. Can also refer to a single sacrament, such as the Rite of Baptism or the Last Rites for the dying.

Roman Pontifical
    A book containing all the prayers and ceremonies exclusively reserved to the office of Bishop, not only the administration of the Sacrament of Holy Orders (clear through the rank of bishop), but other episcopal actions as well such as the blessing of altars, Holy oils, etc.

Saint John Vianney Society
    An order made up of the diocesan priests of the diocese of Campos who were turned out by Bp. Navarro and lead by Bp. de Castro Mayer, and who are now headed by Bp. Licínio Rifan.

Sedevacantism
    The position taken that the Church has no living pope at the moment. An entirely reasonable position after the death of any pope and before the election (and acceptance of office) of the next pope, this position becomes controversial if taken while someone is commonly accepted as a pope and still alive. Such a position means one's belief that despite all appearances, Francis I (or whoever in the years to come) is not really a pope.

Sillonism
    A modernist movement based on the "Rights of Man" and the overall liberalism of the French Revolution, condemned (by name) by Pope Saint Pius X. When Abp. Lefebvre mentioned Sillonism in his episcopal consecration sermon, some reporters misheard him and thought he had said "Zionism" which has nothing to do (in any known or direct way) with Roman Catholicism or the fall of the Vatican organization.

SSPV
    An Acronym for the Society of Saint Pius V. In this book, and in nearly all Traditional Catholic writings, it refers to the priestly society founded in 1983 and run by Fr. Clarence Kelly (now Bp. Kelly), and which continues to function to this day. Increasingly they now use the acronym **CSPV** for Congregation of Saint Pius V.

SSPX

An Acronym for the Society of Saint Pius X. In this book, and in nearly all Traditional Catholic writings, it refers to the priestly society founded in 1970 and run by Marcel Lefebvre until his death, and which continues to function to this day.

Suspension

The forbidding of any cleric to perform the duties of their office. A suspended priest or bishop has been relieved of his duty. If imposed by the Catholic Church, it means that the cleric may only administer sacraments to those who are in immediate danger of death. If imposed by the ex-Catholic Vatican organization, in the sight of God it means precisely nothing.

Thục-line bishop

A bishop who traces his episcopal orders to one of the three Catholic bishops consecrated by Abp. Thục, namely Guérard des Lauriers, Adolpho Zamora Hernandez, or Moises Carmona y Rivera. Sometimes used disparagingly to refer to all other bishops consecrated by Abp. Thục who had Holy Orders tracing from the Old Catholic sect, or of the Palmar de Troya sect. If prefaced by the word "Catholic," it refers exclusively to those Thục-line bishops who trace their orders to des Lauriers, Zamora, or Carmona, and also excluding Bp. Bedingfeld after his defection to the false pope in Canada and any bishops who would trace their orders to him after that point. May also exclude one or two of Bp. Slupski's more irresponsible choices (Webster and /or Petco), and may also include some of the more respectable members through Bp. Datessen.

Traditional

following the Tradition. The Catholic Church has always followed Tradition as a part of revelation from God. To be traditional is to adhere to all of the Church's traditions whether significant or insignificant, despite all pressure, persecution, or coercion to persuade us to do otherwise and no matter what direction it comes from.

Tridentine

pertaining to the Council of Trent. When worship is spoken of as being Tridentine, such as the Tridentine mass, that refers to a mass which is either identical or very close to the mass as codified by Pope Pius V back in the sixteenth century. This codification and canonization did not change the mass other than to smooth out certain local variations and provide a uniformity of rite throughout the Western Church.

Trinating

Ordinarily a priest will say only one or two Masses on any given day, but on rare occasions (Christmas and Easter most usually) he may say as many as three Masses within a single day. Many traditional priests "trinate," saying three Masses, often in different cities, because they are so few.

Validity

Used in reference to sacraments, validity means whether or not a sacrament "worked," namely do we now have the Body and Blood of our Lord or is it still just bread and wine?

Vatican organization

The organization, or society run from Vatican City, located in Rome, Italy. This consists of both the physical facilities, buildings, lands, churches, estates, and other material assets, as well as personnel in charge of various offices, both in Vatican City and abroad, from the leader, who was formerly also the Roman Catholic Pope, down to the presiders of the local assemblies, who were formerly Catholic priests. Prior to Vatican II, the Vatican organization was identical to the Roman Catholic Church, but afterwards (and in most references to it in this book) it is a distinct entity from the Roman Catholic Church.

Vigil

A staying awake on the evening before a feast which involved special prayers.

SOURCES AND BIBLIOGRAPHY

For further details regarding the history and reasons for the stand taken by the Traditional Catholics, the following ten volumes are particularly recommended as supplemental reading to the present volume. Listing in this Bibliography, even in this particularly recommended section (some of which are now out of print but still worth tracking down), does not constitute agreement with all of their contents. The directory recommended here is included, not as a reading source, but because it is helpful for putting the reader in touch with the nearest traditional priest, Mass center, or parish:

White, Dr. David Allen *The Mouth of the Lion*. Kansas City: Angelus Press, 1993.

———*Priest, Where Is Thy Mass? Mass, Where Is Thy Priest? Sixteen Priests Tell Why They Celebrate the Latin Mass*. Kansas City: Angelus Press, 2004.

———*Is Tradition Excommunicated? A Collection of Independent Studies*. Kansas City: Angelus Press, 1993.

Lefebvre, Archbishop Marcel *Open Letter to Confused Catholics*. Kansas City: Angelus Press, 1987.

Ottaviani, Cardinal Alfredo, Cardinal Antonio Bacci, and A Group of Roman Theologians *The Ottaviani intervention: Short Critical Study of the New Order of Mass*. Translated and edited by Fr. Anthony Cekada. Rockford: TAN Books and Publishers, 1992.

Coomaraswamy, Dr. Rama P. *The Problems With the New Mass*. Rockford: TAN Books and Publishers, 1990.

Cekada, Fr. Anthony *The Problems With the Prayers of the Modern Mass*. Rockford: TAN Books and Publishers, 1991.

Le Roux, Abbe Daniel *Peter, Lovest Thou Me?* Gladysdale Victoria: Instauratio Press, 1989.

Radecki, Frs. Francisco and Dominic *What Has Happened to the Catholic Church?* Ontario: The Aylmer Express, 1994.

Morrison, Fr. M. *Official Catholic Directory of Traditional Latin Masses* Santa Monica: Veritas Press, Yearly, 1994-Current.

(Note: Printed copies of the current edition of this directory can no longer be ordered. The most current electronic copy can be found at http://www.traditio.com/nat.htm Donations to defray cost of maintaining this directory would be appreciated.)

A number of other books and periodicals have also served to provide information about various aspects the Traditional Catholic cause and are therefore also of interest as sources of liturgical, historical, and canonical details:

Lefebvre, Archbishop Marcel *Pastoral Letters—1947–1968*. Kansas City: Angelus Press, 1992.

Lefebvre, Archbishop Marcel *I Accuse the Council!* Dickinson: Angelus Press, 1982.

Lefebvre, Archbishop Marcel *A Bishop Speaks.* Dickinson: Angelus Press, 1976.

Lefebvre, Archbishop Marcel *Liberalism.* Dickinson: Angelus Press, 1980.

Lefebvre, Archbishop Marcel *Collected Works Volume 1, 2, and 3.* Dickenson: Angelus Press, 1985.

Lefebvre, Archbishop Marcel *Luther's Reforms and the Modern Mass.* Commack: CTC Books, 1993.

Lefebvre, Archbishop Marcel *They Have Uncrowned Him.* Kansas City: Angelus Press, 1988.

———*Archbishop Lefebvre and the Vatican.* Compiled and edited by Fr. François Laisney. Dickinson: Angelus Press, 1989.

Lefebvre, Archbishop Marcel *Spiritual Journey.* Dickinson: Angelus Press, 1991.

Lefebvre, Archbishop Marcel *Against the Heresies.* Kansas City: Angelus Press, 1997.

Wiltgen, Fr. Ralph M., S.V.D. *The Rhine flows into the Tiber: A history of Vatican II.* Rockford: TAN Books and Publishers, 1967 (1985 TAN Books Edition).

Hildebrand, Dietrich von *The Trojan Horse in the City of God.* Chicago: Franciscan Herald Press, 1965.

Hildebrand, Dietrich von *The Devastated Vineyard.* Translated from the German by John Crosby, Ph.D., and Fred Teichert, Ph.D. Harrison: Roman Catholic Books, 1973 (1985 Edition).

Davies, Michael *Liturgical Revolution, Volume 1: Cranmer's Godly Order.* Devon: Augustine Publishing Co., 1976 (1995 Roman Catholic Books Edition).

Davies, Michael *Liturgical Revolution, Volume 2: Pope John's Council.* Devon: Augustine Publishing Co., 1977.

Davies, Michael *Liturgical Revolution, Volume 3: Pope Paul's New Mass.* Dickinson: Angelus Press, 1980.

Davies, Michael *The Second Vatican Council and Religious Liberty.* Long Prairie: The Neumann Press, 1992.

Davies, Michael *Apologia Pro Marcel Lefebvre, Part One.* Dickinson: Angelus Press, 1979.

Davies, Michael *Apologia Pro Marcel Lefebvre, Part Two.* Dickinson: Angelus Press, 1983.

Davies, Michael *Apologia Pro Marcel Lefebvre, Part Three.* Dickinson: Angelus Press, 1985.

Tissier de Mallerais, Bernard *The Biography of Marcel Lefebvre.* Translated from the French by Brian Sudlow, M.A. with additional material from Rev. Sebastian Wall of the Society of Saint Pius X. Kansas City: Angelus Press, 2004.

Davies, Michael *The New Mass.* Devon: Augustine Publishing Co., 1977 (1985 Edition).

Davies, Michael *The Tridentine Mass.* Devon: Augustine Publishing Co., 1985.

Davies, Michael *Communion Under Both Kinds: An Ecumenical Surrender.* Rockford: TAN Books and Publishers, 1980.

Davies, Michael *An Open Lesson to a Bishop: On the Development of the Roman Rite.* Rockford: TAN Books and Publishers, 1980.

Davies, Michael *Archbishop Lefebvre and Religious Liberty.* Rockford: TAN Books and Publishers, 1980.

Davies, Michael *The Liturgical Revolution.* Dickinson: Angelus Press, 1983.

Davies, Michael *The Barbarians Have Taken Over.* Dickinson: Angelus Press, 1985.

Davies, Michael *On Communion in the Hand and Similar Frauds.* St. Paul: The Remnant Press, Undated.

Davies, Michael *The Roman Rite Destroyed.* Devon: Augustine Publishing Co., 1978 (1992 Angelus Press Edition).

Davies, Michael *Liturgical Shipwreck: 25 Years of the New Mass.* Rockford: TAN Books and Publishers, 1995.

Davies, Michael *A Short History of the Roman Mass.* Rockford: TAN Books and Publishers, 1997.

Davies, Michael *The Catholic Sanctuary and the Second Vatican Council.* Rockford: TAN Books and Publishers, 1997.

Schmidberger, Fr. Franz *The Catholic Church & Vatican II.* Kansas City: Angelus Press, 1996.

De Nantes, Fr. Georges *Books of Accusation.* https://crc-resurrection.org/further-information/liber-accusationis.html

Omlor, Patrick Henry *Questioning The Validity of the Masses using The New All-English Canon.* Reno: Athanasius Press, 1968.

Omlor, Patrick Henry *The Robber Church.* Greenacres: Catholic Research Institute (reprint), Undated.

Omlor, Patrick Henry *Res Sacramenti*. Greenacres: Catholic Research Institute (reprint), Undated.

Omlor, Patrick Henry *Why The "Short Form" Cannot Possibly Suffice*. Greenacres: Catholic Research Institute, 1997.

Omlor, Patrick Henry *Has The Church The Right?* Greenacres: Catholic Research Institute (reprint), 1969.

Omlor, Patrick Henry *The Ecumenist Heresy*. Greenacres: Catholic Research Institute, Undated.

Omlor, Patrick Henry *Insights Into heresy*. Greenacres: Catholic Research Institute (reprint), 1970.

Omlor, Patrick Henry *No Mystery of Faith: No Mass*. Greenacres: Catholic Research Institute (reprint), 1994.

Omlor, Patrick Henry *The Ventriloquists*. Greenacres: Catholic Research Institute (reprint), Undated.

Omlor, Patrick Henry *The Casualties of Thirty-Five Years of Warfare*. Greenacres: Catholic Research Institute, 1997.

Omlor, Patrick Henry *Five Flaws Found*. Greenacres: Catholic Research Institute (reprint), 1970.

Coomaraswamy, Dr. Rama P. *The Destruction of the Christian Tradition*. Bloomington: World Wisdom, Inc., 1981, 2006.

Martin, Malachi *Catholicism Overturned*. Toronto, Triumph Communications, 2003.

Wathen, Fr. James F. *The Great Sacrilege*. Rockford: TAN Books and publishers, 1971.

Gamber, Msgr. Klaus *The Reform of the Roman Liturgy: Its Problems and Background*. Translated from the Original German by Klaus D. Grimm. San Juan Capistrano: Una Voce Press, 1993.

Guimarães, Atila Sinke, Michael J. Matt, John Vennari, and Marian Therese Horvat, Ph.D. *We Resist You To The Face.* Los Angeles: Tradition In Action, Inc., 2000.

Ferrara, Christopher A, and Thomas E. Woods, Jr. *The Great Facade: Vatican II and the Regime of Novelty in the Roman Catholic Church.* Saint Paul: The Remnant Press, 2002, 2015.

Amerio, Romano. *Iota Unum: A Study of Changes in the Catholic Church in the XXth Century.* Translated from the Second Italian Edition by Rev. Fr. John P. Parsons. Kansas City: Sarto House, 1996.

Guimarães, Atila Sinke *Eli, Eli, Lamma Sabacthani? (Michael Saint Amand Interviews Atila Sinke Guimarães On the Vatican II Collection.* Los Angeles: Tradition In Action, Inc., 2017.

Guimarães, Atila Sinke *Eli, Eli, Lamma Sabacthani? (My God, my God, why hast Thou forsaken me?) Michael Saint Amand interviews Atila Sinke Guimarães on the Vatican II Collection.* Los Angeles: Tradition In Action, Inc., 2017.

Guimarães, Atila Sinke *In The Murky Waters of Vatican II: Eli, Eli, Lamma Sabacthani Volume I.* Dallas: Tradition In Action, Inc., 1997, 1999.

Guimarães, Atila Sinke *Animus Injuriandi—I (Desire to Offend): Eli, Eli, Lamma Sabacthani Volume II.* Los Angeles: Tradition In Action, Inc., 2010.

Guimarães, Atila Sinke *Animus Injuriandi—II (Desire to Offend): Eli, Eli, Lamma Sabacthani Volume III.* Los Angeles: Tradition In Action, Inc., 2011.

Guimarães, Atila Sinke *Animus Delendi—I (Desire to Destroy): Eli, Eli, Lamma Sabacthani Volume IV.* Los Angeles: Tradition In Action, Inc., 2000.

Guimarães, Atila Sinke *Animus Delendi—II (Desire to Destroy): Eli, Eli, Lamma Sabacthani Volume V.* Los Angeles: Tradition In Action, Inc., 2002.

Guimarães, Atila Sinke *Inveniet Fidem? (Will He Find Faith?): Eli, Eli, Lamma Sabacthani Volume VI.* Los Angeles: Tradition In Action, Inc., 2007.

Guimarães, Atila Sinke *Destructio Dei (Destruction of God): Eli, Eli, Lamma Sabacthani Volume VII.* Los Angeles: Tradition In Action, Inc., 2012.

Guimarães, Atila Sinke *Fumus Satanae (The Smoke of Satan): Eli, Eli, Lamma Sabacthani Volume VIII*. Los Angeles: Tradition In Action, Inc., 2015.

Guimarães, Atila Sinke *Creatio (Creation): Eli, Eli, Lamma Sabacthani Volume IX*. Los Angeles: Tradition In Action, Inc., 2016.

Guimarães, Atila Sinke *Peccato—Redemptio (Sin—Redemption): Eli, Eli, Lamma Sabacthani Volume X*. Los Angeles: Tradition In Action, Inc., 2017.

Guimarães, Atila Sinke *Ecclesia (The Church): Eli, Eli, Lamma Sabacthani Volume XI*. Los Angeles: Tradition In Action, Inc., 2009.

Da Silveira, Arnaldo Vidigal Xavier. *Can The Pope Go Bad?* Translated by John Russell Spann. Greenacres: Catholic Research Institute, 1998.

Gibson, Hutton *Is the Pope Catholic?* Victoria: Australian Alliance for Catholic Tradition, 1979.

Dormann, Fr. Johannes *Pope John Paul II's Theological Journey to the Prayer Meeting of Religions in Assisi: Part I*. Kansas City: Angelus Press, 1994.

Dormann, Fr. Johannes *Pope John Paul II's Theological Journey to the Prayer Meeting of Religions in Assisi: Part II, Volume 1*. Kansas City: Angelus Press, 1996.

Dormann, Fr. Johannes *Pope John Paul II's Theological Journey to the Prayer Meeting of Religions in Assisi: Part II, Volume 2*. Kansas City: Angelus Press, 1998.

Dormann, Fr. Johannes *Pope John Paul II's Theological Journey to the Prayer Meeting of Religions in Assisi: Part II, Volume 3*. Kansas City: Angelus Press, 2003.

Davies, Michael *The Order of Melchisedech: A Defense of the Catholic Priesthood*. Harrison: Roman Catholic Books, 1979 (1993 Edition).

Depuis, Michael, Keith Roscoe, and John Thomson *Are Today's Seminaries Catholic?* Dickinson: Angelus Press, 1990.

Rose, Michael S. *Goodbye, Good Men* Washington D. C.: Regnery Publishing, Inc., 2002.

Kramer, Fr. Paul L. *A Theological Vindication of Roman Catholic Traditionalism.* Thomaiyar Puram: Apostle Publications, Undated.

Davies, Michael *The Legal Status of the Tridentine Mass.* Dickinson: Angelus Press, 1982.

Schmidberger, Fr. Franz *The Episcopal Consecrations of 30 June 1988.* London: Angelus Press, 1989.

Nemeth, Charles P. Esquire *The Case of Archbishop Marcel Lefebvre: Trial by Canon Law.* Kansas City: Angelus Press, 1994.

Pivert, Fr. Francois *Schism or Not?* Kansas City: Angelus Press, 1995.

Fathers of Holy Cross Seminary *Most Asked Questions about the Society of Saint Pius X.* Kansas City: Angelus Press, 1997.

Kelly, Most Reverend Clarence *The Sacred and the Profane.* Round Top: Seminary Press, 1997.

Cekada, Fr. Anthony *Work of Human Hands: A Theological Critique of the Mass of Paul VI.* West Chester: Philothea Press, 2010.

Cekada, Fr. Anthony *Welcome to the Traditional Latin Mass.* Troy: Catholic Restoration, 1995.

Cekada, Fr. Anthony *Traditionalists Infallibility & the Pope.* Troy: Catholic Restoration, 1995.

Fulham, Rev. Terence R. I.H.M. *Corona Spinarum: A Biography and defense of Archbishop Pierre-Martin Ngô Đình Thục.* http://www.olfatima.com/ (no longer hosted there).

Cain, Michael *Tower of Trent Hall of Honor: Archbishop Pierre-Martin Ngô Đình Thục.* http://www.dailycatholic.org/jul26ttt.htm, 2006.

Vaillancourt, Fr. Kevin *The Answers.* Spokane: OLG Press, 2006.

Derksen, Mario, M.A. *An Open Letter to Bishop Clarence Kelly on the "Thuc Bishops" and the Errors in The Sacred and the Profane* http://www.thucbishops.com/, 2011.

Radecki, Fr. Fransisco, and Fr. Dominic Radecki *Tumultuous Times: The Twenty General Councils of them Catholic Church and Vatican II and its Aftermath.* Wayne: St. Joseph's Media, 2004.

Radecki, Fr. Fransisco, and Fr. Dominic Radecki *Vatican II Exposed as Counterfeit Catholicism.* Wayne: St. Joseph's Media, 2019.

Daly, John S. *Michael Davies – An Evaluation, 2nd Edition.* Saint-Sauveur-de-Meilhan: Tradibooks, 2015.

Birtz, P. Dr. Mircea Remus, and Dr. Manfred Kierein-Kuenring *Voices From Ecclesia Militans In Czechoslovakia.* Claudiopolis: NAPOCA STAR Publishing House, 2011

———*On Tracing the River Back to Its Source: An Interview with Gerry Matatics.* Winchester: The Crusade of Saint Benedict Center, 1995.

Wickens, Fr. Paul A. *Christ Denied.* Rockford: TAN Books and Publishers, 1982.

———*Early Day Documents of the Traditionalist Catholic Church.* Compiled by Brother Hermenegild TOSF. San Bernardino: ———, 2016.

Vaillancourt, Fr. Kevin *The Answers.* Spokane: OLG Press, 2006.

Carre, Marie *AA-1025: The Memoirs of an Anti-Apostle.* Sherbrooke: Editions Saint-Raphael, 1972 (1991 TAN Books Edition).

———*Palmar de Troya: The Light for the Church and for the World Vol. I (December, 1959 – March, 1977).* Belfast: Gregorian Publications, 1979

———*Sursum Corda.* Fort Collins: Foundation for Catholic Reform, Quarterly, 1996–2001.

———*The Latin Mass.* Fort Collins: Foundation for Catholic Reform, Bimonthly, Quarterly, 1992-Current.

———*The Remnant.* St. Paul: The Remnant Press, Biweekly, 1967-Current.

———*Catholic Family News.* Niagara Falls: Catholic Family News Press, Monthly, 1994-Current.

———*The Angelus.* Dickinson, Kansas City: Angelus Press, Monthly, 1978-Current.

———*Catholic Family.* Rockdale: Angelus Press, Bimonthly, 1990–1997.

———*The Roman Catholic.* Oyster Bay Cove, Cincinnati, Norwood: Roman Catholic Association, Monthly, Bimonthly, Irregular, Quarterly, 1979-Current.

———*Catholic Restoration.* Madison Heights, Troy: Instauratio Catholica, Bimonthly, Quarterly, 1991–1995, 2002-Current.

———*Sacerdotium.* Madison Heights, Troy: Instauratio Catholica, Quarterly, Semiannual, 1991–1996.

———*Fortes in Fide.* Saint Louis: William F. J. CHRISTIAN, Irregular, 1975–1993.

———*The Reign of Mary.* Spokane: Mary Immaculate Queen Press, Bimonthly, Quarterly, 1977-Current.

———*Salve Regina.* Coeur d'Alene, Spokane: Mary Immaculate Queen Press, Bimonthly, Quarterly, 1967–1992.

———*The Catholic Voice.* Charlotte: The Society of Traditional Roman Catholics, Bimonthly, 1985-2013.

———*Speculum.* Cullman: The Abbey of Christ the King, Triannual, 1992-2011.

———*The Four Marks.* Molt: Northern Light Publications, 2005-Current.

———*The Catholic Inquisitor.* Hampstead: Inquisitor Press, 2018-Current.

In addition, a number of other periodicals, most in foreign languages (publication history unknown to this writer), might also be of value for those conversant in those languages and/or able to obtain them: *Itinéraires*, *Le Combat de la Foi catholique* by Fr. Louis Coache, *La Contre-Réforme catholique au XXe siecle* by Fr. George De Nantes, *Forts dans la Foi* by Fr. Noël Barbara, *Una Voce*, *Das Zeichen Mariens* by Paul Schenker in Switzerland, *Einsicht* by Dr. E. Heller, *La Voie* in Paris by Myra Davidoglou, and *Triumph* in English.

Such a volume as this would be incomplete without listing the primary opposition literature, what little there is, along with a volume about Traditional Catholicism written by a sociologist, another as a history by a secular journalist with a chip on his shoulder about how the Church raises money for such projects as a new seminary in a missionary land, and even several books by traditionalists attempting to incite division between other traditionalists. The periodicals listed here contain many good and even insightful articles regarding various devotional, apologetic, social, political, economic, and moral issues; it is only in what few places they have dealt with the Traditional Catholic movement (particularly the portions outside the Vatican institution) that they would qualify as opposition literature:

Likoudis, James, and Kenneth D. Whitehead *The Pope, the Council, and the Mass.* W. Hanover: The Christopher Publishing House, 1981, 2006.

Cuneo, Michael W. *The Smoke of Satan*. New York: Oxford University Press, 1997.

Jarvis, Edward, *Sede Vacante: The Life and Legacy of Archbishop Thục*. Berkeley: The Apocryphile Press, 2018.

Davies, Michael *I Am With You Always*. Long Prairie: The Neumann Press, 1997.

Hand, Stephen *Tradition, Traditionalists, and Private Judgment*. Saint Paul: The Wanderer Press, 2000.

———*Sedevacantism: A False Solution to a Real Problem*. Kansas City: Angelus Press, 2003.

Salza, John, and Robert Siscoe *True or False Pope?* Winona: STAS Editions, 2015.

Pontrello, John C. *The Sedevacantist Delusion*. North Charleston: CreateSpace Independent Publishing Platform, 2015.

———*The Wanderer*. Saint Paul: The Wanderer Press, Weekly, 1867-Current.

———*Fidelity*. South Bend: Ultramontane Associates, Monthly, 1982–1998.

———*Crib X Cross X Crown*. Mt. Morris: Franciscan Friars of Mary Immaculate, Quarterly, 1984-Current.

———*This Rock*. San Diego: Catholic Answers, Monthly, 1990-Current.

The following references have provided this writer with a true and deep understanding and appreciation of the Roman Catholic Faith. Those of this category alone are accepted uncritically and unconditionally by this author, except those ideas presented within as being of theoretical or doubtful certitude, and collectively define the theological outlook of this book, especially with reference to the issues discussed in it:

Jesuit Fathers of St. Mary's College *The Church Teaches.* St. Louis: B. Herder Book Co., 1955 (1973 TAN Books Edition).

Denzinger, Henry *The Sources of Catholic Dogma.* Translated by Roy J. Deferrari from the Thirtieth Edition of Henry Denzinger's *Enchiridion Symbolorum.* St. Louis: B. Herder Book Co., 1957.

Bellarmine, Saint Robert De Controversiis: *On the Roman Pontiff Vol. I: Books 1 & 2.* Translated by Ryan Grant. Post Falls: Mediatrix Press, 2015.

Bellarmine, Saint Robert De Controversiis: *On the Roman Pontiff Vol. II: Books 3-5.* Translated by Ryan Grant. Post Falls: Mediatrix Press, 2016.

Bellarmine, Saint Robert De Controversiis: *On Councils: Their Nature and Authority.* Translated by Ryan Grant. Post Falls: Mediatrix Press, 2017.

Bellarmine, Saint Robert De Controversiis: *On the Church Militant.* Translated by Ryan Grant. Post Falls: Mediatrix Press, 2016.

Bellarmine, Saint Robert De Controversiis: *On the Marks of the Church.* Translated by Ryan Grant. Post Falls: Mediatrix Press, 2015.

Ott, Dr. Ludwig *Fundamentals of Catholic Dogma.* Translated from the German by Patrick Lynch, Ph.D. and Edited in English by James Canon Bastible, D.D. Cork: Mercier Press Ltd., 1955 (1974 TAN Books Edition).

Berry, Fr. E. Sylvester, STD *The Church of Christ: An Apologetic and Dogmatic Treatise.* Eugene: Wipf and Stock Publishers (originally Mount Saint Mary's Seminary), 1955.

Noort, Msgr. G. Van *Dogmatic Theology Volume I: The True Religion.* Translated and revised by John J. Castelot, S.S. and William R. Murphy, S.S. Westminster: The Newman Press, 1957.

Noort, Msgr. G. Van *Dogmatic Theology Volume II: Christ's Church.* Translated and revised by John J. Castelot, S.S. and William R. Murphy, S.S. Westminster: The Newman Press, 1957.

Noort, Msgr. G. Van *Dogmatic Theology Volume III: The Sources of Revelation; Divine Faith.* Translated and revised by John J. Castelot, S.S. and William R. Murphy, S.S. Westminster: The Newman Press, 1961.

Journet, Msgr. Charles *The Church of the Word Incarnate: An Essay in Speculative Theology, Volume One: The Apostolic Hierarchy.* New York: Sheed and Ward, 1955.

Trent *The Canons and Decrees of the Council of Trent.* Translated by Fr. H. J. Schroeder, O. P. St. Louis: B. Herder Book Co., 1941 (1978 TAN Books Edition).

Trent, Vatican Council I *Dogmatic Canons and Decrees.* New York: Devin-Adair Company, 1912 (TAN Books Edition).

Paul IV *Cum Ex Apostolatus Officio.* February 15, 1559.

Pius V *Quo Primum Decree of Pope St. Pius V on the Roman Missal.* July 14, 1570.

Pius V *De Defectibus Decree of St. Pius V in the Roman Missal.*

Clement VIII *The Brief: Cum Sanctissimum.* July 7, 1604.

Urban VIII *The Brief: Si Quid Est.* September 2, 1634.

Benedict XIV *De Sacrosanctae Missae Sacrificio.*

Gregory XVI *Mirari Vos: On Liberalism and Religious Indifferentism.* August 15, 1832.

Pius IX *Quanta cura.* December 8, 1864

Pius IX *The Syllabus of Errors: A Collection of the Principal Errors of Our Day Which Have Been Condemned in the Papal Letters of Pope Pius IX.*

Leo XIII *Diuturnum: On Civil Government.* June 29, 1881.

Leo XIII *Humanum Genus: On Freemasonry.* April 20, 1884.

Leo XIII *Immortale Dei: The Christian Constitution of States.* November 1, 1885.

Leo XIII *Libertas Praestantissimum: On Human Liberty.* June 20, 1888.

Leo XIII *Exeunte Jam Anno: On The Right Ordering of Christian Life.* December 25, 1888.

Leo XIII *Rerum Novarum: On the Condition of the Working Classes.* May 15, 1891.

Leo XIII *Providentissimus Deus: On the Study of Sacred Scripture.* November 18, 1893.

Leo XIII *Apostolicae Curae: On the Nullity of Anglican Orders.* September 15, 1896.

Leo XIII *Annum Sacrum: On the Holy Year, 1900.* May 25, 1899.

Leo XIII *"A Light in the Heavens": The Great Encyclical Letters of Pope Leo XIII.* New York: Benziger Brothers, 1903 (1995 TAN Books Edition).

Pius X *Lamentabili Sane: Syllabus Condemning the Errors of the Modernists.* July 3, 1907.

Pius X *Pascendi Dominici Gregis: On the Doctrines of the Modernists.* September 8, 1907.

Pius X *Apostolic Exhortation of Pope Pius X on the Priesthood.* August 4, 1908.

Pius X *Our Apostolic Mandate: On the "Sillon."* August 25, 1910.

Pius X *The Apostolic Constitution: Divino Afflatu.* November 1, 1911.

Pius XI *Quas Primas: On the Kingship of Christ.* December 11, 1925.

Pius XI *Mortalium Animos: On Fostering True Religious Unity.* January 6, 1928.

Pius XI *Casti Connubii: On Christian Marriage.* December 31, 1930.

Pius XI *Quadragesimo Anno: On Reconstructing the Social Order.* May 15, 1931.

Pius XI *The Church In Germany.* March 14, 1937.

Pius XI *Divini Redemptoris: On Atheistic Communism.* March 19, 1937.

Pius XII *Summi Pontificatus: On Function of State in Modern World.* October 20, 1939.

Pius XII *Mystici Corporis: On the Mystical Body of Christ and Our Union in it With Christ.* June 29, 1943.

Pius XII *Mediator Dei: On the Sacred Liturgy.* November 20, 1947.

Pius XII *Sacramentum Ordinis: On the Priesthood.* November 30, 1947.

Pius XII *Humani Generis: On Evolution and Other Modern Errors.* August 12, 1950.

Pius XII *Apostolic Constitution of Pope Pius XII on the Assumption of the Blessed Virgin Mary.* November 1, 1950.

Pius XII *Evangelii Praecones: Heralds of the Gospel.* June 2, 1951.

Pius XII *Fulgens Corona: On the Marian Year and the Dogma of the Immaculate Conception.* September 8, 1953.

Pius XII *Ad Caeli Reginam: On the Queenship of Mary.* October 11, 1954.

Ottaviani, Cardinal Alfredo *Duties of the Catholic State in Regard to Religion.* Translated by Fr. Denis Fahey, C. S. Sp. Tipperary: "The Tipperary Star," Thurles, Co., 1953 (1993 Edition).

Jone, Fr. Heribert *Moral Theology.* Translated and adapted by Fr. Urban Adelman, O. F. M. Cap., J. C. D. Westminster: The Newman Press, 1962 (1993 TAN Books Edition).

———*The 1917 Pio-Benedictine Code of Canon Law.* English Translation by Dr. Edward N. Peters, Curator. San Francisco: Ignatius Press, 2001.

Bouscaren, T. Lincoln, S.J., S.T.D., LL.B. *The Canon Law Digest: Volume I. Officially Published Documents Affecting the Code of Canon Law 1917-1933.* Milwaukee: The Bruce Publishing Company, 1934.

Bouscaren, T. Lincoln, S.J., S.T.D., LL.B. *The Canon Law Digest: Volume II. Officially Published Documents Affecting the Code of Canon Law 1933-1942.* Milwaukee: The Bruce Publishing Company, 1949.

Bouscaren, T. Lincoln, S.J., S.T.D., LL.B. *The Canon Law Digest: Volume III. Officially Published Documents Affecting the Code of Canon Law 1942-1953.* Milwaukee: The Bruce Publishing Company, 1954.

Bouscaren, T. Lincoln, S.J., S.T.D., LL.B. *The Canon Law Digest: Volume IV. Officially Published Documents Affecting the Code of Canon Law 1953-1957.* Milwaukee: The Bruce Publishing Company, 1958.

Miaskiewicz, Francis Sigismund, J.C.L. *Supplied Jurisdiction According to Canon 209.* Boston: ——, 1940 (2014 Reprint).

The Third Plenary Council of Baltimore *Baltimore Catechism Nos. 1–3.* New York: Benziger Brothers, 1885, 1898, and 1933 (1977 TAN Books Edition).

Kinkead, Fr. Thomas L. *An Explanation of the Baltimore Catechism of Christian Doctrine for the use of Sunday-School Teachers and Advanced Classes No. 4.* New York: Benziger Brothers, 1891 and 1921 (1988 TAN Books Edition).

Spiago, Fr. Francis *The Catechism Explained: An Exhaustive Exposition of the Catholic Religion.* Edited by Fr. Richard F Clarke, S. J. New York: Benziger Brothers, 1899 and 1921 (1993 TAN Books Edition).

Morrow, Bp. Louis LaRavoire *My Catholic Faith: A Catechism in Pictures.* Kenosha: My Mission House, 1949, 1952, and 1954 (1992 Sangre De Cristo Products, Inc. Edition).

The Catechism Committee *The Roman Catechism: The Catechism of the Council of Trent.* Translated into English with notes by John A. McHugh, O. P., S. T. M., Litt. D. And Charles J. Callan, O. P., S. T. M., Litt. D. South Bend: Marian Publications, 1976 (1982 TAN Books Edition).

Pius X *The Compendium of Christian Doctrine.* Translated and edited by Right Rev. John Hagen into three parts entitled *Compendium of Catechetical Instruction.* Rome: Irish College, 1911 (1993 Instauratio Press Edition, Victoria).

The following references are general apologetic works which are recommended for those who have doubts about whether the Catholic Church is really the Church which Christ founded:

Keating, Karl *Catholicism and Fundamentalism: The Attack on "Romanism" by "Bible Christians."* San Francisco: Ignatius Press, 1988.

Sales, St. Francis de *The Catholic Controversy.* Translated by Fr. Henry Benedict Mackey, O. S. B. Under the direction of Bp. John Cuthbert Hedley, O. S. B. and reprinted by TAN Books and Publishers. Rockford: TAN, 1989.

Roberts, Fr. Kenneth *Father Roberts Answers Jimmy Swaggart.* West Covina: St. Joseph's Catholic Tapes and Books, 1986.

Graham, Bp. Henry G. *Where We Got The Bible: Our Debt to the Catholic Church.* St. Louis: B. Herder Book Co., 1911 (1987 TAN Books Edition).

Sardá Y Salvany, Don Félex *What Is Liberalism?* Translated and adapted for American readership by Condé B. Pallen, Ph.D., LL.D. St. Louis: B. Herder Book Co., 1899 (1993 TAN Books Edition, titled *Liberalism Is A Sin*).

Nelson, Thomas A. *Which Bible Should You Read?* Rockford: TAN Books and Publishers, 2001.

Rumble, Rev. Dr. Leslie *Radio Replies In Defense of Religion.* Sydney: Missionaries of the Sacred Heart, 1934.

Rumble, Fr. Leslie, and Fr. Charles Mortimer Carty *Radio Replies Volume 2.* St. Paul: Radio Replies Press, 1940.

Rumble, Fr. Leslie, and Fr. Charles Mortimer Carty *Radio Replies Volume 3.* St. Paul: Radio Replies Press, 1942.

Rumble, Fr. Leslie, and Fr. Charles Mortimer Carty *That Catholic Church.* St. Paul: Radio Replies Press, 1954.

Rumble, Fr. Leslie. *Questions People Ask About the Catholic Church.* Kensington: Missionaries of the Sacred Heart, 1972.

Forrest, Rev. M. D. *Chats With Converts.* St. Paul: Radio Replies Press, 1943 (1978 TAN Books Edition).

Ripley, Francis J. *This Is the Faith.* St. Paul: Catechetical Guild Educational Society, 1951.

Alexander, Fr. Anthony F. *College Apologetics: Proof of The Truth of The Catholic Faith.* Chicago: Henry Regnary Company, 1954 (1994 TAN Books Edition).

Finally, the following post-Catholic documents of the Vatican institution have also been of use. Of the encyclicals of John XXIII and those coming after him I have listed here only those few in which their "golden moments" are contained:

Vatican Council II *The Conciliar and Post Conciliar Documents.* Translated and edited by Austin Flannery, O. P. Northport: Costello Publishing Company, 1988.

John XXIII *Veterum Sapientia: Apostolic Constitution On Promoting the Study of Latin.* February 22, 1962.

Paul VI *Humanae Vitae: Of Human Life.* July 25, 1968.

John Paul II *Ecclesia Dei: The Church of God.* July 2, 1988.

John Paul II *Apostolic Constitution Fidei Depositum: On the Publication of the Catechism of the Catholic Church Prepared Following the Second Vatican Ecumenical Council.* October 11, 1992.

John Paul II *Veritatis Splendor: The Splendor of Truth.* August 6, 1993.

John Paul II *Ordinatio Sacerdotalis: On reserving Priestly Ordination to Men Alone.* May 22, 1994.

John Paul II *Evangelium Vitae: The Gospel of Life.* March 25, 1995.

Benedict XVI *Summorum Pontificum: On the Use of The Roman Liturgy Prior To The Reform Of 1970.* July 7, 2007

United States Catholic Conference *Catechism of the Catholic Church.* St. Paul: The Wanderer Press, 1994, 1997.

Altemose, Sr. Charlene *Handbook for Today's Catholic: Beliefs Practices Prayers.* Liguori: Liguori Publications, 1978 and 1991.

Hardon, John A. S. J. *The Catholic Catechism: A Contemporary Catechism of the Teachings of the Catholic Church.* Garden City: Doubleday & Company, 1975.

The Theology behind this *Resurrection* Book!

These two volumes document the Catholic Theology which undergirds this *Resurrection* book. The only thing that can truly bring all real Catholics together is a knowledge of the teachings of the Catholic Church regarding itself, its Ecclesiology.

Sede Vacante! Part One: Dogmatic Ecclesiology Applied to Our Times by Griff Ruby

Part One breaks down all the relevant theological principles into sixteen basic categories, documenting each in detail as discussed by the theologians, and then making deductions based on their content, and the events of recent history pertaining to the Church to show and prove the truth of our situation.

Sede Vacante! Part Two: The *Lumen Gentium* Theory About Our Present Ecclesial Circumstance by Griff Ruby

Part Two explores the theoretical aspects of our present situation, positing the theory most known to be fully in accord with all of the doctrinal findings of Volume One, and then addressing the various theories as have been advanced over the years by concerned Catholics.

Both volumes available on Amazon or direct from the publisher:

iUniverse
1663 Liberty Drive
Bloomington, IN 47403
www.iuniverse.com
1-800-Authors (1-800-288-4677)

You've read the book.
Now, see the movie!

What We have Lost…and the Road to Restoration
A critical look at the changes in the Catholic Church

This is the widely acclaimed video, produced by the "In the Spirit of Chartres" Committee (ISOC), documenting the destruction of Christ's One, Holy, Catholic, and Apostolic Church. A great introduction for those new to Traditional Catholicism and desiring to learn and see the differences between the authentic Catholicism of all ages past (now practiced and believed only by Traditional Catholics) versus that new "Novus Ordo" religion served up as a "Catholicism" commonly seen and mistakenly accepted as such today. The video presently exists and can be viewed free of charge in several places on YouTube:

https://www.youtube.com/watch?v=gkE3hqdlxB8

https://www.youtube.com/watch?v=hJxM7Lo2URw

https://www.youtube.com/watch?v=TR5XT2Vtn6w

Or one may do a search on "What We Have Lost and the Road to Restoration"

Copies of this video on DVD are still available while they last, featuring three foreign language audio tracks (English, Spanish, Lithuanian) and seven different language subtitle tracks (English, Spanish, Lithuanian, French, German, Polish, and Portuguese). Each version includes an introduction and commentary by Dr. David Allen White (always audibly given in English though subtitles will be in the selected language), a foremost Traditional Catholic speaker, author, and educator.

See the changes in architecture, art, music, the Sacramental rites, and most especially the Holy Sacrifice of the Mass. Changes in the Church that have so protestantized it, it is nearly unrecognizable. Changes that are damaging to the faith of almost a billion people. See what we've lost, but more importantly, how we can get it back.

Endorsed Nationally and Internationally by Leading Traditionalists—Clergy and Laymen Alike.

The perfect gift for your Catholic relatives, friends, favorite priest, or bishop.

DVD regular price—**$15.00** (plus $2.50 Shipping and Handling)

Obtain from: https://isoc.ws/

Or by mail:

ISOC,
P. O. Box 87, Glenelg, MD 21737

Made in the USA
Las Vegas, NV
21 August 2022